CHAPMAN & HALL/CRC
Monographs and Surveys in
Pure and Applied Mathematics   143

STABLE SOLUTIONS OF

ELLIPTIC PARTIAL

DIFFERENTIAL

EQUATIONS

CHAPMAN & HALL/CRC
Monographs and Surveys in Pure and Applied Mathematics

CHAPMAN & HALL/CRC
Monographs and Surveys in
Pure and Applied Mathematics   143

STABLE SOLUTIONS OF

ELLIPTIC PARTIAL

DIFFERENTIAL

EQUATIONS

Louis Dupaigne

CRC Press
Taylor & Francis Group
Boca Raton London New York

CRC Press is an imprint of the
Taylor & Francis Group, an **informa** business
A CHAPMAN & HALL BOOK

CRC Press
Taylor & Francis Group
6000 Broken Sound Parkway NW, Suite 300
Boca Raton, FL 33487-2742

First issued in paperback 2019

© 2011 by Taylor & Francis Group, LLC
CRC Press is an imprint of Taylor & Francis Group, an Informa business

No claim to original U.S. Government works

ISBN-13: 978-1-4200-6654-8 (hbk)
ISBN-13: 978-0-367-38297-1 (pbk)

Library of Congress Cataloging-in-Publication Data

Dupaigne, Louis.
    Stable solutions of elliptic partial differential equations / Louis Dupaigne.
        p. cm. -- (Chapman & Hall/CRC monographs and surveys in pure and applied
        mathematics)
    Includes bibliographical references and index.
    ISBN 978-1-4200-6654-8 (hardcover : alk. paper)
    1. Differential equations, Elliptic. I. Title.

QA377.D78 2011
515'.3533--dc23
                                     2011021602

Visit the Taylor & Francis Web site at
http://www.taylorandfrancis.com

and the CRC Press Web site at
http://www.crcpress.com

A mon père,
à mon fils.

Contents

Preface

On the shelves of the library where you may have picked this book, there were probably many more volumes dealing with nonlinear elliptic partial differential equations (PDEs). Thank you for opening this one.

What you will find here is a look at the subject through the loophole of stability. I have tried to present the main ideas on the most simple-looking equations. One could even argue that only one equation is treated, this one:

$$-\Delta u = \lambda e^u \tag{P.1}$$

Chapter 1

What is a stable solution exactly? Oftentimes in physics, one is interested in finding the state of a system of lowest energy E. Such a state is stable because it is trapped at the bottom of an energy well. Through calculus, this leads to solving an equality, $E' = 0$ (critical points), and then an inequality, $E'' \geq 0$.

Accordingly, thinking of solutions of PDEs as critical points of an energy functional, we say that a solution is stable when the second variation of energy is nonnegative. Equivalently, the operator associated with the linearized equation must have a nonnegative spectrum. Yet another way of thinking of stable solutions is through their dynamical properties (asymptotic stability). Using these different points of view, we give examples of several classes of solutions that are stable, such as local minimizers, minimal solutions, and monotone solutions, and we study their basic properties. This being done, we introduce stability outside a compact set, a notion general enough to encompass *all* (classical) solutions defined on, say, bounded domains.

Chapter 2

The PDE (P.1), posed on the unit ball with a homogeneous Dirichlet boundary condition, is our model problem. We construct all *radial* solutions and classify them according to their stability properties (their Morse index). The resulting picture, Figure P.1, is the object of this chapter and will serve as a paradigm throughout the book. As you can see, the picture depends very much on the

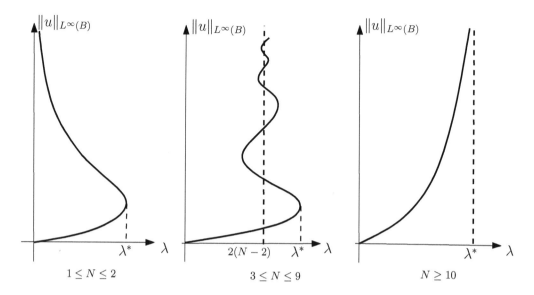

Figure P.1: Bifurcation diagram for the Gelfand problem.

space dimension N. Look, for example, at stable solutions: for $1 \leq N \leq 9$, the stable branch is the lowest part of the curve, before it bends back to the left. In dimension $N \geq 10$, the stable branch is the whole curve. Also observe this: stable solutions are uniformly bounded in dimension $1 \leq N \leq 9$, while they are not for $N \geq 10$.

Chapter 3

In fact, there still exists a weak (unbounded) solution to the equation for $N \geq 10$ and $\lambda = \lambda^*$. In this chapter, we study such stable weak solutions to equations of the type

$$-\Delta u = f(u) \tag{P.2}$$

Under fairly general assumptions, we show that they are unique and arise as limits of classical stable solutions, much as in Figure P.1. For this reason, they earn the name of extremal solutions. We further discuss qualitative properties of the stable branch, in the context of (P.2), except for one major issue—determining when the extremal solution is bounded—which we save for the next chapter.

Chapter 4

We treat the issue of regularity under three different restrictive assumptions: the solution is known to be radial, the nonlinearity has some specific scaling features, such as $f(u) = e^u$, or the space dimension is small. In the latter case, we introduce a restatement of stability in terms of the level sets of solutions, known as the geometric Poincaré formula. In the last section of the chapter, we address the question of regularity for unstable solutions. This is done through a blow-up argument and taking advantage of a Liouville-type theorem, the proof of which we postpone to Chapter 6.

Chapter 5

In the previous chapter, we learned that in low dimensions, at least, stable solutions must be smooth. What happens in large dimensions? Figure P.1 shows that for $N \geq 10$, there exists a singular stable solution to (P.1) on the unit ball. In this chapter, we prove that this remains true on any domain that is sufficiently close to the unit ball, for example, a round ellipsoid. On the contrary, if the ellipsoid is very flat, the extremal solution is smooth. We conclude this chapter by estimating the Hausdorff dimension of the singular set of stable solutions. Such a result is presented for the Lane-Emden equation, $-\Delta u = u^p$, $p > 1$, taking advantage of a monotonicity formula for an appropriate rescaled energy functional.

Chapter 6

In Chapter 4, we derived a regularity theory for solutions to (P.1) in terms of their Morse index. To do this, we used a Liouville-type theorem for solutions to the same equation, posed in the entire Euclidean space. In this chapter, we review the available classification results for stable solutions to (P.2), first in the radial case, then in the context of (P.1), and finally in the general context of (P.2). In the last part of the chapter, we classify solutions to (P.1) that are stable outside a compact set and show in an example that such a result cannot hold for general nonlinearities.

Chapter 7

Chapter 7 is dedicated to a specific class of stable solutions, the monotone solutions, which arise naturally in the context of phase transition phenomena. We discuss the heuristics behind a celebrated conjecture of De Giorgi, its relation to the theory of minimal surfaces, and present the proof of the conjecture in dimensions $N = 2$ and $N = 3$.

Chapter 8

The final chapter contains few proofs. I have tried instead to list some of the directions of current research on stable solutions: classification results in half-spaces and other geometries, partial symmetry of solutions of low Morse index, advanced topics on bifurcation diagrams, extensions to quasi-linear, higher-order and nonlocal equations, and stable solutions on manifolds.

Appendices
If you are new to the subject, Appendices A, B, and C1 are intended to walk you through the fine versions of the maximum principle, the standard regularity theory for linear elliptic equations, and the fundamental functional inequalities used in this book. Two additional topics, the inverse-square potential and some background material on submanifolds of Euclidean space, are also included.

I wish to thank Haim Brezis for giving me the opportunity to write this book. I am indebted to Xavier Cabré, Juan Dávila, and Alberto Farina for the many discussions we have had on the subject. The material presented here shows in itself the great influence they have had on my research. Thanks also to Jean Dolbeault, Filippo Gazzola, and Enrico Valdinoci, for their sharp answers to my picky questions. I feel lucky to be a member of the Laboratoire Amiénois de Mathématiques Fondamentales et Appliquées (LAMFA), Unité Mixte de Recherche Centre National de la Recherche Scientifique (UMR CNRS) 6140. Thanks to Olivier Goubet for creating such a great work environment and for revising this manuscript.

Bergamote, pour le dire brièvement, merci de tes encouragements de chaque instant.

If you find mistakes or typos, please notify me at louis.dupaigne@math.cnrs.fr.

Chapter 1

Defining stability

1.1 Stability and the variations of energy

Intuitively, a system is in a stable state if it can recover from perturbations; a small change will not prevent the system from returning to equilibrium. Place a marble at the center of a smooth bowl and tap it slighty. After some rolling back and forth, the marble will return to its stable position. If instead you turned the bowl over and put the marble carefully on top at the center, then it would be in a rather unstable equilibrium; a slight breeze would suffice to make it fall.

Such phenomena can be understood by looking at the *variations* of the energy of the system.

1.1.1 Potential wells

Consider a smooth function $E : \mathbb{R} \to \mathbb{R}$ and think of it as the potential energy of a system, a quantity that varies with respect to a relevant physical parameter, a quantity that the system tends to minimize. A potential well is a neighborhood of a local minimum of E: potential energy captured in it is unable to convert to another type of energy. As such, a potential well is a *stable* state of this system.

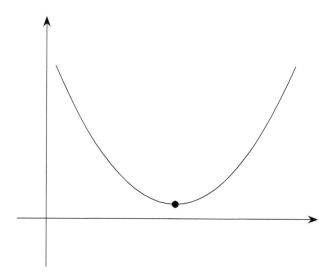

Figure 1.1: A potential well.

From the mathematical point of view, it is clear that $t_0 \in \mathbb{R}$ is a point of local minimum of E as soon as $E'(t_0) = 0$ and $E''(t_0) > 0$. Relaxing the inequality, we obtain the following definition:

Definition 1.1.1 *Let I denote an open interval of $\mathbb{R}$ and $E : I \to \mathbb{R}$ a function of class C^2. $t_0 \in I$ is a stable* [1] *critical point of E if*

$$E'(t_0) = 0 \text{ and } E''(t_0) \geq 0.$$

Exercise 1.1.1 Prove that a point t_0 of local minimum of E is a stable critical point. Is the converse true?

For the marble rolling inside a bowl, the potential energy of the system can be effectively described as a function of one parameter, the distance of the marble to the center of the bowl. The graph of E is then a parabola pointing up, as in Figure 1.1 (or pointing down, in the case where the bowl is turned over). In most situations, however, the state of a system cannot be described

---

[1] A point t_0 satisfying Definition 1.1.1 is usually called *semi*-stable in the literature, while it is said to be stable if the strict inequality $E''(t_0) > 0$ holds. Since the theory we are about to develop does not use the strict inequality, we prefer to make no distinction between semi-stable and stable solutions.

by a single parameter $t \in \mathbb{R}$. Instead, we shall work in a context where the energy is defined on a suitable functional space X. Consider for example a bounded region Ω of the Euclidean space $\mathbb{R}^N$, with smooth boundary $\partial\Omega$, and let $Y = C^2(\overline{\Omega})$. Among the functions in Y, we select those that vanish on the boundary $\partial\Omega$ and denote by $X = C_0^2(\overline{\Omega})$ the subspace thus formed. Now consider the energy functional $\mathscr{E}_\Omega : X \to \mathbb{R}$ defined for $u \in X$ by

$$\mathscr{E}_\Omega(u) = \frac{1}{2} \int_\Omega |\nabla u|^2 \, dx - \int_\Omega F(u) \, dx, \qquad (1.1)$$

where $F : \mathbb{R} \to \mathbb{R}$ is a given function of class C^2. Fixing $u \in X$, we study the variations of the energy along a given direction[2] $\varphi \in X \setminus \{0\}$, that is, we consider the function $E : \mathbb{R} \to \mathbb{R}$ defined for $t \in \mathbb{R}$ by

$$E(t) = \mathscr{E}_\Omega(u + t\varphi). \qquad (1.2)$$

When is 0 a stable critical point of E? To find out, compute the difference quotient of E at 0:

$$\frac{E(t) - E(0)}{t} = \int_\Omega \nabla u \nabla \varphi \, dx + \frac{t}{2} \int_\Omega |\nabla\varphi|^2 \, dx - \int_\Omega \frac{F(u + t\varphi) - F(u)}{t} \, dx.$$

Taking $|t| \leq 1$ and writing $a = \|u\|_{L^\infty(\Omega)} + \|\varphi\|_{L^\infty(\Omega)}$, the mean-value theorem implies that $\left| (F(u + t\varphi) - F(u))/t \right| \leq \|f\|_{L^\infty([-a,a])}$, where $f = F'$. By the dominated convergence theorem, we may pass to the limit as $t \to 0$: $E'(0) = 0$ if and only if

$$\int_\Omega \nabla u \nabla \varphi \, dx = \int_\Omega f(u)\varphi \, dx.$$

Integrating by parts, we see that

$$\int_\Omega (-\Delta u - f(u))\varphi \, dx = 0.$$

If u belongs to X and the above equality holds for arbitrary $\varphi \in X$, then we readily deduce (using, for example, [25, Lemma IV.2]) that u solves the following semilinear partial differential equation (PDE), also known as the Euler-Lagrange equation for the energy functional (1.1):

$$\begin{cases} -\Delta u = f(u) & \text{in } \Omega, \\ u = 0 & \text{on } \partial\Omega. \end{cases} \qquad (1.3)$$

---

[2] Given $\lambda > 0$, the variations of E_λ defined by $E_\lambda(t) := E(\lambda t)$ can be readily deduced from those of E. So, we may always restrict to the case where φ belongs to the unit sphere of X (we then say that φ is a direction).

This holds in particular if u is a critical point of $\mathscr{E}_\Omega$, that is if $\mathscr{E}_\Omega$ is differentiable at u and its Fréchet derivative satisfies $D\mathscr{E}_\Omega(u) = 0$ (see, for example, [51] for Fréchet derivatives and calculus in Banach spaces).

Exercise 1.1.2 Prove that $\mathscr{E}_\Omega$ defined by (1.1) is differentiable in X and prove that for every $u \in X$, the Fréchet derivative of $\mathscr{E}_\Omega$ at u is given by

$$D\mathscr{E}_\Omega(u).\varphi = \int_\Omega \nabla u \nabla \varphi \, dx - \int_\Omega f(u)\varphi \, dx, \qquad \text{for every } \varphi \in X.$$

We have just identified solutions $u \in X$ of (1.3) with critical points of $\mathscr{E}_\Omega$. We summarize this result in the following proposition.

Proposition 1.1.1 *Let Ω denote a smoothly bounded domain of $\mathbb{R}^N$, $N \geq 1$. Let $f : \mathbb{R} \to \mathbb{R}$ denote a function of class C^1 and F an antiderivative of f. Let $X = C_0^2(\overline{\Omega}) = \{u \in C^2(\overline{\Omega}) : u(x) = 0 \text{ for all } x \in \partial\Omega\}$ endowed with its natural norm $\|\cdot\|_{C^2(\overline{\Omega})}$. Let $\mathscr{E}_\Omega : X \to \mathbb{R}$ defined by (1.1). Then, $u \in C^2(\overline{\Omega})$ is a solution to (1.3) if and only if u is a critical point of $\mathscr{E}_\Omega$.*

Which of these solutions should be called stable? To find out, we compute $E''(0)$:

$$\frac{E'(t) - E'(0)}{t} = \frac{E'(t)}{t} = \frac{D\mathscr{E}_\Omega(u + t\varphi).\varphi}{t}$$

$$= \frac{1}{t} \int_\Omega \nabla u \nabla \varphi \, dx + \int_\Omega |\nabla \varphi|^2 \, dx - \int_\Omega \frac{1}{t} f(u + t\varphi)\varphi \, dx.$$

$$= \int_\Omega |\nabla \varphi|^2 \, dx - \int_\Omega \frac{f(u + t\varphi) - f(u)}{t} \varphi \, dx.$$

By a standard dominated convergence argument,

$$E''(0) = \int_\Omega |\nabla \varphi|^2 \, dx - \int_\Omega f'(u)\varphi^2 \, dx.$$

Recalling Definition 1.1.1, it is natural to define stability as follows.

Definition 1.1.2 *Let $f \in C^1(\mathbb{R})$ and let Ω denote an open set of $\mathbb{R}^N$, $N \geq 1$. A solution $u \in C^2(\Omega)$ of*

$$-\Delta u = f(u) \qquad \text{in } \Omega \qquad (1.4)$$

is stable if

$$Q_u(\varphi) := \int_\Omega |\nabla \varphi|^2 \, dx - \int_\Omega f'(u)\varphi^2 \, dx \geq 0, \quad \forall \, \varphi \in C_c^1(\Omega). \qquad (1.5)$$

Remark 1.1.1

- *The quadratic form Q_u is called the second variation of the energy $\mathcal{E}_\Omega$.*

- *u is stable in Ω if and only if u is stable in every subdomain $\omega \subset\subset \Omega$. In particular, the notion of stability is relevant also for unbounded and/or nonsmooth domains Ω.*

- *However, if u is stable in two given domains ω_1, ω_2, u need not be stable in $\omega_1 \cup \omega_2$. See Proposition 1.5.1.*

- *Using a standard approximation argument, if Ω is bounded or merely if $f'(u)^-$ is bounded in Ω, one can take $\varphi \in H_0^1(\Omega)$ in the above definition[3].*

1.1.2   Examples of stable solutions

In the following list of examples, we assume again that Ω is smoothly bounded.

Example 1.1.1 Local minimizers of the energy are stable.

Definition 1.1.3 *Let X denote a Banach space of functions, with norm $\|\cdot\|$. Assume X contains $C_c^1(\Omega)$ and let $\mathcal{E}_\Omega : X \to \mathbb{R}$, a functional. $u \in X$ is a local minimizer of $\mathcal{E}_\Omega$ if there exists $t_0 > 0$ such that*

$$\mathcal{E}_\Omega(u) \le \mathcal{E}_\Omega(u + \varphi) \qquad \text{for all } \varphi \in C_c^1(\Omega) \text{ s.t. } \|\varphi\| < t_0.$$

In particular, assume u is a local minimizer of $\mathcal{E}_\Omega$ defined by (1.1). Then 0 is a point of local minimum of E defined by (1.2). We deduce that $E'(0) = 0$ and $E''(0) \ge 0$, and so u is a stable solution to (1.3).

Example 1.1.2 ([35]) Critical points that minimize the energy from one side are stable.

Definition 1.1.4 *Let X denote a Banach space of functions containing $C_c^1(\Omega)$ and $\mathcal{E}_\Omega : X \to \mathbb{R}$, a functional. $u \in X$ is a one-sided local minimizer of $\mathcal{E}_\Omega$ if there exists $t_0 > 0$ and $\varepsilon \in \{-1, +1\}$ such that*

$$\mathcal{E}_\Omega(u) \le \mathcal{E}_\Omega(u + \varepsilon \varphi) \qquad \text{for all } \varphi \in C_c^1(\Omega) \text{ s.t. } \varphi \ge 0 \text{ and } \|\varphi\| < t_0.$$

---

[3]See, for example, [25] for an introduction to Sobolev spaces such as $H_0^1(\Omega)$.

Assume that u is both a critical point and a one-sided minimizer of $\mathscr{E}_\Omega$ defined by (1.1) and assume for example that $\varepsilon = +1$. Then, given $\varphi \geq 0$, 0 is a point of local minimum of $E|_{[0,t_0)}$, where E is defined by (1.2). We deduce that $E'(0) = 0$ and $E''(0) \geq 0$, and so (1.5) holds for $\varphi \geq 0$. Now take an arbitrary test function $\varphi \in C_c^1(\Omega)$ and split it into its positive and negative parts: $\varphi = \varphi^+ - \varphi^-$. Since Ω is bounded, by a standard approximation argument, (1.5) remains valid for the test functions φ^+, φ^-. Then, $Q_u(\varphi) = Q_u(\varphi^+) + Q_u(\varphi^-) \geq 0$ and so u is stable.

Example 1.1.3 ([193]) Minimal solutions are stable.

A function $\underline{u} \in C^2(\overline{\Omega})$ satisfying

$$\begin{cases} -\Delta \underline{u} \leq f(\underline{u}) & \text{in } \Omega, \\ \underline{u} \leq 0 & \text{on } \partial\Omega, \end{cases}$$

is called a subsolution to (1.3). Similarly, a function $\overline{u} \in C^2(\overline{\Omega})$ satisfying the reverse set of inequalities is called a supersolution to (1.3). Assume such a sub- and a supersolution exist and assume in addition that they are distinct and ordered: $\underline{u} \leq \overline{u}$ in Ω. By the strong maximum principle (Proposition A.8.1), it turns out that the inequality is always strict:

$$\underline{u} < \overline{u} \quad \text{in } \Omega. \tag{1.6}$$

Lemma 1.1.1 *Let Ω denote a smoothly bounded domain of $\mathbb{R}^N$, $N \geq 1$ and let $f \in C^1(\mathbb{R})$. Assume that there exists $\underline{u}, \overline{u} \in C^2(\overline{\Omega})$ a sub- and a supersolution to (1.3). In addition, assume (1.6). Then, there exists a unique solution u of (1.3) having the following properties: Given any supersolution $\overline{u}_2 \in C^2(\overline{\Omega})$ of (1.3) such that $\overline{u}_2 \geq \underline{u}$,*

1. *$\underline{u} \leq u \leq \overline{u}_2$ and*

2. *u minimizes $\mathscr{E}_\Omega$ among all functions $v \in C^2(\overline{\Omega})$ such that $\underline{u} \leq v \leq \overline{u}_2$.*

u is called the minimal solution to (1.3) relative to $\underline{u}$.

Assume temporarily that Lemma 1.1.1 is established. Let u denote the minimal solution relative to $\underline{u}$. The strong maximum principle (Proposition A.8.1) implies that either $u \in \{\underline{u}, \overline{u}\}$ or $\underline{u} < u < \overline{u}$. Say $u = \underline{u}$ (all other cases can be treated similarly). Then, by (1.6), given $\varphi \in C_c^1(\Omega)$, $\varphi \geq 0$, there exists $t_0 > 0$ such that $\underline{u} \leq u + t\varphi \leq \overline{u}$ for $0 \leq t < t_0$. By item 2 of the lemma, we deduce that 0 is a point of minimum of $E|_{[0,t_0)}$. Since u solves (1.3), $E'(0) = 0$. So, we must have $E''(0) \geq 0$ and (1.5) holds for $\varphi \geq 0$. Decomposing an arbitrary test

6

function φ into its positive and negative part as in Example 1.1.2, we deduce that u is stable. It remains to prove Lemma 1.1.1.

Proof. The proof we present here is not standard and a bit technical (we follow this strategy to prove point 2 of the lemma). If you are unfamiliar with the method of sub- and supersolutions, we recommend you first read [193] or simply do Exercise 1.2.1. Define the truncated nonlinearity $g(x,u)$ for $x \in \Omega$ and $u \in \mathbb{R}$ by

$$g(x,u) = \begin{cases} f(\underline{u}(x)) & \text{if } u < \underline{u}(x), \\ f(u) & \text{if } \underline{u}(x) \le u \le \overline{u}(x), \\ f(\overline{u}(x)) & \text{if } u > \overline{u}(x). \end{cases} \tag{1.7}$$

Let $G(x,u) = \int_0^u g(x,t)\,dt$ and define $\mathscr{E}_b : H_0^1(\Omega) \to \mathbb{R}$ by

$$\mathscr{E}_b(u) = \frac{1}{2}\int_\Omega |\nabla u|^2\,dx - \int_\Omega G(x,u(x))\,dx. \tag{1.8}$$

Then, $\mathscr{E}_b$ is bounded below (see Exercise 1.1.3). So, there exists a minimizing sequence (u_n) of $\mathscr{E}_b$. Letting $a = \min_{\overline{\Omega}} \underline{u}$ and $b = \max_{\overline{\Omega}} \overline{u}$, there exists constants $C_1, C_2, C_3 > 0$ such that

$$\frac{1}{2}\int_\Omega |\nabla u_n|^2\,dx \le \mathscr{E}_b(u_n) + \left| \int_\Omega G(x,u_n(x))\,dx \right|$$

$$\le C_1 + \|f\|_{L^\infty(a,b)} \int_\Omega |u_n|\,dx$$

$$\le C_1 + C_2\|u_n\|_{L^2(\Omega)}$$

$$\le C_1 + C_3\|\nabla u_n\|_{L^2(\Omega)},$$

where we used the mean-value theorem in the second inequality, Hölder's inequality in the third, and Poincaré's inequality in the last. Solving the previous (quadratic) inequality for $\|\nabla u_n\|_{L^2(\Omega)}$, we deduce that (u_n) is bounded in $H_0^1(\Omega)$. It follows that a subsequence (u_{k_n}) converges to a function $u \in H_0^1(\Omega)$ weakly in $H_0^1(\Omega)$, strongly in $L^2(\Omega)$ and almost everywhere (a.e.) in Ω. Since $\left| G(x,u_{k_n}(x)) \right| \le C \left| u_{k_n}(x) \right|$, we deduce that $\int_\Omega G(x,u_{k_n}(x))\,dx \to \int_\Omega G(x,u(x))\,dx$ as $n \to \infty$. Since $u \mapsto \int_\Omega |\nabla u|^2\,dx$ is weakly lower-semicontinuous in $H^1(\Omega)$, it follows that

$$\mathscr{E}_b(u) \le \liminf_{n \to \infty} \mathscr{E}_b(u_{k_n}) = \inf_{u \in H_0^1(\Omega)} \mathscr{E}_b(u).$$

So, u minimizes $\mathcal{E}_b$. In particular, u is a weak solution to

$$\begin{cases} -\Delta u = g(x,u) & \text{in } \Omega, \\ \quad u = 0 & \text{on } \partial\Omega, \end{cases} \tag{1.9}$$

that is, $u \in H_0^1(\Omega)$ and for all $\varphi \in C_c^1(\Omega)$,

$$\int_\Omega \nabla u \nabla \varphi \, dx = \int_\Omega g(x,u)\varphi \, dx.$$

In addition, $u \in C^{1,\alpha}(\overline{\Omega})$ (see Exercise 1.1.4).

Observe that *any* solution to (1.9) satisfies $\underline{u} \leq u \leq \overline{u}$ and solves (1.3). Indeed, assume by contradiction that $\omega = \{x \in \Omega \ : \ u(x) < \underline{u}(x)\}$ is non-empty. Then, $v = u - \underline{u}$ is harmonic in ω and $v \geq 0$ on $\partial\omega$. By the maximum principle (Proposition A.2.2), $v \geq 0$ in ω, which is absurd by definition of ω. So $u \geq \underline{u}$. Similarly, $u \leq \overline{u}$. By (1.7) and (1.9), u is a solution to (1.3).

Let $\overline{u}_2$ denote another supersolution to (1.3) such that $\overline{u}_2 \geq \underline{u}$. Define the truncation g^2 of g by

$$g^2(x,u) = \begin{cases} g(x,\underline{u}(x)) & \text{if } u < \underline{u}(x), \\ g(x,u) & \text{if } \underline{u}(x) \leq u \leq \overline{u}_2(x), \\ g(x,\overline{u}_2(x)) & \text{if } u > \overline{u}_2(x). \end{cases}$$

Working as above, we can construct a solution u^2 of (1.9), satisfying $\underline{u} \leq u^2 \leq \overline{u}_2$. As observed earlier, u^2 then solves (1.3) and $u^2 \leq \overline{u}$.

Next, take a finite family of supersolutions $i = \{\overline{u}, \overline{u}_2, \ldots, \overline{u}_n\}$ such that $\overline{u}_k \geq \underline{u}$ for $k = 2, \ldots, n$. Also let I denote the set of all such families, ordered by inclusion. Repeating the truncation process inductively, we obtain a solution u_i of (1.3) such that $\underline{u} \leq u_i \leq \overline{u}, \overline{u}_2, \ldots, \overline{u}_n$. Note that given $i \in I$, there may be more than one such solution u_i. Nevertheless, using the axiom of choice on the set of all such solutions, we can construct a well-defined generalized sequence $(u_i)_{i \in I}$, contained in the set K of all solutions u satisfying $\underline{u} \leq u \leq \overline{u}$. By standard elliptic estimates, K is a compact subset of $C^2(\overline{\Omega})$ so there exists a generalized subsequence $(u_{\phi(j)})_{j \in J}$ converging to a solution u of (1.3). Now choose an arbitrary supersolution $\overline{v} \geq \underline{u}$ and let $i_1 := \{\overline{v}, \overline{u}\} \in I$. Given $\epsilon > 0$, let $j_0 \in J$ such that $j > j_0 \implies \|u_{\phi(j)} - u\|_\infty < \epsilon$. Also choose $j_1 \in J$ such that $j > j_1 \implies \phi(j) > i_1$. Finally, pick $j_2 > j_0, j_1$. Then, for $j > j_2$,

$$u \leq \|u_{\phi(j)} - u\|_\infty + u_{\phi(j)} \leq \epsilon + \overline{v}.$$

Letting $\epsilon \to 0$, we conclude that $u \leq \overline{v}$ for any supersolution $\overline{v} \geq \underline{u}$, which proves item 1 of the lemma. Since each u_i was obtained as a global minimizer of a truncated energy, $u_{\phi(j)}$, $j > j_1$ minimizes $\mathscr{E}_\Omega$ among all functions $v \in C^2(\overline{\Omega})$ such that $\underline{u} \leq v \leq \overline{v}$. Passing to the limit, so does u. $\qquad\square$

Exercise 1.1.3

- Prove that g defined by (1.7) is a continuous map and that the family of functions $(g_x)_{x \in \Omega}$ defined by $g_x(u) := g(x, u)$ for $u \in \mathbb{R}$ is uniformly Lipschitz continuous on any compact interval $[a, b]$.

- Prove that $\mathscr{E}_b$ defined by (1.8) is bounded below.

Exercise 1.1.4

- By looking at the variations of $E_b(t) = \mathscr{E}_b(u + t\varphi)$, prove that a minimizer u of $\mathscr{E}_b$ is a weak solution to (1.9).

- Prove that $u \in C^{1,\alpha}(\overline{\Omega})$.

Exercise 1.1.5 Let u denote the minimal solution to (1.3) relative to $\underline{u}$. Let ω denote a subdomain of Ω. We say that $\overline{u}_2 \in C^2(\overline{\omega})$ is a local supersolution to (1.3) if $-\Delta \overline{u}_2 \geq f(\overline{u}_2)$ in ω, $\overline{u}_2 \geq u$ on $\partial \omega$ and $\overline{u}_2 \geq \underline{u}$ in ω. Prove that $\overline{u}_2 \geq u$ in ω for any local supersolution $\overline{u}_2$.

1.2 Linearized stability

1.2.1 Principal eigenvalue of the linearized operator

Take a second look at the definition of stability, Inequality (1.5). As observed in Remark 1.1.1, if Ω is bounded, $Q_u \geq 0$ on all of $H_0^1(\Omega)$, which we can rephrase as

$$\inf_{\varphi \in H_0^1(\Omega), \|\varphi\|_{L^2(\Omega)}=1} \left(\int_\Omega |\nabla \varphi|^2 \, dx - \int_\Omega f'(u) \varphi^2 \, dx \right) \geq 0. \qquad (1.10)$$

The left-hand side of the above inequality coincides with the principal eigenvalue $\lambda_1(-\Delta - f'(u); \Omega)$ of the elliptic operator $-\Delta - f'(u)$ with Dirichlet boundary conditions (see Theorem A.6.1). By Theorem A.6.1, we also have $\lambda_1(-\Delta - f'(u); \Omega) \geq 0$, if and only if there exists $v \in C^2(\overline{\Omega})$ such that $v > 0$ in Ω and

$$-\Delta v - f'(u)v \geq 0 \qquad \text{in } \Omega. \qquad (1.11)$$

Recalling that $\omega \mapsto \lambda_1(-\Delta - f'(u); \omega)$ is nonincreasing, we obtain the following equivalent definition of stability (which remains valid even if Ω is unbounded):

Proposition 1.2.1 *Let $f \in C^{1,\alpha}(\mathbb{R})$ and let Ω denote a domain of $\mathbb{R}^N$, $N \geq 1$. A solution $u \in C^2(\Omega)$ of (1.4) is stable if and only if either of the following conditions hold*

- *$\lambda_1(-\Delta - f'(u); \omega) \geq 0$ for every bounded subdomain $\omega \subset\subset \Omega$, or*

- *there exists $v \in C^2(\Omega)$ such that $v > 0$ in Ω and (1.11) holds.*

Remark 1.2.1 *When Ω is bounded, the first condition simply reduces to $\lambda_1(-\Delta - f'(u); \Omega) \geq 0$.*

Proof. Assume first that (1.5) holds and fix $\omega \subset\subset \Omega$. Then, $Q_u(\varphi) \geq 0$ for all $\varphi \in H_0^1(\omega)$, that is, $\lambda_1(-\Delta - f'(u); \omega) \geq 0$.

Assume next that $\lambda_1(-\Delta - f'(u); \omega) \geq 0$ for every subdomain ω. Let (ω_n) denote an increasing sequence of bounded subdomains covering Ω and $x_0 \in \cap \omega_n$. Denote $\varphi_{1,n} > 0$ the associated eigenfunction normalized by $\varphi_{1,n}(x_0) = 1$. Using Harnack's inequality (see Proposition A.3.2), $(\varphi_{1,n})$ is uniformly bounded on compact subsets of Ω. By standard elliptic regularity, a subsequence (φ_{1,k_n}) converges uniformly on compact subsets to some function $v \in C^2(\Omega)$ such that $v \geq 0$ and $-\Delta v - f'(u)v = \lambda v \geq 0$ in Ω, where $\lambda = \lim_{n\to\infty} \lambda_1(-\Delta - f'(u); \omega_n)$. By the strong maximum principle, either $v \equiv 0$, which is excluded by the normalization $\varphi_{1,n}(x_0) = 1$, or $v > 0$ as requested in Definition 1.2.1.

Assume at last that there exists $v \in C^2(\Omega)$ such that $v > 0$ in Ω and (1.11) holds. Take $\varphi \in C_c^1(\Omega)$ and multiply (1.11) by φ^2/v:

$$
\begin{aligned}
0 &\leq \int_\Omega (-\Delta v - f'(u)v) \frac{\varphi^2}{v}\, dx \\
&\leq \int_\Omega \left(\nabla v \cdot \nabla \frac{\varphi^2}{v} - f'(u)\varphi^2 \right) dx \\
&\leq -\int_\Omega \frac{\varphi^2}{v^2} |\nabla v|^2\, dx + 2\int_\Omega \frac{\varphi}{v} \nabla v \cdot \nabla \varphi\, dx - \int_\Omega f'(u)\varphi^2\, dx \\
&\leq \int_\Omega |\nabla \varphi|^2\, dx - \int_\Omega f'(u)\varphi^2\, dx,
\end{aligned}
$$

where we used Young's inequality in the last inequality. Equation (1.5) follows. $\square$

Exercise 1.2.1 Let $f \in C^1(\mathbb{R})$ and let Ω denote a smoothly bounded domain of $\mathbb{R}^N$, $N \geq 1$. Assume that there exists $\underline{u}, \overline{u} \in C^2(\overline{\Omega})$, a sub- and a supersolution to (1.3), such that $\underline{u} < \bar{u}$. Let $u_0 = \underline{u}$ and for $n \in \mathbb{N}^*$, let $u_n \in C^2(\overline{\Omega})$ denote the unique solution to

$$\begin{cases} -\Delta u_n + \Lambda u_n = f(u_{n-1}) + \Lambda u_{n-1} & \text{in } \Omega, \\ \qquad\qquad u_n = 0 & \text{on } \partial\Omega, \end{cases} \qquad (1.12)$$

where $\Lambda = \|f'\|_{L^\infty(a,b)}$, $a = \min_{\overline{\Omega}} \underline{u}$, $b = \max_{\overline{\Omega}} \bar{u}$.

- Prove that u_n is well defined and that $\underline{u} \le u_n \le u_{n+1} \le \bar{u}$ for all $n \in \mathbb{N}$.

- Prove that $u : \Omega \to \mathbb{R}$ defined by $u(x) := \lim_{n \to +\infty} u_n(x)$ for all $x \in \overline{\Omega}$ is a classical solution to (1.3).

- Prove that u is the minimal solution to (1.3) relative to $\underline{u}$.

- Let $\bar{u}_\varepsilon = u - \varepsilon\varphi_1$, where $\varepsilon > 0$ and $\varphi_1 > 0$ is an eigenfunction associated to $\lambda_1 = \lambda_1(-\Delta - f'(u); \Omega)$. Assume by contradiction that $\lambda_1 < 0$. Prove that for small $\varepsilon > 0$, $\bar{u}_\varepsilon$ is a supersolution to (1.3) such that $\bar{u}_\varepsilon > \underline{u}$, and obtain a contradiction. Conclude that u is stable.

1.2.2 New examples of stable solutions

Using Definition 1.2.1, we can produce new examples of stable solutions.

Example 1.2.1 Monotone solutions are stable.

Definition 1.2.1 *A solution $u \in C^2(\Omega)$ of (1.4) is monotone if up to a rotation of space*

$$\frac{\partial u}{\partial x_N} > 0 \qquad in \ \Omega.$$

Differentiate (1.4) with respect to x_N: if u is monotone then $v = \partial u/\partial x_N$ is positive and solves

$$-\Delta v - f'(u)v = 0 \qquad in \ \Omega.$$

So, u is stable, by Proposition 1.2.1.

Example 1.2.2 ([97]) Positive solutions on coercive epigraphs are stable.

Definition 1.2.2 *A domain Ω is a coercive epigraph if, up to a rotation of space, there exists $g \in C(\mathbb{R}^{N-1}, \mathbb{R})$ such that $\Omega = \{x = (x', x_N) \in \mathbb{R}^{N-1} \times \mathbb{R} : x_N > g(x')\}$ and $\lim_{|x'| \to +\infty} g(x') = +\infty$.*

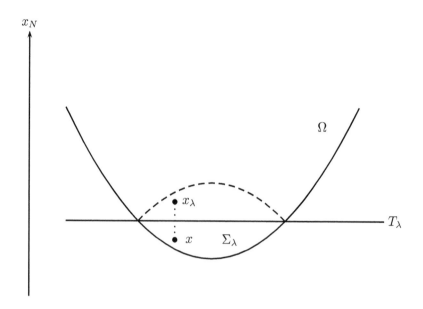

Figure 1.2: The moving-plane device.

Lemma 1.2.1 ([92]) *Let Ω denote a coercive epigraph. If $u > 0$ solves (1.3), then u is monotone, hence stable.*

Proof. The argument relies on the celebrated moving-plane device: given $\lambda > 0$, consider a hyperplane $T_\lambda = \{(x', x_N) \in \mathbb{R}^{N-1} \times \mathbb{R} : x_N = \lambda\}$ and $\Sigma_\lambda = \{(x', x_N) \in \Omega : x_N < \lambda\}$, the region below T_λ. Reflect Σ_λ through the hyperplane T_λ: for $x \in \Sigma_\lambda$, the reflection of x is given by $\hat{x}_\lambda = (x', 2\lambda - x_N)$. Observe that since Ω is an epigraph, $x \in \Sigma_\lambda \Rightarrow \hat{x}_\lambda \in \Omega$ and we may define $\hat{u}_\lambda(x) = u(\hat{x}_\lambda)$ for $x \in \Sigma_\lambda$, see Figure 1.2. Also let $w_\lambda = \hat{u}_\lambda - u$. We shall prove that given any $\lambda > 0$,

$$w_\lambda > 0 \quad \text{in } \Sigma_\lambda. \tag{1.13}$$

We establish (1.13) first for small $\lambda > 0$. Let

$$V_\lambda(x) = \frac{f(\hat{u}_\lambda) - f(u)}{\hat{u}_\lambda - u}(x), \qquad \text{whenever } u(x) \neq \hat{u}_\lambda(x),$$

and $V_\lambda(x) = f'(u(x))$ otherwise. Then,

$$-\Delta w_\lambda - V_\lambda(x)w_\lambda = 0 \quad \text{in } \Sigma_\lambda \tag{1.14}$$

12

and $w_\lambda \geq 0$ on $\partial \Sigma_\lambda$. The maximum principle for small domains (Proposition A.7.1) implies that $w_\lambda \geq 0$ in Σ_λ, provided $\lambda > 0$ is small. Since $u > 0$ in Ω and $u = 0$ on $\partial \Omega$, $w_\lambda > 0$ on $\partial \Omega \cap \overline{\Sigma_\lambda}$. By the strong maximum principle applied to (1.14), $w_\lambda > 0$ in Σ_λ, that is, (1.13) holds for all small values of $\lambda > 0$.

Now let λ_0 denote the supremum of all $\mu > 0$ such that (1.13) holds for all $0 < \lambda \leq \mu$. Assume by contradiction that $\lambda_0 < \infty$. By continuity, we have $w_{\lambda_0} \geq 0$ in Σ_{λ_0}. By the strong maximum principle, the inequality is strict: $w_{\lambda_0} > 0$ in Σ_{λ_0}. Fix $\delta > 0$ small and a compact set $K \subset\subset \Sigma_{\lambda_0}$ such that $|\Sigma_{\lambda_0} \setminus K| \leq \delta/2$. Since K is compact, $w_{\lambda_0} \geq \eta > 0$ in K for some constant $\eta > 0$. Choosing $\epsilon > 0$ sufficiently small, it follows that $w_{\lambda_0 + \varepsilon} > 0$ in K, while $|\Sigma_{\lambda_0 + \varepsilon} \setminus K| \leq \delta$. So,

$$-\Delta w_{\lambda_0 + \varepsilon} + V_{\lambda_0 + \varepsilon} w_{\lambda_0 + \varepsilon} = 0 \qquad \text{in } \Sigma_{\lambda_0 + \varepsilon} \setminus K,$$
$$w_{\lambda_0 + \varepsilon} \geq 0 \qquad \text{on } \partial \left(\Sigma_{\lambda_0 + \varepsilon} \setminus K \right).$$

By the maximum principle for small domains, we conclude that $w_{\lambda_0 + \varepsilon} \geq 0$ in $\Sigma_{\lambda_0 + \varepsilon} \setminus K$. Since we already had $w_{\lambda_0 + \varepsilon} > 0$ in K, the strong maximum principle then implies that (1.13) holds for $\lambda = \lambda_0 + \varepsilon$, contradicting the definition of λ_0. So, (1.13) holds for all $\lambda > 0$.

We want to prove that u is monotone in the x_N direction, that is, that $\partial u / \partial x_N (x', \lambda) > 0$ for all $(x', \lambda) \in T_\lambda \cap \Omega$ and all $\lambda > 0$. To see this, fix $\lambda > 0$ and observe that w_λ is a positive solution to (1.14) in Σ_λ and that $w_\lambda = 0$ on $T_\lambda \cap \Omega$. By the boundary point lemma (Lemma A.5.1), we deduce that

$$2 \frac{\partial u}{\partial x_N}(x', \lambda) = -\frac{\partial w_\lambda}{\partial x_N}(x', \lambda) > 0, \qquad \text{for all } (x', \lambda) \in T_\lambda \cap \Omega.$$

$\square$

Exercise 1.2.2 ([123]) Let $f \in C^1(\mathbb{R}, \mathbb{R})$ and let B denote the unit ball in $\mathbb{R}^N$, $N \geq 2$. Let $u \in C^2(\overline{B})$ denote a positive solution to (1.18) below. We want to prove that u is radially symmetric.

- Check that it suffices to prove that u is an even function of the variable x_N.

- Check that it suffices to prove that $u(x', x_N) \leq u(x', -x_N)$ for every $x = (x', x_N) \in B$ such that $x_N > 0$.

- For $\lambda \in (0, 1)$, define $T_\lambda = \{(x', x_N) \in \mathbb{R}^{N-1} \times \mathbb{R} : x_N = \lambda\}$, $\Sigma_\lambda = \{(x', x_N) \in B : \lambda < x_N < 1\}$, $\hat{x}_\lambda = (x', 2\lambda - x_N)$, $\hat{u}_\lambda(x) = u(\hat{x}_\lambda)$, and

$w_\lambda(x) = \hat{u}_\lambda(x) - u(x)$, for $x = (x', x_N) \in \Sigma_\lambda$. Prove that w_λ is well defined and

$$0 < w_\lambda \quad \text{in } x \in \Sigma_\lambda, \tag{1.15}$$

for all $\lambda \in (0, 1)$. Conclude that u is radially symmetric.

- Prove that $\frac{\partial u}{\partial r} < 0$ in $B \setminus \{0\}$, where $r = |x|$.

Remark 1.2.2 *Lemma 1.2.1 remains valid if $\Omega = \mathbb{R}^N_+$ and either $N = 2$ or $N \geq 3$, $f(0) \geq 0$, and f globally Lipschitz, or $N \geq 3$, $f(0) \geq 0$, and u bounded. See [16] and [70]. Unfortunately, positivity does not imply stability for a general domain Ω, as the following examples show.*

Example 1.2.3 ([97]) Let $I = (0, 1)$, $p > 1$ and $v \in C^2([0, 1])$, $v > 0$ be a solution to

$$\begin{cases} -v'' = v^p & \text{in } (0, 1), \\ v(0) = v(1) = 0. \end{cases} \tag{1.16}$$

Multiplying the above equation by v and integrating by parts and recalling Remark 1.1.1, we easily see that $Q_v(v) < 0$, that is, v is unstable. So positive solutions defined on a bounded domain need not be stable.

Since Lemma 1.2.1 relied on the existence of a privileged direction in which the solution u was monotone, one may hope that positivity implies stability for a larger class of unbounded domains, say for cylinders. Unfortunately, this is still not the case, for example, take $N \geq 2$ and $\Omega = I \times \mathbb{R}^{N-1} \subset \mathbb{R}^N$. If v is a solution to (1.16), then $u(x_1, \ldots, x_N) = v(x_1)$ is a positive solution to (1.3) with $f(u) = u^p$. However, u is not stable. Take $R > 0$, the $N - 1$ dimensional ball $\omega_R = \{x' = (x_1, \ldots, x_{N-1}) \in \mathbb{R}^{N-1} : \sum_{j=1}^{N-1} |x_j|^2 < R\}$ and φ_R an eigenfunction associated to the principal eigenvalue $\lambda_1(-\Delta; \omega_R)$ of the Laplace operator on ω_R. Normalize φ_R by $\|\varphi_R\|_{L^2(\omega_R)} = 1$ and set $\psi_R = u\varphi_R$. Then,

$$Q_u(\psi_R) = \int_\Omega \nabla u \nabla(u\varphi_R^2)\, dx + \int_\Omega u^2 |\nabla \varphi_R|^2\, dx - p\int_\Omega u^{p+1}\varphi_R^2\, dx$$

$$= \int_\Omega -\Delta u\, u\varphi_R^2\, dx + \int_\Omega u^2 |\nabla \varphi_R|^2\, dx - p\int_\Omega u^{p+1}\varphi_R^2\, dx$$

$$= (1-p)\int_\Omega u^{p+1}\varphi_R^2\, dx + \int_\Omega u^2 |\nabla \varphi_R|^2\, dx$$

$$= (1-p)\int_0^1 v^{p+1}\, dt + \lambda_1(-\Delta; \omega_R)\int_0^1 v^2\, dt.$$

By a direct scaling argument, $\lambda_1(-\Delta; \omega_R) = \lambda_1(-\Delta; \omega_1)\frac{1}{R^2}$. So, choosing $R > 0$ large, we deduce that $Q_u(\psi_R) < 0$. By Remark 1.1.1, u is not stable. □

Exercise 1.2.3 Let $\mathscr{E}(v) = \frac{1}{2}\int_0^1 \left(\frac{dv}{dx}\right)^2 dx$ and let F denote the set of all functions $v \in H_0^1(0,1)$ such that $\int_0^1 |v|^{p+1}\, dx = 1$. Prove that $\min_F \mathscr{E}$ is achieved. What PDE does the minimizer solve? Deduce that there exists a positive solution to (1.16).

1.3 Elementary properties of stable solutions

1.3.1 Uniqueness

If f is nonincreasing and Ω bounded, the maximum principle implies that (1.3) has a unique solution.

Exercise 1.3.1 Prove it.

Otherwise, the equation can have many solutions. But perhaps uniqueness holds in the class of stable solutions. This is indeed the case when the nonlinearity is convex as the following proposition shows.

Proposition 1.3.1 ([32]) *Let $N \geq 1$ and $\Omega \subset \mathbb{R}^N$ denote a smoothly bounded domain. Let $\lambda_1 = \lambda_1(-\Delta; \Omega) > 0$ denote the principal eigenvalue of the Dirichlet Laplacian on Ω. Assume that $f \in C^1(\mathbb{R})$ is convex. Let $u_1, u_2 \in C^2(\overline{\Omega})$ denote two stable solutions of (1.3). Then, either $u_1 = u_2$ or $f(u) = \lambda_1 u$ on the ranges of u_1 and u_2.*

Proof. Let u_1, u_2 denote two stable solutions of (1.3). Then, $w = u_2 - u_1$ solves

$$-\Delta w = f(u_2) - f(u_1) \qquad \text{in } \Omega.$$

Multiply the above equality by w^+, the positive part of w, and integrate by parts:

$$\int_\Omega \left|\nabla w^+\right|^2 dx = \int_\Omega (f(u_2) - f(u_1))w_+ \, dx.$$

Since u_2 is stable, we also have

$$\int_\Omega \left|\nabla w^+\right|^2 dx \geq \int_\Omega f'(u_2)(w^+)^2 \, dx,$$

15

hence

$$\int_\Omega \left(f(u_2) - f(u_1) - f'(u_2)w^+ \right) w^+ \, dx \geq 0.$$

By convexity, the integrand in the above inequality is nonpositive. If f is strictly convex, we readily deduce that $w_+ = 0$, that is, $u_2 \leq u_1$. The reverse inequality is obtained by exchanging the roles of u_1 and u_2. Otherwise, f is linear on the union of all intervals with end points $u_1(x), u_2(x)$, $x \in \Omega$, hence on the whole ranges of u_1 and u_2. So, u_1, u_2 are stable solutions of a linear problem, which is possible only if they belong to the eigenspace associated to the principal eigenvalue λ_1. $\qquad\square$

Exercise 1.3.2 Prove Proposition 1.3.1 for concave f.

Assume that there exists a minimal solution to (1.3), denoted u. It follows from the above uniqueness result that if f is strictly convex, then u is the unique stable solution to (1.3). By minimality, u lies below all other solutions of the equation. Can all solutions of the equation be ordered in such a way? When f is strictly convex, the answer is always negative. Indeed, if u_1, u_2 denote two solutions both different from u, then necessarily they must cross as the following proposition shows.

Proposition 1.3.2 ([111, 112, 139]) *Let Ω denote a smoothly bounded domain of $\mathbb{R}^N$. Assume that $f \in C^1(\mathbb{R})$ is strictly convex. Let $u_1, u_2 \in C^2(\overline{\Omega})$ denote two distinct solutions of (1.3) such that*

$$u_1 \leq u_2 \qquad \text{in } \Omega.$$

Then, u_1 is the unique stable solution to (1.3).

Proof. Let $w = u_2 - u_1$. Then, $w \geq 0$ in Ω and using convexity,

$$-\Delta w = f(u_2) - f(u_1) \geq f'(u_1)w \qquad \text{in } \Omega.$$

By the strong maximum principle (Corollary A.5.1), either $w \equiv 0$, which we have excluded or $w > 0$ in Ω, hence u_1 is stable (by Proposition 1.2.1). By Proposition 1.3.1, we deduce that u_1 is the unique stable solution to (1.3). $\qquad\square$

1.3.2 Nonuniqueness

The convexity assumption cannot be dropped in Proposition 1.3.1. A counter-example is given by the Allen-Cahn nonlinearity $f(u) = u - u^3$:

Proposition 1.3.3 *Let Ω denote a smoothly bounded domain of $\mathbb{R}^N$ and $\lambda_1 = \lambda_1(-\Delta; \Omega) > 0$ denote the principal eigenvalue of the Dirichlet Laplacian on Ω. Let $f(u) = u - u^3$ and $\lambda \geq 0$. Consider the equation*

$$\begin{cases} -\Delta u = \lambda f(u) & \text{in } \Omega, \\ u = 0 & \text{on } \partial\Omega. \end{cases} \tag{1.17}$$

- *If $\lambda \leq \lambda_1$, then $u = 0$ is the unique solution to (1.17) and it is stable.*

- *If $\lambda > \lambda_1$, then there exists at least two stable nontrivial solutions of (1.17).*

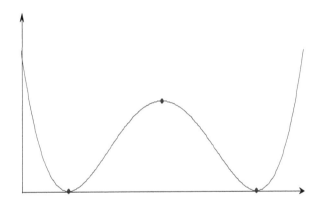

Figure 1.3: $W(u) = \frac{1}{4}(u^2 - 1)^2$ is a double-well potential, indicating that at least two stable solutions may exist.

Proof. We begin with the case $\lambda \leq \lambda_1$. Let $X = H_0^1(\Omega) \cap L^\infty(\Omega)$. Then, any (classical) solution to (1.17) is a critical point of the energy $\mathscr{E} : X \to \mathbb{R}$ defined for $u \in X$ by

$$\mathscr{E}_\Omega(u) = \frac{1}{2}\int_\Omega |\nabla u|^2 \, dx + \frac{\lambda}{4}\int_\Omega (u^2 - 1)^2 \, dx.$$

We are going to prove that $\mathscr{E}_\Omega$ is strictly convex and bounded below by $\lambda/4$, which implies that $u = 0$ is the only critical point of $\mathscr{E}_\Omega$. Since $u = 0$ is a global minimizer (or using a direct computation), we easily obtain that it is stable. We rewrite $\mathscr{E}_\Omega$ as $\mathscr{E}_\Omega = \mathscr{E}_1 + \mathscr{E}_2$, where

$$\mathscr{E}_1(u) = \frac{1}{2}\left(\int_\Omega |\nabla u|^2 \, dx - \lambda \int_\Omega u^2 \, dx\right) \quad \text{and} \quad \mathscr{E}_2(u) = \frac{\lambda}{4}\int_\Omega \left(u^4 + 1\right) \, dx.$$

Clearly, $\mathscr{E}_1 \geq 0$ if $\lambda \leq \lambda_1$ and $\mathscr{E}_2 \geq \lambda/4$ is strictly convex. So, it suffices to prove that $\mathscr{E}_1$ is convex. Now, since $\lambda \leq \lambda_1$, $\mathscr{E}_1$ is a positive semidefinite quadratic form. Let $\mathbb{B}_1$ denote its associated symmetric bilinear form. Then, given $t \in [0,1]$, $u, v \in X$,

$$\begin{aligned}
\mathscr{E}_1(tu + (1-t)v) &= t^2\mathscr{E}_1(u) + (1-t)^2\mathscr{E}_1(v) + 2t(1-t)\mathbb{B}_1(u,v) \\
&\leq t^2\mathscr{E}_1(u) + (1-t)^2\mathscr{E}_1(v) + 2t(1-t)\mathscr{E}_1(u)^{1/2}\mathscr{E}_1(v)^{1/2} \\
&\leq \left(t\mathscr{E}_1(u)^{1/2} + (1-t)\mathscr{E}_1(v)^{1/2}\right)^2 \\
&\leq t\mathscr{E}_1(u) + (1-t)\mathscr{E}_1(v),
\end{aligned}$$

where we used the Cauchy-Schwarz inequality in the first inequality and the convexity of $x \mapsto x^2$ in the last. We turn next to the case $\lambda > \lambda_1$. $\underline{u} = 0$ and $\overline{u} = 1$ are ordered sub- and supersolutions of the equation, so by Lemma 1.1.1, there exists u, the minimal solution (relative to $\underline{u} = 0$) of (1.17). Since u is minimal, u is stable. Since $\lambda > \lambda_1$, 0 is unstable and so $u \not\equiv 0$. By the strong maximum principle, $0 < u < 1$. By a direct computation, $-u$ is another stable solution to the problem. $\qquad\square$

1.3.3   Symmetry

When an equation exhibits a certain symmetry, it is natural to ask whether its solutions are also symmetric. For example, one can ask whether any solution u of (1.3) in $\Omega = B$, the unit ball of $\mathbb{R}^N$, is radially symmetric. This is indeed the case if u is stable, as the following proposition shows.

Proposition 1.3.4 ([4]) *Let $f \in C^1(\mathbb{R})$. Also let B denote the unit ball in $\mathbb{R}^N$, $N \geq 2$ and $u \in C^2(\overline{B})$ denote a stable solution to*

$$\begin{cases} -\Delta u = f(u) & \text{in } B, \\ \quad u = 0 & \text{on } \partial B. \end{cases} \tag{1.18}$$

Then u is radially symmetric. Moreover, $r \mapsto u(r)$ is either constant, increasing or decreasing in $(0,1)$.

Remark 1.3.1 *A celebrated result of Gidas, Ni, and Nirenberg* [123] *states that every positive solution to* (1.18) *is radially symmetric and decreasing. See Exercise 1.2.2.*

However, unstable sign-changing solutions of (1.18) *can be nonradial. This is the case for example if* $f(u) = \lambda_2 u$, *where* λ_2 *is the second eigenvalue of the Dirichlet Laplace operator (see paragraph V.5.5 in* [62]) *or if* $f(u) = |u|^{p-1}u$, $p > 1$ *(see* [2]*).*

Proof. We first show that u is radial. It suffices to prove that any tangential derivative $v = x_i\partial_j u - x_j\partial_i u$, $i,j = 1,\dots,N$ is identically zero. Integration by parts implies on the one hand that

$$\int_B v\,dx = \int_{\partial B}\left(x_i un_j - x_j un_i\right)\,d\sigma = 0, \tag{1.19}$$

since $n_i = x_i$ is the i-th component of the normal unit vector to ∂B pointing outwards. On the other hand, a direct calculation shows that v solves the linearized equation

$$\begin{cases} -\Delta v = f'(u)v & \text{in } B, \\ \quad v = 0 & \text{on } \partial B, \end{cases} \tag{1.20}$$

where the boundary condition follows from the facts that u is constant on ∂B and that v is a tangential derivative. Multiplying (1.20) by v and integrating by parts, it follows that

$$\int_B |\nabla v|^2\,dx - \int_B f'(u)v^2\,dx = 0.$$

Since u is stable, it follows that if $v \not\equiv 0$, v minimizes the Rayleigh quotient (1.10). Hence, the linearized operator $-\Delta - f'(u)$ has principal eigenvalue $\lambda_1 = 0$ and v is an eigenfunction, so it must be everywhere positive or everywhere negative, contradicting (1.19). So, $v \equiv 0$ and u is radial.

We prove next that $u(r)$ is constant or monotone. If u is not constant, $v = du/dr$ is nontrivial. Assume by contradiction that $v(r_0) = 0$ for some $r_0 \in (0,1]$. By Remark 1.1.1, u is stable in B_{r_0} and we may use v as a test function in (1.5). We obtain

$$\int_{B_{r_0}} |\nabla v|^2\,dx - \int_{B_{r_0}} f'(u)v^2\,dx \geq 0. \tag{1.21}$$

Differentiating (1.18) with respect to r we also have

$$\begin{cases} -\Delta v + \dfrac{N-1}{r^2}v = f'(u)v & \text{in } B_{r_0}, \\ \qquad\qquad v = 0 & \text{on } \partial B_{r_0}. \end{cases}$$

Multiplying the above equation by v and integrating by parts, we get

$$\int_{B_{r_0}} |\nabla v|^2 \, dx - \int_{B_{r_0}} f'(u)v^2 \, dx = -\int_{B_{r_0}} \frac{N-1}{r^2} v^2 \, dx < 0,$$

which contradicts (1.21). □

Exercise 1.3.3 Prove that Proposition 1.3.4 remains valid if assumption (1.5) is replaced by $\lambda_2(-\Delta - f'(u); \Omega) > 0$. What if $\lambda_2(-\Delta - f'(u); \Omega) = 0$?

1.4 Dynamical stability

At the beginning of this chapter, we introduced stability as the ability of a system to return to equilibrium after a small perturbation. This intuitive description can be stated mathematically as follows.

Definition 1.4.1 *Let Ω denote a smoothly bounded domain of $\mathbb{R}^N$, $N \geq 1$. A solution $u \in C^2(\overline{\Omega})$ of (1.3) is asymptotically stable if there exists $\epsilon > 0$ such that for all $u_0 \in C^2(\overline{\Omega})$ with $\|u - u_0\|_{L^\infty(\Omega)} < \epsilon$, the solution $v \in C^2(\overline{\Omega} \times [0, T])$ of*

$$\begin{cases} \dfrac{\partial v}{\partial t} - \Delta v = f(v) & \text{in } \Omega \times (0, T), \\ \qquad v = 0 & \text{on } \partial\Omega \times (0, T), \\ \quad v(x, 0) = u_0(x) & \text{for } x \in \Omega \end{cases} \qquad (1.22)$$

is defined for all times $T > 0$ and

$$\lim_{t \to +\infty} \|v(t) - u\|_{L^\infty(\Omega)} = 0. \qquad (1.23)$$

Unfortunately, the notion of stability, as introduced in (1.5), is not equivalent to that of dynamical stability. Before looking at counter-examples, let us first state a positive result.

Proposition 1.4.1 *Let Ω denote a smoothly bounded domain of $\mathbb{R}^N$, $N \geq 1$ and let $u \in C^2(\overline{\Omega})$ denote a solution to (1.3) such that the linearized operator is positive, that is,*

$$\lambda_1(-\Delta - f'(u); \Omega) > 0.$$

Then, u is asymptotically stable.

Proof. We follow [193] and [28]. We start out by constructing a subsolution $\underline{u}_n$ and a supersolution $\bar{u}_n$ of (1.3) such that $\underline{u}_n < u < \bar{u}_n$ and $\|\bar{u}_n - \underline{u}_n\|_{L^\infty(\Omega)} \to 0$ as $n \to \infty$. To do so, let $v \in C^2(\overline{\Omega})$ denote the solution to

$$\begin{cases} -\Delta v - f'(u)v = |f'(u)| + 1 & \text{in } \Omega, \\ v = 0 & \text{on } \partial\Omega. \end{cases}$$

Note that since $\lambda_1 = \lambda_1(-\Delta - f'(u); \Omega) > 0$, v is well defined and unique by the Lax-Milgram lemma, and $v \in C^2(\overline{\Omega})$ by standard elliptic regularity. Let $\bar{u}_n = u + \frac{1}{n}(v+1)$. Then,

$$-\Delta\bar{u}_n - f(\bar{u}_n) = -\Delta u + \frac{1}{n}(-\Delta v) - f\left(u + \frac{1}{n}(v+1)\right)$$

$$= f(u) + \frac{1}{n}\left(f'(u)v + |f'(u)| + 1\right) - f\left(u + \frac{1}{n}(v+1)\right)$$

$$= f(u) + \frac{1}{n}\left(f'(u)v + |f'(u)| + 1\right)$$

$$- \left(f(u) + \frac{1}{n}f'(u)(v+1) + o\left(\frac{v+1}{n}\right)\right)$$

$$\geq \frac{1}{n} + o(1/n) \geq 0,$$

where the last inequality holds for n sufficiently large. Similar inequalities prove that $\underline{u}_n = u - \frac{1}{n}(v+1)$ is a subsolution to the problem.

We are left with proving that u is asymptotically stable. To do so, we remark that if $\varepsilon > 0$ is chosen small enough, then any initial datum $u_0 \in C^2(\overline{\Omega})$ such that $\|u - u_0\|_{L^\infty(\Omega)} < \epsilon$ lies in the interval $(\underline{u}_{n_0}, \bar{u}_{n_0})$ for some fixed $n_0 \in \mathbb{N}$. By the method of sub- and supersolutions, the solution v of (1.22) is well defined for all times and $\underline{u}_{n_0} \leq v \leq \bar{u}_{n_0}$. In particular, for all $t > 0$

$$\|v(t) - u\|_{L^\infty(\Omega)} \leq \frac{1}{n_0}(\|v\|_{L^\infty(\Omega)} + 1). \tag{1.24}$$

Fix $\delta \in (0, \lambda_1)$ small. Since f is C^1, the inequality

$$\left|f(z) - f(u) - f'(u)(z - u)\right| \leq \delta |z - u|$$

holds for small enough $|z - u|$. Choosing n_0 sufficiently large (which is allowed if $\varepsilon > 0$ is small), we conclude that for all time $t > 0$,

$$\left|f(v) - f(u) - f'(u)(v - u)\right| \leq \delta |v - u|.$$

So, $w := v - u$ satisfies

$$\frac{\partial w}{\partial t} - \Delta w - (f'(u) + \delta)w = f(v) - f(u) - (f'(u) + \delta)(v - u) \leq 0.$$

The principal eigenvalue μ_1 of the operator $-\Delta - (f'(u) + \delta)$ is given by

$$\mu_1 = \lambda_1 - \delta > 0.$$

By comparison, we deduce that $w(t) \leq Ce^{-\mu_1 t}$. Similar arguments lead to $-w(t) \leq Ce^{-\mu_1 t}$, so that u is asymptotically stable. $\qquad\square$

Conversely, if a solution u of (1.3) is asymptotically stable, then necessarily it must be stable, that is, (1.5) holds. This is the object of the following proposition.

Proposition 1.4.2 *Let Ω denote a smoothly bounded domain of $\mathbb{R}^N$, $N \geq 1$ and let $u \in C^2(\overline{\Omega})$ denote a solution to (1.3) such that*

$$\lambda_1(-\Delta - f'(u); \Omega) < 0.$$

Then, u is not asymptotically stable.

Proof. Take $\varepsilon > 0$ and consider the function $\overline{u} = u - \varepsilon\varphi_1$, where $\varphi_1 > 0$ is an eigenfunction associated to $\lambda_1 = \lambda_1(-\Delta - f'(u); \Omega)$. Then,

$$
\begin{aligned}
-\Delta\overline{u} - f(\overline{u}) &= -\Delta u - \varepsilon\left(-\Delta\varphi_1\right) - f\left(u - \varepsilon\varphi_1\right) \\
&= f(u) - \varepsilon\left(f'(u) + \lambda_1\right)\varphi_1 - f\left(u - \varepsilon\varphi_1\right) \\
&= f(u) - \varepsilon\left(f'(u) + \lambda_1\right)\varphi_1 - \left(f(u) - \varepsilon f'(u)\varphi_1 + o(\varepsilon\varphi_1)\right) \\
&= -\varepsilon\lambda_1\varphi_1 + o(\varepsilon\varphi_1) \geq 0,
\end{aligned}
$$

if $\varepsilon > 0$ is sufficiently small. So, $\overline{u}$ is a supersolution to (1.3). It follows that the solution v of (1.22) with initial datum $u_0 = \overline{u}$ is a nonincreasing function of time. So, for all times $t > 0$ for which v is well defined,

$$u - v(t) \geq u - u_0 = \varepsilon\varphi_1,$$

contradicting $\lim_{t \to \infty} \|u - v(t)\|_{L^\infty(\Omega)} = 0$. $\qquad\square$

There is still one scenario that we haven't considered: the case $\lambda_1(-\Delta - f'(u)) = 0$. As we are about to see, examples show that no conclusion can be drawn in full generality.

Proposition 1.4.3 *Let Ω denote a smoothly bounded domain of $\mathbb{R}^N$, $N \geq 3$, let $\lambda_1 = \lambda_1(-\Delta; \Omega) > 0$ and let $\varphi_1 > 0$ denote a corresponding eigenvector. Take $\varepsilon > 0$, $1 < p < (N+2)/(N-2)$ and consider the equation*

$$\begin{cases} \dfrac{\partial v}{\partial t} - \Delta v = \lambda_1 v + |v|^{p-1} v & \text{in } \Omega \times (0, T), \\ \qquad\quad v = 0 & \text{on } \partial\Omega \times (0, T), \\ \quad v(x, 0) = \pm\varepsilon\varphi_1(x) & \text{for } x \in \Omega. \end{cases} \tag{1.25}$$

Then, $u = 0$ is a solution to (1.3) with $f(u) = \lambda_1 u + |u|^{p-1} u$ such that $\lambda_1(-\Delta - f'(0); \Omega) = 0$. Yet, $u = 0$ is not asymptotically stable.

Proof. Let $\mathscr{E}_\Omega : H_0^1(\Omega) \to \mathbb{R}$ denote the usual energy functional defined by

$$\mathscr{E}_\Omega(u) = \frac{1}{2} \int_\Omega |\nabla u|^2 \, dx - \frac{\lambda_1}{2} \int_\Omega u^2 \, dx - \frac{1}{p+1} \int_\Omega |u|^{p+1} \, dx.$$

Note that $\mathscr{E}_\Omega(0) = 0$, while $\mathscr{E}_\Omega(\pm\varepsilon\varphi_1) < 0$. Let v denote the solution to (1.25).

$$\frac{d}{dt} \mathscr{E}_\Omega(v) = D\mathscr{E}_\Omega(v).\frac{\partial v}{\partial t} = -\int_\Omega \left(\frac{\partial v}{\partial t} \right)^2 dx \leq 0.$$

In particular, $\mathscr{E}_\Omega(v) \leq \mathscr{E}_\Omega(\pm\varepsilon\varphi_1) < 0$. So, even if v were defined for all times $t > 0$, v could not converge to $u = 0$, which has zero energy. $\qquad\square$

The reverse conclusion can also occur.

Proposition 1.4.4 *Let Ω denote a smoothly bounded domain of $\mathbb{R}^N$, $N \geq 1$ and let $\lambda_1 = \lambda_1(-\Delta; \Omega) > 0$. Take $\varepsilon > 0$, $1 < p$ and consider the equation*

$$\begin{cases} \dfrac{\partial v}{\partial t} - \Delta v = \lambda_1 v - |v|^{p-1} v & \text{in } \Omega \times (0, T), \\ \qquad\quad v = 0 & \text{on } \partial\Omega \times (0, T), \\ \quad v(x, 0) = u_0(x) & \text{for } x \in \Omega. \end{cases} \tag{1.26}$$

Then, $u = 0$ is a solution to (1.3) with $f(u) = \lambda_1 u - |u|^{p-1} u$ such that $\lambda_1(-\Delta - f'(0); \Omega) = 0$ and $u = 0$ is asymptotically stable.

Proof. Let $\varphi_1 > 0$ denote a eigenvector associated to λ_1. One may easily check that $\underline{u}_n = -\frac{1}{n}\varphi_1$ and $\bar{u}_n = \frac{1}{n}\varphi_1$ are respectively a sub- and a supersolution to (1.26) provided u_0 is small enough. We can then argue as in the proof of Proposition 1.4.1. $\qquad\square$

1.5 Stability outside a compact set

In this section, following [94], we generalize the notion of stability by requiring that (1.5) holds only for test functions supported away from a given compact set.

Definition 1.5.1 *Let $f \in C^1(\mathbb{R})$, let Ω denote an open set of $\mathbb{R}^N$, $N \geq 1$, and let $K \subset\subset \Omega$ denote a compact subset of Ω. A solution $u \in C^2(\Omega)$ of (1.4) is stable outside the compact set K if*

$$Q_u(\varphi) := \int_\Omega |\nabla\varphi|^2 \, dx - \int_\Omega f'(u)\varphi^2 \, dx \geq 0, \quad \forall \, \varphi \in C_c^1(\Omega \setminus K). \quad (1.27)$$

Definition 1.5.1 encompasses the following fundamental class of solutions.

Example 1.5.1 Solutions of finite Morse index are stable outside a compact set.

Definition 1.5.2 *Let $f \in C^1(\mathbb{R})$, let Ω denote a domain of $\mathbb{R}^N$, $N \geq 1$. A solution $u \in C^2(\Omega)$ of (1.4) has Morse index $k \geq 1$ if k is the maximal dimension of a subspace X_k of $C_c^1(\Omega)$ such that*

$$Q_u(\varphi) < 0 \qquad \forall \varphi \in X_k \setminus \{0\}.$$

We then write $k = \text{ind}(u)$.

Remark 1.5.1 *If u is stable, we write $0 = \text{ind}(u)$.*

Let us check that every solution of finite Morse index is stable outside a compact set. Assume $\text{ind}(u) = k \in \mathbb{N}^*$ and take k independent vectors $\varphi_1, \ldots, \varphi_k \in C_c^1(\Omega)$ spanning the associated space X_k. Each φ_i has compact support so that $K = \cup_{i=1}^k \text{supp} \, \varphi_i$ is a compact subset of Ω. Then, $Q_u(\varphi) \geq 0$ for all $\varphi \in C_c^1(\Omega \setminus K)$ for otherwise there would exist a function φ with $\text{supp} \, \varphi \not\subset K$ and $Q_u(\varphi) < 0$. In particular, $\varphi, \varphi_1, \ldots, \varphi_k$ would be linearly independent, hence $\text{ind}(u) \geq k+1$, a contradiction. $\qquad \square$

If we assume that Ω is a bounded domain, then *every* solution $u \in C^2(\bar\Omega)$ of (1.3) has finite Morse index and is thus stable outside a compact set.

Proposition 1.5.1 *Let $f \in C^1(\mathbb{R})$ and let Ω denote a smoothly bounded domain of $\mathbb{R}^N$, $N \geq 1$. Every solution $u \in C^2(\bar\Omega)$ of (1.4) has finite Morse index. Furthermore, $k = \text{ind}(u)$ if and only if the linearized operator $-\Delta - f'(u)$ (with Dirichlet boundary conditions) has exactly k strictly negative eigenvalues.*

Remark 1.5.2 *In the above statement, eigenvalues are repeated according to their geometric multiplicity.*

Proof. We just need to prove that the Morse index k of u is equal to the number $\tilde{k}$ of negative eigenvalues of the operator.

Step 1. $\tilde{k} \geq k$.

If $k = 0$, we refer the reader to Definition 1.2.1. If $k = 1$, then for any $\psi \in X_k \setminus \{0\}$, $Q_u(\psi) < 0$. In particular, $\lambda_1 = \lambda_1(-\Delta - f'(u); \Omega) < 0$, since λ_1 is given by (1.10). If $k \geq 2$, the k-th eigenvalue $\lambda_k = \lambda_k(-\Delta - f'(u); \Omega)$ of the linearized operator is given by

$$
\lambda_k = \inf \left\{ \frac{\int_\Omega |\nabla \varphi|^2 \, dx - \int_\Omega f'(u)\varphi^2 \, dx}{\int_\Omega \varphi^2 \, dx} : \varphi \in H_0^1(\Omega) \setminus \{0\} \text{ s.t.} \right.
$$

$$
\left. \int_\Omega \varphi \varphi_j \, dx = 0 \quad \forall \, j = 1, \ldots, k-1 \right\}, \quad (1.28)
$$

where the functions φ_j, $j = 1, \ldots, k-1$ are linearly independent eigenvectors associated to the eigenvalues $\lambda_j < \lambda_k$ (repeated according to their geometric multiplicity). Consider the linear map $\Lambda : X_k \to \mathbb{R}^{k-1}$ defined for $\varphi \in X_k$ by $\Lambda \varphi = (\int_\Omega \varphi \varphi_1 dx, \ldots, \int_\Omega \varphi \varphi_{k-1} dx)$. Since X_k is k-dimensional, the kernel of Λ is nontrivial and there exists $\varphi \in X_k \setminus \{0\}$ such that $\int_\Omega \varphi \varphi_j \, dx = 0$ for all $j = 1, \ldots, k-1$. We deduce that $\lambda_k \leq Q_u(\varphi) < 0$.

Step 2. $\tilde{k} \leq k$.

Using the notation of Step 1, $Q_u(\varphi_j) < 0$ for all $j = 1, \ldots, \tilde{k}$. Since $\varphi_j \in H_0^1(\Omega)$, there exists $\psi_j \in C_c^1(\Omega)$ such that $Q_u(\psi_j) < 0$ for $j = 1, \ldots, \tilde{k}$. Furthermore, the functions (ψ_j) can be chosen linearly independant, for otherwise (φ_j) would be linearly dependent. So, $\mathrm{ind}\,(u) \geq \tilde{k}$. $\qquad\square$

Solutions of finite Morse index have the following additional stability property.

Proposition 1.5.1 *Let $f \in C^1(\mathbb{R})$ and let Ω denote an open set of $\mathbb{R}^N$, $N \geq 2$. Let u be a solution to (1.4) with finite Morse index. Then, for every $x_0 \in \Omega$, there exists $r_0 > 0$ such that u is stable in $B(x_0, r_0)$.*

Proof. We may always assume that $B(0, 1) \subset \Omega$ and it suffices to prove that u is stable near the origin. Assume first that $\mathrm{ind}\,u = 1$. Either u is stable in $B(0, 1/n)$ for some $n \geq 2$ and we are done. Or, for all $n \geq 2$, there exists a direction $\varphi_n \in C_c^1(B(0, 1/n))$ such that $Q_u(\varphi_n) < 0$. Since $\mathrm{ind}\,u = 1$, this implies that u is stable in $B(0,1) \setminus \overline{B(0,1/n)}$. This being true for all $n \geq 2$, we deduce that u is stable in $B(0,1) \setminus \{0\}$. In fact, since $N \geq 2$, u is stable in $B(0,1)$. Take an arbitrary test function $\varphi \in C_c^1(B(0,1))$ and let $\varphi_n = \varphi \zeta_n$, where

$$\zeta_n(x) = \begin{cases} 0 & \text{if } |x| < 1/n^2, \\[2mm] 2 - \dfrac{\ln|x|}{\ln 1/n} & \text{if } 1/n^2 \leq |x| < 1/n, \\[2mm] 1 & \text{if } |x| \geq 1/n, \end{cases}$$

if $N = 2$, and $\zeta_n(x) = \zeta(nx)$, with $\zeta \in C^1(\mathbb{R}^N)$ such that $\zeta \equiv 0$ in B_1 and $\zeta \equiv 1$ in $\mathbb{R}^N \setminus B_2$, if $N \geq 3$. Then, $\varphi_n \in C_c^{0,1}(B(0,1) \setminus \{0\})$ and so $Q_u(\varphi_n) \geq 0$. In addition, one easily verifies that $\varphi_n \to \varphi$ pointwise and in $H^1(B(0,1))$, and $|\varphi_n| \leq |\varphi|$. Hence, $Q_u(\varphi) \geq 0$. So, every solution of index $\operatorname{ind} u = 1$ is stable in a neighborhood of 0. Now take a solution u of index k. Working exactly as above, we deduce that u has index $k - 1$ in some ball $B(0, r_1)$. Working inductively on k, we deduce that u is stable in some ball $B(0, r_k)$.

$\square$

Exercise 1.5.1 When working on unbounded domains, the notion of stability outside a compact is still insufficient to encompass all solutions. Check that the function u constructed in Example 1.2.3 is unstable outside every compact set of $\Omega = I \times \mathbb{R}^{N-1}$.

1.6   Resolving an ambiguity

Consider again (1.4) in the special case $\Omega = \mathbb{R}$, that is,

$$-\frac{d^2 u}{dx^2} = f(u) \qquad \text{in } \mathbb{R}. \tag{1.29}$$

Assume that $u = 0$ is a stable solution to the above equation. Then, we must have $f'(0) \leq 0$. To see this, simply apply stability with test function $\varphi_\varepsilon(x) = \varphi(\varepsilon x)$, where $\varphi \in C_c^1(\mathbb{R}^N) \not\equiv 0$, and let $\varepsilon \to 0$. Assume further that $f'(0) < 0$. By Proposition 1.4.1, $u = 0$ must be asymptotically stable for the corresponding parabolic equation, at least on any bounded open interval of $\mathbb{R}$ (or even just for the ODE $\frac{du}{dt} = f(u)$).

Now, we could have also thought of (1.29) as an ODE, where x is now interpreted as the time variable. For this interpretation, our notion of stability is opposite to the notion of asymptotic stability:

Proposition 1.6.1 ([214]) *Let $f \in C^1(\mathbb{R})$ such that $f(0) = 0$ and $f'(0) < 0$. In particular, $u = 0$ is a stable solution to (1.29). However, u is asymptotically unstable for the ode*

$$-\frac{d^2 u}{dt^2} = f(u). \tag{1.30}$$

Proof. Assume to the contrary that 0 is an asymptotically stable equilibrium of (1.30). Then, for $\varepsilon > 0$ sufficiently small, the solution $u(t)$ to (1.30) with $u'(0) = 0$ and $u(0) = \varepsilon$, is defined for all times $t > 0$, and

$$\lim_{t \to +\infty} u(t) = 0,$$

and so

$$\liminf_{t \to +\infty} |u'(t)| = 0.$$

Multiplying (1.30) by u' and integrating between 0 and t, we obtain the following law of conservation of energy:

$$\frac{1}{2} u'(t)^2 + F(u(t)) = F(\varepsilon),$$

where $F(t) = \int_0^t f(s)\, ds$. Passing to the liminf as $t \to +\infty$, we deduce that

$$F(\varepsilon) = F(0).$$

This being true for all ε small, we deduce that $f \equiv 0$ in a neighborhood of 0, contradicting $f'(0) < 0$. $\qquad\square$

Chapter 2

The Gelfand problem

In Chapter 1, we gained familiarity with the notion of stability for semilinear elliptic equations of the form (1.3). It is time to look at a concrete example. This chapter is devoted to the study of the following problem:

$$\begin{cases} -\Delta u = \lambda e^u & \text{in } B, \\ \quad u = 0 & \text{on } \partial B, \end{cases} \tag{2.1}$$

where $\lambda > 0$ is a parameter and B is the unit ball of $\mathbb{R}^N$, $N \geq 1$. Equation (2.1) bears many names: Barenblatt, Bratu, Emden, Fowler, Frank-Kamenetskii, Gelfand, and Liouville are some of the famous scientists to whom the equation has been attributed. For short, we call (2.1) the Gelfand problem.[1] It arises as a (crude) model in a number of interesting physical contexts,[2] one of which we outline next.

2.1 Motivation

In dimension $N = 1, 2, 3$, Equation (2.1) can be derived from the thermal self-ignition model. The full model describes the reaction process in a combustible material during what is referred to as the ignition period. A solution u of (2.1) represents a dimensionless temperature inside a cylindrical vessel (which walls are ideally conducting), when the system has reached an intermediate-asymptotic steady state. The underlying space variable $x \in B$

---

[1]One may argue that our choice is not historically sound. See [26], [131].

[2]We warn the reader that the considerations presented in this book have little impact on the understanding of the underlying physical phenomenon. Instead, we shall *use* physics (well, truly heuristics) to gain intuition on the mathematical analysis of (2.1).

should be thought of as dimensionless, that is, the vessel's size has been normalized. We refer the interested reader to [13] and Chapters VI-VII of [110] for the detailed derivation of the model.

Now, take a second look at Equation (2.1). On the left-hand side, there is a diffusion operator, $-\Delta$, accounting for the diffusion of heat from the hot reactants to the cold boundary. On the right-hand side, we have a reaction term, e^u. The exponential nonlinearity has to do with the so-called Arrhenius law. More precisely, it is an approximation of the nonlinear term given in this (empirical) law, which truly takes the form $f(u) = e^{\frac{u}{1+\varepsilon u}}$. This term models the production of heat induced by the chemical reaction. The diffusion operator and the reaction nonlinearity compete. In one kind of reaction, the produced heat does not have time to be carried away through the walls of the vessel: either the combustible rarifies and the reaction dies out, or there is so much combustible that a thermal explosion occurs. Either way, no steady-state, that is, no solution to (2.1), should be expected. In another kind of reaction, on the contrary, an equilibrium between the produced and the diffused heat quickly occurs, so that existence of solutions of (2.1) should hold.

The balance between diffusion and reaction is quantified by the parameter $\lambda > 0$. This parameter is sometimes referred to as the Frank-Kamenetskii constant. According to our previous discussion, we should expect no solution to (2.1) if λ is large, while solutions should exist for small λ. This is indeed the case, as the following proposition demonstrates.

Proposition 2.1.1 *Assume $N \geq 1$. Then, every solution $u \in C^2(\overline{B})$ of (2.1) is radial. Furthermore, there exists $\lambda^* = \lambda^*(N) > 0$ such that*

- *For $0 < \lambda < \lambda^*$, there exists the minimal solution $u_\lambda \in C^2(\overline{B})$ of (2.1). In particular, u_λ is stable.*

- *For $\lambda > \lambda^*$, there exists no solution $u \in C^2(\overline{B})$ of (2.1).*

We shall prove Proposition 2.1.1 for $N = 1$ in Proposition 2.2.1, $N = 2$ in Proposition 2.3.1, and $N \geq 3$ in Proposition 2.4.1. See also Proposition 3.3.1 for a more general and streamlined argument.

2.2 Dimension $N = 1$

For $N = 1$, (2.1) reads

$$-u'' = \lambda e^u, \qquad u(-1) = u(1) = 0. \tag{2.2}$$

We completely characterize solutions to the equation as follows.

Proposition 2.2.1 *There exists $\lambda^* > 0$ such that*

- *For $0 < \lambda < \lambda^*$, there exists exactly two solutions $u \in C^2([-1,1])$ of (2.2). One of them, denoted u_λ, is minimal, hence stable. The other one, denoted U_λ, has Morse index 1. Both solutions are positive, even, strictly decreasing on $(0,1]$, and uniquely determined by their value at $x = 0$. In addition, the curve*

$$\begin{cases} (0, \lambda^*) \to C^2([-1,1]) \times C^2([-1,1]), \\ \quad \lambda \mapsto (u_\lambda, U_\lambda) \end{cases}$$

is smooth and for all $x \in (-1,1)$,

$$\lim_{\lambda \to 0^+} (u_\lambda(x), U_\lambda(x)) = (0, +\infty), \qquad (2.3)$$

$$\lim_{\lambda \to \lambda^*} (u_\lambda(x), U_\lambda(x)) = (u^*(x), u^*(x)), \qquad (2.4)$$

where $u^* \in C^2([-1,1])$ is the unique solution to (2.2) for $\lambda = \lambda^*$.

- *For $\lambda > \lambda^*$, there exists no solution $u \in C^2([-1,1])$ to (2.1).*

Plotting λ on the x-axis and the norm $\|u\|_\infty = u(0)$ on the y-axis for each solution $u \in \{u_\lambda, U_\lambda\}$, we obtain the following *bifurcation diagram* summarizing Proposition 2.2.1.

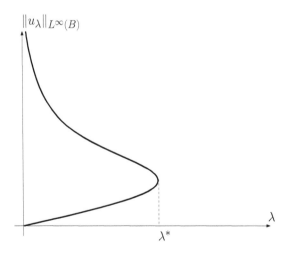

Figure 2.1: Bifurcation diagram for the Gelfand problem in dimension $N = 1$.

Proof.
Step 1. Every solution is positive, even, radially decreasing and characterized by its value at 0.

Let u denote a solution to (2.2). Since $-u'' > 0$ in $(-1,1)$, u cannot achieve an interior point of minimum. In particular, $u > 0$ in $(-1,1)$. Also, u is even and $u'(r) < 0$ for $r \in (0,1)$. Indeed, assume by contradiction that u achieves its maximum at some $x_0 \in (-1,1) \setminus \{0\}$. Without loss of generality, $x_0 \in (0,1)$. Since $v(t) = u(x_0 + t)$ and $\tilde{v}(t) = u(x_0 - t)$ satisfy the same initial value problem with initial conditions $v(0) = u(x_0)$ and $v'(0) = 0$, they must coincide, that is, v is even. In particular, $u(2x_0 - 1) = u(1) = 0$, contradicting $u > 0$ in $(-1,1)$. So, u achieves its unique point of maximum at 0. It follows that u is even and that $u'(r) < 0$ for $r \in (0,1)$. Now multiply (2.2) by u' and integrate between 0 and $r \in (0,1)$:

$$-\frac{u'^2}{2} = \lambda \left(e^u - e^{u_0} \right), \tag{2.5}$$

where $u_0 = u(0)$, which we rewrite as

$$\frac{-u'}{\sqrt{2\lambda \left(e^{u_0} - e^u \right)}} = 1.$$

Integrating once more between 0 and 1, it follows that every solution to (2.2) satisfies

$$\int_0^{u_0} \frac{dt}{\sqrt{2\lambda(e^{u_0} - e^t)}} = 1. \tag{2.6}$$

Conversely, every time there exists $u_0 > 0$ such that (2.6) holds, the even function u defined for $r \in (0,1)$ by

$$\int_{u(r)}^{u_0} \frac{dt}{\sqrt{2\lambda(e^{u_0} - e^t)}} = r$$

solves (2.2). This completes Step 1.

Step 2. For some $\lambda^* > 0$, there are exactly two solutions for $\lambda \in (0, \lambda^*)$, one for $\lambda = \lambda^*$, and none for $\lambda > \lambda^*$. In addition, (2.3) and (2.4) hold.

According to Step 1, there exists a solution such that $u(0) = u_0 > 0$ if and only if (2.6) holds. Using standard calculus, one easily sees that the real-valued function $I : \mathbb{R}_+^* \to \mathbb{R}$ defined for $u_0 > 0$ by

$$I(u_0) = \int_0^{u_0} \frac{dt}{\sqrt{e^{u_0} - e^t}}$$

takes values in some bounded interval $(0, M^*]$, achieves its maximum M^* at a unique point u_0^*, and satisfies $\lim_{u_0 \to 0^+} I(u_0) = \lim_{u_0 \to +\infty} I(u_0) = 0$. Step 2 follows.

Step 3. Any solution u is stable outside the compact set $K = \{0\}$. To see this, recalling that u is even, it suffices to prove that

$$Q_u(\varphi) \geq 0 \qquad \text{for every } \varphi \in C_c^1(0, 1).$$

Since $u' < 0$ in $(0, 1)$, every test function $\varphi \in C_c^1(0, 1)$ can be written in the form $\varphi = u'\psi$, where $\psi \in C_c^1(0, 1)$. So,

$$
\begin{aligned}
Q_u(\varphi) &= \int_0^1 \left(\left(\frac{d(u'\psi)}{dr} \right)^2 - \lambda e^u (u'\psi)^2 \right) dr \\
&= \int_0^1 \left((u'')^2 \psi^2 + 2u'\psi u''\psi' + (u')^2(\psi')^2 - \lambda e^u (u')^2 \psi^2 \right) dr \\
&= \int_0^1 \left(u''(u'\psi^2)' + (u')^2(\psi')^2 - \lambda e^u (u')^2 \psi^2 \right) dr. \qquad (2.7)
\end{aligned}
$$

Now, differentiating (2.2), we obtain $-u''' = \lambda e^u u'$. Multiplying by $u'\psi^2$ and integrating, it follows that

$$\int_0^1 u''(u'\psi^2)' \, dr = \lambda \int_0^1 e^u (u')^2 \psi^2 \, dr.$$

Using this in (2.7), we finally obtain

$$Q_u(\varphi) = \int_0^1 (u')^2(\psi')^2 \, dr \geq 0.$$

Step 4. U_λ has Morse index 1.

First observe that by Proposition 1.3.2, U_λ must have nonzero Morse index. Take a direction $\psi \in C_c^1(-1, 1)$ such that $Q_{U_\lambda}(\psi) < 0$. By Step 3 and by density, $Q_{U_\lambda}(\varphi) \geq 0$, for every $\varphi \in C_c^1(-1, 1)$ such that $\varphi(0) = 0$. In particular, $\psi(0) \neq 0$ and we may as well assume that $\psi(0) = 1$. Now take any $\varphi \in C_c^1(-1, 1)$. Then, $\tilde{\varphi} = \varphi - \varphi(0)\psi$ vanishes at 0 and so $Q_{U_\lambda}(\tilde{\varphi}) \geq 0$. It follows that the Morse index of U_λ is at most 1.

□

2.3   Dimension $N = 2$

In dimension $N = 2$, the solutions to the Gelfand problem (2.1) can be explicitly computed.

Proposition 2.3.1 ([147]) *Let $N = 2$ and $\lambda^* = 2$. Then,*

- *For $0 < \lambda < \lambda^*$, there exists exactly two solutions, $u_\lambda, U_\lambda \in C^2(\overline{B})$ to (2.1). u_λ is minimal, hence stable, while U_λ is unstable. Both solutions are radial and explicitly given by*

$$u_\lambda(r) = \ln \frac{8b_-}{(1 + \lambda b_- r^2)^2}, \qquad U_\lambda(r) = \ln \frac{8b_+}{(1 + \lambda b_+ r^2)^2} \qquad (2.8)$$

where $b_\pm = \frac{4 - \lambda \pm \sqrt{16 - 8\lambda}}{\lambda^2}$, $r \in [0, 1]$.

- *For $\lambda = \lambda^*$, there exists a unique solution given by*

$$u(r) = \ln \frac{4}{(1 + r^2)^2}, \qquad for\ r \in [0, 1]. \qquad (2.9)$$

- *For $\lambda > \lambda^*$, there exists no solution $u \in C^2(\overline{B})$ to (2.1).*

Plotting λ on the x-axis and the norm $\|u\|_\infty = u(0)$ on the y-axis for each solution $u \in \{u_\lambda, U_\lambda\}$, we obtain a bifurcation diagram (Figure 2.2) summarizing Proposition 2.3.1.

Remark 2.3.1 *The bifurcation diagrams in dimension $N = 1$ and $N = 2$ are very similar. We point out one important difference. For $N = 1$, as $\lambda \to 0$, the unstable solution U_λ blows up at every point $x \in (-1, 1)$, by (2.4). For $N = 2$, U_λ blows up only at the origin. In fact, $U_\lambda(r) \to 4 \ln \frac{1}{r}$ for $r \in (0, 1)$, as follows from (2.8).*

Proof. By the maximum principle, every solution is positive. By the Gidas-Ni-Nirenberg symmetry result (see Exercise 1.2.2), every solution is radial. One can verify by direct inspection that any function of the form (2.8) is a solution. Conversely, since there is at most one solution to the ODE $-u'' - \frac{N-1}{r}u' = \lambda e^u$ with the prescribed initial value $u'(0) = 0$ and $u(0) = a$, every solution must be of the form (2.8) for some $b \in \mathbb{R}$. Since solutions must also satisfy the boundary condition $u(1) = 0$, we must have

$$8b = (1 + \lambda b)^2.$$

This equation is quadratic in b and has respectively 2, 1, or 0 solutions if $\lambda < \lambda^* = 2$, $\lambda = \lambda^*$, and $\lambda > \lambda^*$, respectively. $\qquad\square$

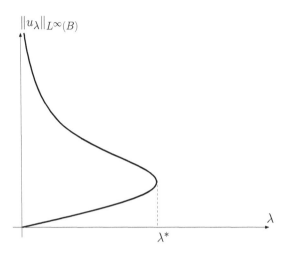

Figure 2.2: Bifurcation diagram for the Gelfand problem in dimension $N = 2$.

2.4   Dimension $N \geq 3$

The structure of the solution set to (2.1) in dimension $N \geq 3$ is radically different. In this section, we establish two bifurcation diagrams (Figures 2.3 and 2.5), and calculate the Morse index of solutions.

Proposition 2.4.1 ([132]) *Let $3 \leq N \leq 9$. There exists $\lambda^* > 2(N - 2)$ such that*

- *For $0 < \lambda < \lambda^*$, $\lambda \neq 2(N - 2)$, there exists finitely many solutions $u \in C^2(\overline{B})$ to (2.1).*

- *Given any $k \in \mathbb{N}$, there exists $\epsilon > 0$ such that for $|\lambda - 2(N - 2)| < \epsilon$, there exists at least k solutions.*

- *For $\lambda = 2(N - 2)$, there exists infinitely many solutions.*

- *For $\lambda = \lambda^*$, there exists a unique solution.*

- *For $\lambda > \lambda^*$, there exists no solution.*

Proposition 2.4.1 is summarized in Figure 2.3.

Proof. By the maximum principle, every solution is positive. By the Gidas-Ni-Nirenberg symmetry result (see Exercise 1.2.2), every solution is radial and

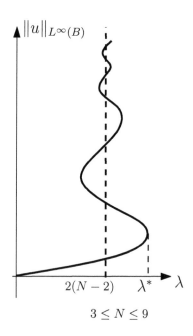

Figure 2.3: Bifurcation diagram in dimension $3 \leq N \leq 9$.

radially decreasing. So, every solution to (2.1) must satisfy the initial value problem

$$\begin{cases} r^{-(N-1)} \left(r^{N-1} u' \right)' + \lambda e^u = 0 \\ u(0) = a \qquad u'(0) = 0, \end{cases} \tag{2.10}$$

for some $a > 0$. Equation (2.10) has a unique maximal solution, obtained using the Picard fixed point theorem applied to the equivalent integral equation

$$u(r) = a - \lambda \int_0^r \int_0^s \left(\frac{t}{s} \right)^{N-1} e^{u(t)} \, dt \, ds.$$

We claim that u is defined for all $r \geq 0$. To see this, we apply the Emden transformation

$$u(r) = w(t) - 2t + a \qquad \text{with} \qquad r = \sqrt{\frac{2(N-2)}{\lambda e^a}} e^t, \tag{2.11}$$

(2.10) is equivalent to the autonomous ODE

$$w'' + (N-2)w' + 2(N-2)(e^w - 1) = 0, \tag{2.12}$$

$$\lim_{t \to -\infty} w(t) - 2t = \lim_{t \to -\infty} e^{-t} \left(w'(t) - 2 \right) = 0.$$

36

Let $v(t) = w(t) - 2t$. Then w solves (2.12) if and only if v solves the integral equation

$$v(t) = -2(N-2) \int_{-\infty}^{t} e^{-(N-2)s} \left(\int_{-\infty}^{s} e^{N\sigma + v(\sigma)} \, d\sigma \right) ds. \qquad (2.13)$$

Applying the Picard fixed point theorem to (2.13), we deduce that there exists a unique solution to (2.12) defined on a maximal interval $(-\infty, T)$. In fact, $T = +\infty$. To see this, consider the Lyapunov function

$$\mathscr{L}(w) = \frac{1}{2}(w')^2 + 2(N-2)(e^w - w).$$

Then, using (2.12),

$$\frac{d\mathscr{L}(w)}{dt} = w''w' + 2(N-2)(e^w - 1)w' = -(N-2)(w')^2 \leq 0.$$

Hence, $\mathscr{L}(w)$ is bounded from above, and so w, w' remain bounded as $t \to T^-$. It follows that $T = +\infty$.

Now rewrite (2.12) as a system

$$\frac{d}{dt} \begin{pmatrix} w \\ w' \end{pmatrix} = \begin{pmatrix} w' \\ -(N-2)(w' + 2(e^w - 1)) \end{pmatrix}$$

and observe that $(0,0)$ is the unique stationary point. To determine its nature, linearize the system at $(0,0)$:

$$\frac{d}{dt} \begin{pmatrix} z \\ z' \end{pmatrix} = \begin{pmatrix} 0 & 1 \\ -2(N-2) & -(N-2) \end{pmatrix} \begin{pmatrix} z \\ z' \end{pmatrix}.$$

The associated eigenvalues are given by

$$\mu_{\pm} = -\frac{1}{2}\left(N - 2 \pm i\sqrt{(N-2)(10-N)} \right)$$

and so $(0,0)$ is a spiral attractor.

Now let w denote the solution to (2.12). By definition, the orbit $\mathscr{O} = \{(w, w') = (w(t), w'(t)) : t \in \mathbb{R}\}$ is asymptotic to the line $w' = 2$ at $t = -\infty$ in the (w, w')-phase plane. Consider the following four regions of this plane

$$\begin{aligned}
\Omega_1 &= \{w' > 0, w' + 2(e^w - 1) > 0\}, \\
\Omega_2 &= \{w' < 0, w' + 2(e^w - 1) > 0\}, \\
\Omega_3 &= \{w' < 0, w' + 2(e^w - 1) < 0\}, \\
\Omega_4 &= \{w' < 0, w' + 2(e^w - 1) < 0\}.
\end{aligned}$$

We prove next that the orbit $\mathcal{O}$ is contained in the half-plane $\{(w, w') : w' < 2\}$, and that, starting in Ω_1, $\mathcal{O}$ then enters successively Ω_i, $i \in \mathbb{Z}/4\mathbb{Z}$, spiraling toward the unique stationary point $(0, 0)$.

Observe that since $u'(r) < 0$ for $r > 0$, $\mathcal{O}$ lies in $\{(w, w') : w' < 2\}$. To see that $\mathcal{O}$ starts in Ω_1, we calculate

$$
\begin{aligned}
\lim_{t \to -\infty} r^{-2} \left(w'(t) + 2(e^{w(t)} - 1) \right) &= \lim_{r \to 0^+} r^{-2} \left(ru'(r) + 2e^{u(r) + 2t - a} \right) \\
&= \lim_{r \to 0^+} \left(u''(0) + 2e^{u(r)} \frac{\lambda}{2(N-2)} \right) \\
&= u''(0) + \frac{\lambda e^a}{N-2} \\
&= \lambda e^a \left(-\frac{1}{N} + \frac{1}{N-2} \right) > 0.
\end{aligned}
$$

So, $\mathcal{O}$ starts in Ω_1. Since $w' > 0$ in Ω_1, (2.12) can be rewritten in this region as

$$
\frac{dw'}{dw} = -\frac{N-2}{w'} (w' + 2(e^w - 1)). \tag{2.14}
$$

Since $0 < w' < 2$ in Ω_1, (2.14) implies that the orbit cannot escape to infinity in Ω_1, that is, $\mathcal{O} \cap \Omega_1 \cap \{(w, w') : w > M\}$ is empty for large enough M. From (2.14), we also deduce that $\mathcal{O}$ goes right and downwards in Ω_1. So, either $\mathcal{O}$ remains in Ω_1, or $\mathcal{O}$ leaves Ω_1 at some time t_1. If the former case occurs, $(w(t), w'(t))$ remains in Ω_1 and converges to the unique stationary point $(0, 0)$ as $t \to +\infty$. This is excluded since $(0, 0)$ is a spiral attractor. So, $\mathcal{O}$ leaves Ω_1 at some time t_1 where $w'(t_1) = 0$ and $e^{w(t_1)} - 1 \geq 0$. In fact, $e^{w(t_1)} - 1 > 0$. Otherwise, it would follow by uniqueness for the second order ODE (2.12) with initial datum $w(t_1) = w'(t_1) = 0$ that $w \equiv 0$. Hence, $\mathcal{O}$ must enter Ω_2 at t_1. A similar analysis shows that thereon, $\mathcal{O}$ enters successively Ω_i, $i \in \mathbb{Z}/4\mathbb{Z}$ and remains in a bounded region of the phase-plane. In particular, w must converge to the unique stationary point $(0, 0)$. We summarize the previous results in Figure 2.4.

Going back to (2.1), the boundary condition $u(1) = 0$ translates to

$$
w(\tau) - 2\tau + a = 0,
$$

where τ satisfies

$$
\sqrt{\frac{2(N-2)}{\lambda e^a}} e^\tau = 1.
$$

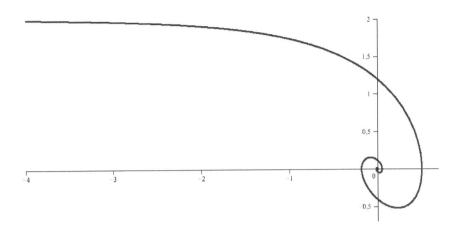

Figure 2.4: Phase portrait of $\mathcal{O}$ in the (w, w') plane.

This is equivalent to asking that

$$w(\tau) = \ln \frac{\lambda}{2(N-2)}.$$

From Figure 2.4, we find infinitely many such values of τ for $\lambda = 2(N-2)$, finitely many for λ in a bounded punctured interval $(0, \lambda^*) \setminus \{2(N-2)\}$, more than k solutions for λ close enough to $2(N-2)$, one solution for $\lambda = \lambda^*$, and none for $\lambda > \lambda^*$. □

Exercise 2.4.1
Let $N \geq 10$ and $\lambda^* = 2(N-2)$. Prove that (2.1) has a unique solution $u \in C^2(\overline{B})$ (which is stable) for $0 < \lambda < \lambda^*$, and no solution $u \in C^2(\overline{B})$ for $\lambda > \lambda^*$. Prove that u_λ converges pointwise to $\ln \frac{1}{|x|^2}$, as $\lambda \to \lambda^*$.

2.4.1 Stability analysis

In Proposition 2.4.1, we have proved that in dimension $3 \leq N \leq 9$, given any $\tau \in \mathbb{R}$, (λ_τ, u_τ) defined by

$$\begin{aligned} u_\tau(r) &= w(t) - 2t - (w(\tau) - 2\tau) \\ &= w(\ln r + \tau) - w(\tau) - 2\ln r, \end{aligned} \tag{2.15}$$

with $r = e^{t-\tau}$ and

$$\lambda_\tau = 2(N-2)e^{w(\tau)} \tag{2.16}$$

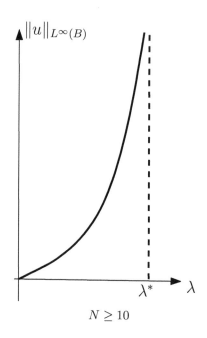

Figure 2.5: Bifurcation diagram in dimension $N \geq 10$.

satisfies (2.1). Conversely, every solution to (2.1) is expressed in the form (2.15) and (2.16). Thus, we can parametrize the solution set $\mathscr{S}$ with $\tau \in \mathbb{R}$. That is,

$$\mathscr{S} = \{(\lambda_\tau, u_\tau) \,:\, \tau \in \mathbb{R}\}.$$

From Figure 2.4, we infer that the orbit $\mathcal{O}$ crosses the w-axis infinitely many times. Let (τ_k) denote the corresponding crossing times, that is, $\tau_1 < \tau_2 < \cdots < \tau_k < \ldots$ and

$$w'(\tau_k) = 0. \tag{2.17}$$

Then,

$$w(\tau_2) < \cdots < w(\tau_{2k}) < w(\tau_{2k-1}) < \cdots < w(\tau_1),$$

and $w(t)$ achieves a local maximum and minimum at $t = \tau_{2k-1}$ and $t = \tau_{2k}$, respectively. $T_k = (\lambda_{\tau_k}, u_{\tau_k})$ is called a turning point.

Proposition 2.4.2 ([165]) *Let $3 \leq N \leq 9$ and $k \geq 1$. The Morse index of solutions to (2.1) belonging to the arc $(T_k, T_{k+1}]$ is at most equal to k.*

Proof. Consider the eigenvalue problem

$$\begin{cases} -\Delta\varphi - \lambda_\tau e^{u_\tau}\varphi = \mu\varphi & \text{in } B, \\ \qquad\qquad\qquad \varphi = 0 & \text{on } \partial B. \end{cases} \tag{2.18}$$

We claim that all eigenfunctions must be radial if $\mu \leq 0$.

Lemma 2.4.1 ([144]) *Assume that $\mu \leq 0$. Then, any solution $\varphi \in C^2(\overline{B})$ to (2.18) is radial.*

Proof. For $k \in \mathbb{N}$, let $\mu_k = k(k + N - 2)$ be the k-th eigenvalue of the Laplace-Beltrami operator $-\Delta_{\mathscr{S}^{N-1}}$ on the sphere (see Theorem C.4.1). Also let $\{\varphi_{k,l} : k \geq 0, l = 1, \ldots, m_k\}$ be an orthonormal system in $L^2(\mathscr{S}^{N-1})$ formed with eigenfunctions associated to each μ_k. Take a solution φ of (2.18) and set

$$a_{k,l}(r) = \int_{\mathscr{S}^{N-1}} \varphi \varphi_{k,l} d\sigma.$$

For $k \geq 1$, $a_{k,l}(0) = a_{k,l}(1) = 0$ and $a_{k,l}$ solves

$$a_{k,l}'' + \frac{N-1}{r}a_{k,l}' + \left[\lambda_\tau e^{u_\tau} - \frac{\mu_k}{r^2}\right]a_{k,l} + \mu a_{k,l} = 0. \tag{2.19}$$

We need to prove that $a_{k,l} \equiv 0$ for $k \geq 1$. Suppose to the contrary that $a_{k,l} \not\equiv 0$. Let r_0 be the first zero of $a_{k,l}$. Then, we may assume without loss of generality that $a_{k,l} > 0$ in $(0, r_0)$. Recall that $u = u_\tau$ is radial and radially decreasing. Then, $v := -u' > 0$ in $(0, r_0]$, $v(0) = 0$, and differentiating (2.1) with respect to r,

$$v'' + \frac{N-1}{r}v' + \left(\lambda_\tau e^{u_\tau} - \frac{N-1}{r^2}\right)v = 0, \quad \text{in } (0, r_0). \tag{2.20}$$

Multiplying (2.19) by $r^{N-1}v$, integrating and using (2.20), we deduce that

$$a_{k,l}'(r_0)v(r_0)r_0^{N-1} + \int_0^{r_0} \frac{N-1-\mu_k}{r^2}a_{k,l}(r)v(r)r^{N-1}\,dr =$$
$$-\mu \int_0^{r_0} a_{k,l}(r)v(r)r^{N-1}\,dr.$$

Since $a_{k,l}'(r_0) < 0$, $a_{k,l}, v > 0$ in $(0, r_0)$, and $N - 1 \leq \mu_k$, the left-hand side is negative. Thus, μ must be positive, a contradiction. $\square$

Thanks to the previous lemma, (2.18) reduces to

$$\begin{cases} \varphi'' + \dfrac{N-1}{r}\varphi' + \lambda_\tau e^{u_\tau}\varphi + \mu\varphi = 0 & \text{for } r \in (0,1) \\ \varphi'(0) = 0, \quad \varphi(1) = 0. \end{cases} \tag{2.21}$$

At a turning point T_k, $\mu = 0$ must be one of the eigenvalues of (2.21). To see this, simply observe that $v = \frac{du_\tau}{d\tau}$ is an eigenfunction associated to $\mu = 0$, as follows from differentiating (2.1) with respect to τ and using (2.16) and (2.17). So, 0 is the l-th eigenvalue of the linearized equation (2.21), for some $l \in \mathbb{N}$. We want to prove that $l = k$. For this, we first show that the l-th eigenvalue μ_τ of (2.21) changes sign from $+$ to $-$, as τ increases across τ_k. Note that

$$\frac{d\lambda_\tau}{d\tau}(\tau_k) = 0,$$

by (2.16) and (2.17). Since λ_τ and u_τ are smooth functions of τ, and since μ_τ is a simple eigenvalue, it follows from the implicit function theorem that μ_τ is a smooth function of τ. So, letting $v = \frac{du_\tau}{d\tau}$ and differentiating (2.1) in the τ variable at $u = u_\tau$, we get

$$\begin{cases} -\Delta v = \lambda_\tau e^{u_\tau} v + \dfrac{d\lambda_\tau}{d\tau} e^{u_\tau} & \text{in } B, \\ v = 0 & \text{on } \partial B. \end{cases} \tag{2.22}$$

At $\tau = \tau_k$, (2.22) simplifies to

$$\begin{cases} -\Delta v = \lambda_{\tau_k} e^{u_{\tau_k}} v & \text{in } B, \\ v = 0 & \text{on } \partial B. \end{cases} \tag{2.23}$$

That is, $v = du_\tau/d\tau\big|_{\tau=\tau_k}$ is an eigenfunction associated to the l-th eigenvalue 0 at $\tau = \tau_k$. Differentiating again (2.22) with respect to τ, we get for $z = \frac{d^2 u_\tau}{d\tau^2}\big|_{\tau=\tau_k}$,

$$\begin{cases} -\Delta z - \lambda_{\tau_k} e^{u_{\tau_k}} z = \lambda_{\tau_k} e^{u_{\tau_k}} v^2 + \dfrac{d^2\lambda_\tau}{d\tau^2}\bigg|_{\tau=\tau_k} e^{u_{\tau_k}} & \text{in } B, \\ z = 0 & \text{on } \partial B. \end{cases} \tag{2.24}$$

Multiplying by v and integrating, we deduce that

$$\lambda_{\tau_k} \int_B e^{u_{\tau_k}} v^3 \, dx + \frac{d^2\lambda_\tau}{d\tau^2}\bigg|_{\tau=\tau_k} \int_B e^{u_{\tau_k}} v \, dx = 0. \tag{2.25}$$

Integrating (2.23),

$$\lambda_{\tau_k} \int_B e^{u_{\tau_k}} v \, dx = -\int_{\partial B} \frac{\partial v}{\partial n} \, d\sigma = -|\mathscr{S}^{N-1}| v'(1). \tag{2.26}$$

Furthermore, by (2.15),

$$v'(1) = w''(\tau_k)$$

and by (2.16),

$$\frac{d^2\lambda_\tau}{d\tau^2}\bigg|_{\tau=\tau_k} = \lambda_{\tau_k} w''(\tau_k).$$

So, (2.25) becomes

$$\lambda_{\tau_k} \int_B e^{u_{\tau_k}} v^3 \, dx = |\mathscr{S}^{N-1}| w''(\tau_k)^2. \tag{2.27}$$

For any $\tau \in \mathbb{R}$, let φ_τ denote an eigenfunction associated to the l-th eigenvalue, normalized by $\varphi_\tau(0) = du_\tau/d\tau\big|_{\tau=\tau_k}$, that is,

$$\begin{cases} -\Delta\varphi_\tau - \lambda_\tau e^{u_\tau}\varphi_\tau = \mu_\tau\varphi_\tau & \text{in } B, \\ \varphi_\tau = 0 & \text{on } \partial B, \end{cases} \tag{2.28}$$

and

$$\varphi_{\tau_k} = \frac{du_\tau}{d\tau}\bigg|_{\tau=\tau_k}.$$

Differentiate (2.28) with respect to τ. We get for $\psi = \frac{d\varphi_\tau}{d\tau}$,

$$\begin{cases} -\Delta\psi - \lambda_\tau e^{u_\tau}\psi - \mu_\tau\psi = \lambda_\tau e^{u_\tau}\dfrac{du_\tau}{d\tau}\varphi_\tau + \dfrac{d\lambda_\tau}{d\tau}e^{u_\tau}\varphi_\tau + \dfrac{d\mu_\tau}{d\tau}\varphi_\tau & \text{in } B, \\ \psi = 0 & \text{on } \partial B. \end{cases}$$

At $\tau = \tau_k$, the equation reduces to

$$-\Delta\psi - \lambda_{\tau_k} e^{u_{\tau_k}}\psi = \lambda_{\tau_k} e^{u_{\tau_k}} v^2 + \frac{d\mu_\tau}{d\tau}\bigg|_{\tau=\tau_k} v.$$

Multiplying by v and integrating, we deduce that

$$\lambda_{\tau_k} \int_B e^{u_{\tau_k}} v^3 \, dx + \frac{d\mu_\tau}{d\tau}\bigg|_{\tau=\tau_k} \int_B v^2 \, dx = 0. \tag{2.29}$$

Using (2.27), we obtain

$$\frac{d\mu_\tau}{d\tau}\bigg|_{\tau=\tau_k} \int_B v^2 \, dx = -|\mathscr{S}^{N-1}| w''(\tau_k)^2.$$

43

By (2.12) and (2.17), $w''(t)$ never vanishes at $t = \tau_k$. Hence, the l-th eigenvalue to (2.18) decreases from $+$ to $-$ as τ increases through τ_k. Equivalently, the Morse index increases by one, as τ increases through τ_k.

We claim that on the arc $(T_k T_{k+1})$, zero is not an eigenvalue of (2.18) and hence the Morse index never changes there. If this were not the case, there would exist a time $\tau \neq \tau_k$ and an eigenfunction $\varphi \not\equiv 0$, which is radial by Lemma 2.4.1, solving

$$\begin{cases} -\Delta\varphi = \lambda_\tau e^{u_\tau}\varphi & \text{in } B, \\ \varphi = 0 & \text{on } \partial B. \end{cases} \tag{2.30}$$

Integrating the above equation, we obtain on the one hand

$$\lambda_\tau \int_B e^{u_\tau}\varphi \, dx = -|\mathscr{S}^{N-1}|\varphi'(1). \tag{2.31}$$

Multiplying (2.22) by φ and integrating, we have on the other hand

$$\frac{d\lambda_\tau}{d\tau} \int_B e^{u_\tau}\varphi \, dx = 0. \tag{2.32}$$

Since $\tau \neq \tau_k$, it follows from (2.16) that $\frac{d\lambda_\tau}{d\tau} \neq 0$. Collecting (2.31) and (2.32), we deduce that $\varphi'(1) = 0$. By uniqueness for the ode (2.30) with initial value $\varphi(1) = \varphi'(1) = 0$, this forces $\varphi \equiv 0$, a contradiction.

We have just proved that the Morse index remains constant on each arc $(T_k T_{k+1})$ and increases by one through each turning point T_k.

Finally, observe that any solution on the lowest part of the bifurcation diagram (Figure 2.3), that is, any solution of the form $(\lambda, u) = (\lambda_\tau, u_\tau)$, $\tau \in (-\infty, \tau_1]$, is minimal, hence stable. Proposition 2.4.2 follows.

$\square$

Exercise 2.4.2 Let $N = 2$. Prove that the unstable solution U_λ to (2.1) has Morse index $\text{ind}(U_\lambda) = 1$.

2.5 Summary

Let us review the main features of the bifurcation diagrams in Figure 2.6. The curve on in each of these diagrams is the graph

$$G = \{(\lambda, \|u\|_{L^\infty(B)}) : u \text{ is a solution to (2.1) with parameter } \lambda\}.$$

All classical solutions of the equation are represented.

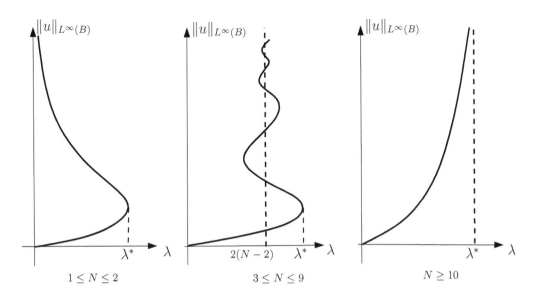

Figure 2.6: Bifurcation diagrams for the Gelfand problem.

1. **Nonexistence.** In any dimension $N \geq 1$, there exists no solution to the equation for λ larger than a certain value $\lambda^* = \lambda^*(N)$, called the extremal parameter. For $1 \leq N \leq 9$, $\lambda^* > 2(N-2)$ (for $N = 1$, $\lambda^* \simeq 0.88$, for $N = 2$, $\lambda^* = 2$ and for $N = 3$, $\lambda^* \simeq 3.32$). For $N \geq 10$, $\lambda^* = 2(N-2)$.

2. **Multiplicity.** Starting from right to left, we see that

   - For $N \geq 10$, there exists a unique solution for each $\lambda \in (0, \lambda^*)$.
   - For $3 \leq N \leq 9$, the problem has
     - a unique solution for λ sufficiently small,
     - finitely many solutions for $\lambda \neq 2(N-2)$,
     - more than any given number of solutions for λ sufficiently close to $2(N-2)$,
     - infinitely many solutions for $\lambda = 2(N-2)$,
     - two solutions for λ close to λ^*, and
     - a unique solution for $\lambda = \lambda^*$.
   - For $1 \leq N \leq 2$, the problem has exactly two solutions for every $\lambda \in (0, \lambda^*)$ and a unique one for $\lambda = \lambda^*$.

3. **The stable branch.** For each $\lambda \in (0, \lambda^*)$, the minimal solution must be the one with the smallest L^∞ norm. Furthermore, it is stable. By

45

Proposition 1.3.1, there exists at most one stable solution to the equation for each λ. So, the lowest piece of the diagrams is the branch of stable solutions.

4. **Boundedness of the stable branch.** In dimensions $1 \leq N \leq 9$, all stable solutions are uniformly bounded, while this fails in dimensions $N \geq 10$.

5. **Turning points.** In dimensions $1 \leq N \leq 9$, the solution curve turns around at $\lambda = \lambda^*$, where we have $\lambda_1(-\Delta - \lambda^* e^{u^*}, B) = 0$. All solutions lying above the stable branch are unstable, so their Morse index is at least equal to one. In dimensions $3 \leq N \leq 9$, the solution curve exhibits infinitely many turning points, accumulating toward $\lambda_s = 2(N-2), u_s = -2 \ln |x|$. At each of these points, the Morse index of solutions increases by one unit.

6. **Boundedness of the branch of index** k. It follows from the above discussion, that all solutions of Morse index at most k are uniformly bounded by a constant depending on k and N only.

7. **Singular solutions.** In dimensions $N \geq 2$, the curve has a vertical asymptote at $\lambda = \lambda_s = 2(N-2)$. Solutions accumulate toward $u_s(x) = -2 \ln |x|$, $\lambda_s = 2(N-2)$. Note that u_s is a solution to (2.1) (in the sense of distributions) only for $N \geq 3$. Note also that by Hardy's inequality (Proposition C.1.1), u_s is stable if $N \geq 10$, while u_s is not even of finite Morse index if $1 \leq N \leq 9$. Other singular solutions of the equation exist (see [186]) but they are not represented here.

Chapter 3

Extremal solutions

In Chapter 2, we obtained a complete picture of the set of classical solutions to the Gelfand problem (2.1) (see Figure 2.6). Such a result was obtained for a very specific equation and one may wonder what happens, say, if the domain is no longer a ball or if the nonlinearity is not the exponential function.

In this chapter, we consider the equation

$$\begin{cases} -\Delta u = \lambda f(u) & \text{in } \Omega, \\ u = 0 & \text{on } \partial\Omega, \end{cases} \tag{3.1}$$

where this time, $\Omega \subset \mathbb{R}^N$, $N \geq 1$, denotes any smoothly bounded domain. Throughout the chapter, the nonlinearity $f \in C^1(\mathbb{R})$ will be required to satisfy the following sign assumption:

$$f(t) \geq 0 \quad \text{for } t \geq 0. \tag{3.2}$$

See, for example, [197, 199] for more general nonlinearities.

Singular solutions play an important role in the study of (3.1), even if one is interested solely in classical solutions. Recall that

$$u_s(x) = -2\ln|x| \tag{3.3}$$

is a singular solution to the Gelfand problem (2.1) in the unit ball $B \subset \mathbb{R}^N$, $N \geq 3$, for $\lambda = \lambda_s := 2(N-2)$. Also recall that, when $N \geq 3$, u_s is the limit of the whole curve of regular solutions to the equation; the curve is said to bifurcate from infinity. Our first task, in the setting of (3.1), is to properly define weak solutions.

3.1 Weak solutions

3.1.1 Defining weak solutions

In this section, we identify the relevant notion of weak solutions to (1.3). In particular, we need to decide whether the equation is satisfied up to the singular set of the solution or only away from it. To answer this question, consider again the function u_s given by (3.3). In dimension $N \geq 3$, u_s solves (2.1) in $\mathscr{D}'(B)$ and is therefore a natural candidate. Now, for $N = 2$, Equation (2.1) with $\lambda = \lambda_s = 0$, simplifies to $\Delta u = 0$, that is, we request that u be harmonic in B. For sure, u_s should be excluded from the solution set. Since u_s is harmonic away from the origin (but $-\Delta u_s = 4\pi\delta_0$ in $\mathscr{D}'(\mathbb{R}^2)$), we shall request that the equation is satisfied *up to* the singular set. For the equation to hold, in the sense of distributions, we must also ask that weak solutions be (locally) integrable. This leads us to the following definition.

Definition 3.1.1 *Let $N \geq 1$, let $\Omega \subset \mathbb{R}^N$ denote a smoothly bounded domain, and let d_Ω denote the distance to the boundary of Ω, that is,*

$$d_\Omega(x) = dist(x, \partial\Omega) \qquad for\ all\ x \in \mathbb{R}^N. \tag{3.4}$$

Let $f \in C(\mathbb{R})$. We say that u is an L^1–weak solution (or simply a weak solution) to (1.3) if $u \in L^1(\Omega)$, $f(u)d_\Omega(x) \in L^1(\Omega)$, and

$$\int_\Omega u\,(-\Delta\varphi)\ dx = \int_\Omega f(u)\varphi\ dx, \tag{3.5}$$

for all $\varphi \in C_0^2(\overline{\Omega})$, where

$$C_0^2(\overline{\Omega}) = \left\{ \varphi \in C^2(\overline{\Omega}) \ : \ \varphi\big|_{\partial\Omega} \equiv 0 \right\}. \tag{3.6}$$

Remark 3.1.1 *Note that (3.5) makes sense for $\varphi \in C_0^2(\overline{\Omega})$. Indeed, by the mean-value theorem, $|\varphi(x)| \leq C d_\Omega(x)$.*

Note also that the boundary condition in (1.3) is encoded in the chosen space $C_0^2(\overline{\Omega})$ of test functions.

Exercise 3.1.1 Prove that if $u \in C^2(\overline{\Omega})$ is a classical solution to (1.3), then u is an L^1-weak solution to (1.3). Prove that if $u \in H_0^1(\Omega)$, $f(u)d_\Omega(x) \in L^1(\Omega)$, and

$$\int_\Omega \nabla u \nabla \varphi\ dx = \int_\Omega f(u)\varphi\ dx$$

for all $\varphi \in C_c^\infty(\Omega)$, then u is a weak solution to (1.3). Conversely, prove that if u is a weak solution, $f \in C^{0,\alpha}(\mathbb{R})$ for some $\alpha \in (0,1)$ and $f(u) \in L^p(\Omega)$ for some $p > N/2$, then in fact $u \in C^2(\overline{\Omega})$ and u solves the equation in the classical sense.

Exercise 3.1.1 shows that L^1-weak solutions are a natural extension of the standard notions of variational and classical solutions. It is however not clear at this stage why Definition 3.1.1 encompasses *all* the reasonable singularities one might encounter when dealing with equations of the form (1.3). The model singular solution given by (3.3) happens to belong to $L^1(B)$, but perhaps this is not general and weaker nonlinearities f might produce nonintegrable singularities. We prove next that singular solutions that do not belong to $L^1(\Omega)$ are rather unlikely (at least under the assumption $f \geq 0$). That is, they cannot be calculated by any reasonable approximation argument.

Definition 3.1.2 *Let $f \in C^1(\mathbb{R})$, $f \geq 0$. Consider an arbitrary nondecreasing sequence (f_n) of bounded continuous functions such that $f_n \nearrow f$ pointwise. Let $u_n \in C^2(\overline{\Omega})$ denote a solution to (1.3) with nonlinearity f_n. If for any choice of such (f_n) and (u_n), there holds*

$$\lim_{n \to +\infty} \frac{u_n(x)}{d_\Omega(x)} = +\infty \qquad \text{uniformly in } x \in \Omega,$$

where d_Ω is given by (3.4), then we say that there is complete blow-up in (1.3).

Remark 3.1.2 *A standard way of approximating f is to use the truncation f_n of f at level n, defined for $t \in \mathbb{R}$ by*

$$f_n(t) = \begin{cases} f(t) & \text{if } f(t) \leq n, \\ n & \text{if } f(t) > n. \end{cases} \tag{3.7}$$

Proposition 3.1.1 ([27]) *Let $N \geq 1$ and let $\Omega \subset \mathbb{R}^N$ denote a smoothly bounded domain of $\mathbb{R}^N$. Assume that $f \in C^1(\mathbb{R})$ satisfies $f \geq 0$. If there exists no weak solution to (1.3), then there is complete blow-up in (1.3).*

Proof. Take a nondecreasing sequence of bounded functions (f_n) converging pointwise to f. Let ζ_0 denote the unique solution to

$$\begin{cases} -\Delta\zeta_0 = 1 & \text{in } \Omega, \\ \zeta_0 = 0 & \text{on } \partial\Omega. \end{cases} \tag{3.8}$$

For each $n \in \mathbb{N}$, since f_n is bounded above by a constant M_n and below by 0, the functions $\underline{u} = 0$ and $\overline{u} = M_n\zeta_0$ form a sub- and a supersolution to the approximated problem (3.9) below. Furthermore, $\underline{u} < \bar{u}$ in Ω. By the method of sub- and supersolutions (Lemma 1.1.1), there exists a minimal nonnegative solution $u_n \in C^2(\overline{\Omega})$ of

$$\begin{cases} -\Delta u_n = f_n(u_n) & \text{in } \Omega, \\ u_n = 0 & \text{on } \partial\Omega. \end{cases} \tag{3.9}$$

We claim that

$$\lim_{n\to+\infty} \int_\Omega f_n(u_n)d_\Omega(x)\,dx = +\infty. \tag{3.10}$$

If not, since (u_n) is a nondecreasing sequence (each u_n being minimal), $\int_\Omega f_n(u_n)d_\Omega(x)\,dx \leq C$. Multiplying (3.9) with the solution ζ_0 of (3.8), we deduce that $\|u_n\|_{L^1(\Omega)} \leq C$. By monotone convergence, (u_n) converges in $L^1(\Omega)$ to a weak solution to (1.3), a contradiction. We have just proved (3.10). We may now apply the boundary point lemma (Proposition A.4.2) to conclude that for some constant $c = c(\Omega) > 0$,

$$u_n \geq c \left(\int_\Omega f_n(u_n)d_\Omega(x)\,dx \right) d_\Omega.$$

So complete blow-up occurs for the sequence of minimal solutions (u_n). Since $\tilde{u}_n \geq u_n$ for any other solution $\tilde{u}_n$ of (3.9), the result follows. $\qquad\square$

Exercise 3.1.2 Let $\Omega = (-1,1)$, $\lambda > 0$, and let $u \in L^2_{loc}(\Omega \setminus \{0\})$. Assume that u solves

$$-x^2 u'' = u^2 + \lambda, \qquad \text{in } \mathcal{D}'(\Omega \setminus \{0\}). \tag{3.11}$$

- Prove that $u \in C^\infty(\Omega \setminus \{0\})$ and that

$$|u(x)| \geq \lambda |\ln x| - C, \quad \text{near } x = 0.$$

- Let $\zeta_n \in C_c^\infty(\Omega \setminus \{0\})$ denote a cutoff function such that $0 \leq \zeta_n \leq 1$ in Ω, $\zeta_n(x) = 0$ for $|x| \leq 1/n$, and $\zeta_n(x) = 1$ for $|x| \geq 2/n$. Multiplying (3.11) by ζ_n^4/x^2, prove that

$$\int_{2/n}^{3/n} \frac{u^2}{x^2}\,dx \leq Cn.$$

- Deduce that there is no $u \in L^2_{loc}(\Omega \setminus \{0\})$, solving (3.11).

- Given $n \in \mathbb{N}^*$, prove that there exists $\lambda_n > 0$ and u_n solution to

$$\begin{cases} -(x^2 + 1/n)u_n'' = u_n^2 + \lambda_n & \text{in } \Omega, \\ u_n(-1) = u_n(1) = 0. \end{cases} \tag{3.12}$$

- Let λ_n^* denote the supremum of all λ_n such that (3.12) has a (classical) solution. Prove that $\lambda_n^* \to 0$, as $n \to +\infty$.

3.2 Stable weak solutions

The notion of stability for weak solutions is naturally defined as follows.

Definition 3.2.1 *Let $N \geq 1$, let Ω denote an open set of $\mathbb{R}^N$ and let $f \in C^1(\mathbb{R})$. Let $u \in L^1_{loc}(\Omega)$ satisfy $f(u) \in L^1_{loc}(\Omega)$ and*

$$-\Delta u = f(u), \qquad in \ \mathscr{D}'(\Omega).$$

We say that u is stable if $f'(u) \in L^1_{\mathrm{loc}}(\Omega)$ and (1.5) holds.

Remark 3.2.1 *By Hardy's inequality (Proposition C.1.1), the function $u_s(x) = -2\ln|x|$ is a stable weak solution to the Gelfand problem (2.1) if and only if $N \geq 10$.*

3.2.1 Uniqueness of stable weak solutions

Recall that for (strictly) convex f, there exists at most one classical stable solution to (1.3), see Proposition 1.3.1. For weak solutions, the situation is more delicate. A first partial answer is provided by the following proposition.

Proposition 3.2.1 ([32]) *Let $N \geq 1$ and $\Omega \subset \mathbb{R}^N$ denote a smoothly bounded domain. Let $\lambda_1 = \lambda_1(-\Delta; \Omega) > 0$ denote the principal eigenvalue of the Dirichlet Laplacian on Ω. Assume that $f \in C^1(\mathbb{R})$ is convex. Let u_1, u_2 denote two stable weak solutions of (1.3). In addition, assume that*

$$u_1, u_2 \in H_0^1(\Omega).$$

Then, either $u_1 = u_2$ almost everywhere (a.e.) or $f(u) = \lambda_1 u$ on the essential ranges of u_1 and u_2. In the latter case, u_1 and u_2 belong to the eigenspace associated to λ_1. In particular, they are colinear.

Proof. Simply repeat the proof of Proposition 1.3.1. $\qquad\qquad\qquad\qquad\square$

When solutions do not belong to the energy space $H_0^1(\Omega)$, uniqueness fails in general, as the following example demonstrates.

Example 3.2.1 ([32]) Let $N \geq 3$ and let $\Omega = B$ denote the unit ball in $\mathbb{R}^N$. Given any p in the range

$$\frac{N}{N-2} < p \leq \tilde{p} = \frac{N+2\sqrt{N-1}}{N-4+2\sqrt{N-1}}, \tag{3.13}$$

let $\lambda_s = \frac{2}{p-1}\left(N - \frac{2p}{p-1}\right) > 0$ and consider

$$\begin{cases} -\Delta u = \lambda_s(1+u)^p & \text{in } B, \\ \quad u = 0 & \text{on } \partial B. \end{cases} \tag{3.14}$$

There exists at least two stable solutions of (3.14). One of them belongs to $H_0^1(B)$, while the other does not and is given by $u_s(x) = |x|^{-\frac{2}{p-1}} - 1$.

Proof. Thanks to Hardy's inequality (Proposition C.1.1), a direct computation shows that u_s is a stable weak solution that does not belong to $H_0^1(B)$, for p in the range (3.13). By the method of sub- and supersolutions, there exists a stable solution u, satisfying $0 \leq u \leq u_s$ in Ω, obtained as the monotone limit of the sequence (u_n) given by $u_0 = 0$ and for $n \geq 1$,

$$\begin{cases} -\Delta u_n = \lambda_s(1+u_{n-1})^p & \text{in } B, \\ \quad u_n = 0 & \text{on } \partial B. \end{cases} \tag{3.15}$$

Since $u_s \notin H_0^1(B)$, it suffices to prove that $u \in H_0^1(B)$ to deduce that $u \neq u_s$. To do so, we observe by an obvious inductive argument, that each u_n is smooth. So, we need only prove that there exists a constant M such that for all $n \in \mathbb{N}$,

$$\|\nabla u_n\|_{L^2(B)} \leq M. \tag{3.16}$$

Multiply (3.15) by u_n and integrate. Then,

$$\int_B |\nabla u_n|^2 \, dx = \lambda_s \int_B (1+u_{n-1})^p u_n \, dx \leq \lambda_s \int_B (1+u_n)^p u_n \, dx. \tag{3.17}$$

Since u_s is stable, we also have

$$\int_B |\nabla u_n|^2 \, dx \geq \lambda_s p \int_B (1+u_s)^{p-1} u_n^2 \, dx \geq \lambda_s p \int_B (1+u_n)^{p-1} u_n^2 \, dx.$$

It follows that

$$p \int_B (1+u_n)^{p-1} u_n^2 \, dx \leq \int_B (1+u_n)^p u_n \, dx,$$

and so,

$$(p-1)\int_B (1+u_n)^p u_n \, dx \le p \int_B (1+u_n)^{p-1} u_n \, dx$$

$$\le p \int_{[\frac{p+1}{p-1} < u_n]} (1+u_n)^{p-1} u_n \, dx + p \int_{[\frac{p+1}{p-1} \ge u_n]} (1+u_n)^{p-1} u_n \, dx$$

$$\le p \int_{[\frac{p+1}{p-1} < u_n]} (1+u_n)^{p-1} u_n \, dx + C(p,N)$$

$$\le \frac{p-1}{2} \int_{[\frac{p+1}{p-1} < u_n]} (1+u_n)^p u_n \, dx + C.$$

It follows that

$$\int_B (1+u_n)^p u_n \, dx \le C.$$

Recalling (3.17), we deduce (3.16). □

3.2.2   Approximation of stable weak solutions

As we have seen in Chapter 2, when $N \ge 10$, the singular solution u_s given by (3.3) is the limit of the branch of stable solutions for the Gelfand problem (2.1). In other words, u_s can be approximated by a curve of smooth stable solutions to the same equation (with a slightly smaller value of the parameter λ). This approximation procedure turns out to hold in a more general setting, as the following two results demonstrate.

Theorem 3.2.1 ([29]) *Let $N \ge 1$ and $\Omega \subset \mathbb{R}^N$ denote a smoothly bounded domain. Assume that $f \in C^1(\mathbb{R})$, $f \ge 0$, and f is convex. Assume that*

$$u \in L^1(\Omega)$$

is a stable weak solution to (1.3). Then, for every $\varepsilon > 0$, there exists a stable solution to

$$\begin{cases} -\Delta u_\varepsilon = (1-\varepsilon)f(u_\varepsilon) & \text{in } \Omega, \\ u_\varepsilon = 0 & \text{on } \partial\Omega, \end{cases} \tag{3.18}$$

which is bounded, hence regular.

Combining this with the uniqueness result Proposition 3.2.1, we obtain the following corollary.

Corollary 3.2.1 *Let $N \geq 1$ and $\Omega \subset \mathbb{R}^N$ denote a smoothly bounded domain. Assume that $f \in C^1(\mathbb{R})$, $f \geq 0$, and f is convex. Assume that*

$$u \in H_0^1(\Omega)$$

is a weak solution to (1.3). Then, $u = \lim_{\varepsilon \to 0} u_\varepsilon$, where u_ε is the classical stable solution to (3.18) and where convergence holds in $H_0^1(\Omega)$.

To prove Theorem 3.2.1, we establish first the following lemma, known as the concave truncation method, or Kato's inequality.

Lemma 3.2.1 ([133], [29]) *Let $N \geq 1$ and $\Omega \subset \mathbb{R}^N$ denote a smoothly bounded domain. Let d_Ω denote the distance to the boundary of Ω, as defined in (3.4). Given $f \in L^1(\Omega, d_\Omega(x)dx)$, let $u \in L^1(\Omega)$ denote the unique weak solution to*

$$\begin{cases} -\Delta u = f & \text{in } \Omega, \\ \quad\ \, u = 0 & \text{on } \partial\Omega, \end{cases}$$

given by Lemma A.9.1. Let $\Phi \in C^1(\mathbb{R})$ denote a concave function such that Φ' is bounded and $\Phi(0) \geq 0$. Then, $\Phi(u) \in L^1(\Omega)$ and

$$-\Delta\Phi(u) \geq \Phi'(u)f \qquad \text{in } \Omega,$$

in the sense that

$$-\int_\Omega \Phi(u)\Delta\varphi \, dx \geq \int_\Omega \Phi'(u)f\varphi \, dx, \qquad \text{for all } \varphi \in C_0^2(\overline{\Omega}), \ \varphi \geq 0.$$

Remark 3.2.2 *The lemma remains valid for $\Phi(u) = -u^+$, that is,*

$$\int_\Omega u^+\Delta\varphi \, dx \geq -\int_{[u\geq 0]} f\varphi \, dx, \qquad \forall \, \varphi \in C_0^2(\overline{\Omega}), \varphi \geq 0. \qquad (3.19)$$

Proof. Take $f_n \in C_c^1(\Omega)$ such that $f_n \to f$ in $L^1(\Omega, d_\Omega(x)dx)$. Let $u_n \in C^2(\overline{\Omega})$ denote the solution to

$$\begin{cases} -\Delta u_n = f_n & \text{in } \Omega, \\ \quad\ \, u_n = 0 & \text{on } \partial\Omega. \end{cases}$$

Then,

$$-\Delta\Phi(u_n) = -\Phi'(u_n)\Delta u_n - \Phi''(u_n)|\nabla u_n|^2 \geq -\Phi'(u_n)\Delta u_n = \Phi'(u_n)f_n.$$

Therefore, given $\varphi \in C_0^2(\overline{\Omega})$, $\varphi \geq 0$,

$$-\int_\Omega \Phi(u_n)\Delta\varphi \, dx \geq \int_\Omega \Phi'(u_n)f_n\varphi \, dx.$$

By Lemma A.9.1, $u_n \to u$ in $L^1(\Omega)$. Since Φ' is bounded,

$$|\Phi(t)| \leq C|t| + \Phi(0),$$

for all $t \in \mathbb{R}$. So, we also have $\Phi(u_n) \to \Phi(u) \in L^1(\Omega)$. The lemma follows letting $n \to +\infty$. To see that Remark 3.2.2 holds, take a sequence of smooth concave functions in $\mathbb{R}$ such that $\Phi_n(t) = -t$ if $t \geq 0$ and $|\Phi_n(t)| \leq 1/n$ if $t < 0$. In particular, $0 \geq \Phi_n' \geq -1$ in $\mathbb{R}$. So, we may apply Lemma 3.2.1: given $\varphi \in C_0^2(\overline{\Omega})$, $\varphi \geq 0$,

$$-\int_\Omega \Phi_n(u)\Delta\varphi \, dx \geq \int_\Omega \Phi_n'(u)f\,\varphi \, dx.$$

Letting $n \to +\infty$, we deduce (3.19). □

Proof of Theorem 3.2.1. Thanks to Lemma 3.2.1, we may construct a bounded supersolution $U \geq 0$ to (3.18) as follows. If $f(t_0) = 0$ for some $t_0 \geq 0$, then t_0 is a supersolution to (3.18). By the method of sub- and super-solutions, we deduce that there exists a bounded stable solution to (3.18). If $f(t) > 0$ for $t \geq 0$, we seek a supersolution to the form $U = \Phi(u)$, where Φ satisfies the assumptions of Lemma 3.2.1. Then, U is a supersolution to (3.18) as soon as

$$\Phi'(u)f(u) \geq (1-\varepsilon)f(\Phi(u)).$$

So, we set Φ to be a solution to the differential equation

$$\Phi'(t)f(t) = (1-\varepsilon)f(\Phi(t)), \qquad \text{for all } t > 0.$$

with initial value $\Phi(0) = 0$. In other words, for $t > 0$, $\Phi(t)$ is the unique real number such that

$$\int_0^{\Phi(t)} \frac{ds}{f(s)} = (1-\varepsilon)\int_0^t \frac{ds}{f(s)}. \tag{3.20}$$

Note that $0 \leq \Phi(t) \leq t$, that Φ is C^2, that $\Phi' \geq 0$, and that

$$
\begin{aligned}
\Phi''(t) &= (1-\varepsilon)\left(\frac{f(\Phi(t))}{f(t)}\right)' \\
&= \frac{1-\varepsilon}{f(t)^2}(f'(\Phi(t))\Phi'(t)f(t) - f(\Phi(t))f'(t)) \\
&= \frac{(1-\varepsilon)f(\Phi(t))}{f(t)^2}((1-\varepsilon)f'(\Phi(t)) - f'(t)) \\
&\leq \frac{(1-\varepsilon)f(\Phi(t))}{f(t)^2}(f'(\Phi(t)) - f'(t)).
\end{aligned}
$$

Since $\Phi(t) \leq t$ and f is convex, we deduce that Φ is concave. In particular, $0 \leq \Phi'(t) \leq \Phi'(0)$ for $t \geq 0$, that is, Φ' is bounded. So, we may apply Lemma 3.2.1 and deduce that U is a supersolution to (3.18). We split the rest of the proof in two cases:

Case 1. Assume that

$$
\int_0^{+\infty} \frac{ds}{f(s)} < +\infty.
$$

Then, it follows from (3.20) that Φ is bounded and so must be U. By the method of sub- and supersolutions, we deduce that there exists a classical stable solution to (3.18).

Case 2. Assume that

$$
\int_0^{+\infty} \frac{ds}{f(s)} = +\infty.
$$

If f is nonincreasing, then f is bounded and $U = u$ provides a bounded supersolution to (3.18). Otherwise, since f is convex, there exists $t_0 \geq 0$ such that $f'(t) > 0$ for $t \geq t_0$. So, the function h defined for $t \geq 0$ by

$$
h(t) = \int_0^t \frac{ds}{f(s)}
$$

is concave in $[t_0, +\infty)$ and for $t_0 \leq U \leq u$,

$$
h(u) \leq h(U) + h'(U)(u - U) = h(U) + \frac{u - U}{f(U)}.
$$

Apply the above inequality with $U = \Phi(u)$. Then, $h(U) = (1-\varepsilon)h(u)$ and

$$
\varepsilon f(U) \leq \frac{u - U}{h(u)} \leq \frac{u}{h(u)} \leq Cu,
$$

56

whenever $u \geq U \geq t_0$. Clearly, we also have $f(U) \leq C$ for $0 \leq U \leq t_0$ and so

$$f(U) \leq C(1+u), \tag{3.21}$$

for all $0 \leq U \leq u$. By the definition of Φ and Lemma 3.2.1, $U(x) = \Phi(u(x))$ is a weak supersolution to

$$\begin{cases} -\Delta u_1 = (1-\varepsilon)f(u_1) & \text{in } \Omega, \\ \quad u_1 = 0 & \text{on } \partial\Omega. \end{cases}$$

By the method of sub- and supersolutions, there exists a stable weak solution u_1 of the above equation, such that $0 \leq u_1 \leq U$. In particular, we have $0 \leq f(u_1) \leq f(U)$. By (3.21) and Corollary A.9.1, we deduce[1] that $u_1 \in L^p(\Omega)$ for all

$$1 \leq p < N/(N-1). \tag{3.22}$$

By the same construction, we find a solution to

$$\begin{cases} -\Delta u_2 = (1-\varepsilon)^2 f(u_2) & \text{in } \Omega, \\ \quad u_2 = 0 & \text{on } \partial\Omega. \end{cases}$$

such that $0 \leq u_2 \leq u_1$ and $f(u_2) \leq C(1+u_1)$. In particular, $f(u_2) \in L^p(\Omega)$ for any p in the range (3.22). This implies that $u_2 \in L^q(\Omega)$ for all $q < \frac{N}{N-3}$. By iteration, the solution u_k to the equation

$$\begin{cases} -\Delta u_k = (1-\varepsilon)^k f(u_k) & \text{in } \Omega, \\ \quad u_k = 0 & \text{on } \partial\Omega, \end{cases}$$

is bounded provided $k \geq [(N+1)/2]+1$. Since $\varepsilon \in (0,1)$, this completes the proof. $\qquad\square$

Proof of Corollary 3.2.1. By Theorem 3.2.1, the stable solution u_ε of (3.18) is classical. u_ε is also minimal, hence $0 \leq u_\varepsilon \leq u_{\varepsilon'} \leq u$ for $\varepsilon' \leq \varepsilon$. By monotone convergence, (u_ε) converges in $L^1(\Omega)$ to a stable weak solution $v \in L^1(\Omega)$ such that $0 \leq v \leq u$. Also, since f is convex, the function $g(t) = f(t) + \Lambda t$, with $\Lambda = f'(0)^-$, is nondecreasing in $\mathbb{R}^+$. Multiplying (3.18) by u_ε and integrating, we obtain

$$\int_\Omega |\nabla u_\varepsilon|^2\, dx = (1-\varepsilon) \int_\Omega f(u_\varepsilon)u_\varepsilon\, dx \leq \int_\Omega g(u_\varepsilon)u_\varepsilon\, dx$$

$$\leq \int_\Omega g(u)u\, dx = \int_\Omega |\nabla u|^2\, dx + \Lambda \int_\Omega u^2\, dx < +\infty. \tag{3.23}$$

---

[1]In fact, we could even take $p < N/(N-2)$, since $f(U)$ is $L^1(\Omega)$ and not merely $L^1(\Omega, d_\Omega\, dx)$ as in Corollary A.9.1.

It follows that (u_ε) is bounded in $H^1_0(\Omega)$, hence $v \in H^1_0(\Omega)$. By Proposition 3.2.1, $u = v$. Returning to (3.23), we deduce from the monotone convergence theorem and the compactness of (u_ε) in $L^2(\Omega)$ that

$$\int_\Omega |\nabla u_\varepsilon|^2 \, dx = (1 - \varepsilon) \int_\Omega f(u_\varepsilon) u_\varepsilon \, dx$$

$$= (1 - \varepsilon) \left(\int_\Omega g(u_\varepsilon) u_\varepsilon \, dx - \Lambda \int_\Omega u_\varepsilon^2 \, dx \right)$$

$$\to \int_\Omega g(u) u \, dx - \Lambda \int_\Omega u^2 \, dx = \int_\Omega f(u) u \, dx$$

$$= \int_\Omega |\nabla u|^2 \, dx,$$

as $\varepsilon \to 0^+$ and so $u_\varepsilon \to u$ in $H^1_0(\Omega)$.

$\square$

Exercise 3.2.1 Assume that there exists a classical solution to

$$\begin{cases} v_t - \Delta v = f(v) & \text{in } \Omega \times (0, +\infty), \\ v = 0 & \text{on } \partial\Omega \times (0, +\infty), \\ v = 0 & \text{in } \Omega \times \{0\}, \end{cases} \tag{3.24}$$

where $f \in C^2(\mathbb{R})$ is such that $f, f', f'' > 0$.

- Prove that there exists a bounded global solution $v = v_\varepsilon$ to

$$\begin{cases} v_t - \Delta v = (1 - \varepsilon) f(v) & \text{in } \Omega \times (0, +\infty), \\ v = 0 & \text{on } \partial\Omega \times (0, +\infty), \\ v = 0 & \text{in } \Omega \times \{0\}. \end{cases}$$

- Prove that $u_\varepsilon = \lim_{t \to +\infty} v_\varepsilon(t)$ solves (3.18).

3.3 The stable branch

In this section, we construct and analyze the branch to stable solutions of (3.1).

Proposition 3.3.1 ([29, 63, 136, 137]) *Assume that $N \geq 1$. Let $\Omega \subset \mathbb{R}^N$ denote a smoothly bounded domain. Assume that $f \in C^1(\mathbb{R})$, $f \geq 0$. Then, there exists $\lambda^* = \lambda^*(\Omega, N, f) \in (0, +\infty]$ such that*

- *For $0 < \lambda < \lambda^*$, there exists the minimal solution $u_\lambda \in C^2(\overline{\Omega})$ of (3.1). In particular, u_λ is stable. If in addition f is convex, then u_λ is the unique classical stable solution to (3.1).*

- *For $\lambda > \lambda^*$, there exists no classical solution $u \in C^2(\overline{\Omega})$ of (3.1). In addition, if f is convex or if Ω is a ball, then there exists no weak solution to (3.1) either.*

Remark 3.3.1 *λ^* is called the extremal parameter of (3.1).*

Remark 3.3.2 *As we have seen in Chapter 1, the minimal solution to a semilinear elliptic problem is defined relatively to a subsolution. Here, since $f \geq 0$, $\underline{u} = 0$ is a subsolution to (3.1) and by the maximum principle (Proposition A.2.2), every solution to (3.1) is positive. So, u_λ is minimal among all solutions of the problem.*

Proof. Let $\zeta_0 \in C^2(\overline{\Omega})$ denote the solution to (3.8). For $\lambda > 0$ sufficiently small, we have $1 \geq \lambda f(\zeta_0)$. So, $\overline{u} = \zeta_0$ is a supersolution to (3.1), while $\underline{u} = 0$ is a subsolution. In addition, $\overline{u} > \underline{u}$. So, we may apply the method of sub- and supersolutions (Lemma 1.1.1) and obtain the minimal solution u_λ to (3.1) for $\lambda > 0$ small. Define

$$\lambda^* = \sup\{\lambda > 0 \; : \; (3.1) \text{ has a classical solution } u \in C^2(\overline{\Omega})\}.$$

We claim that there exists solutions to (3.1) for all $\lambda \in (0, \lambda^*)$. Fix such a λ. By definition of λ^*, there exists $\mu \in (\lambda, \lambda^*)$ and a function $u_\mu \in C^2(\overline{\Omega})$ solving

$$\begin{cases} -\Delta u_\mu = \mu f(u_\mu) & \text{in } \Omega, \\ \quad u_\mu = 0 & \text{on } \partial\Omega. \end{cases}$$

In particular, since $\mu > \lambda$, $\overline{u} = u_\mu$ is a supersolution to (3.1), while $\underline{u} = 0$ is a subsolution, and $\overline{u} \geq \underline{u}$. By the method of sub- and supersolutions, there exists the minimal solution $u_\lambda \in C^2(\overline{\Omega})$ to (3.1). By Proposition 1.3.1, if f is convex, then u_λ is the unique stable solution to (3.1).

It remains to prove that there exists no weak solution to (3.1) for $\lambda > \lambda^*$, whenever f is convex or Ω is a ball. Assume by contradiction that there exists a weak solution $u \in L^1(\Omega)$ of (3.1) for some $\lambda > \lambda^*$. By the method of sub- and supersolutions, we may always assume that u is the minimal solution to (3.1).

59

If f is convex, Theorem 3.2.1 implies that there exists a classical solution to the problem for all $\mu \in [0, \lambda)$, which contradicts the definition of λ^*. Assume now that Ω is a ball B. Assume for simplicity that B is the unit ball centered at the origin. Since u is minimal, u must be radial. Indeed, given any rotation of space R, $v(x) = u(Rx)$ is a weak solution to (3.1). Since u is minimal, $u(x) \leq v(x) = u(Rx)$ for $x \in B$. Applying the inequality at $y = R^{-1}x$, we deduce that $u(x) = u(Rx)$ for all $x \in B$ and so u is radial. In particular, u must be regular away from the origin and $\partial u / \partial r < 0$ in $B \setminus \{0\}$. Given $\varepsilon > 0$, set $V(x) = V(r) := u(r + \varepsilon)$, for $r = |x| \leq 1 - \varepsilon$. Then, V is a bounded function and for $r \in (0, 1 - \varepsilon)$, we have

$$-V'' - \frac{N-1}{r}V' = -V'' - \frac{N-1}{r+\varepsilon}V' - (N-1)\left(\frac{1}{r} - \frac{1}{r+\varepsilon}\right)V' \geq \lambda f(V).$$

Take a test function $\varphi \in C_0^2(\overline{B_{1-\varepsilon}})$, $\varphi \geq 0$ and $\delta > 0$. Since V is bounded, it follows that

$$\int_{B_{1-\varepsilon}\setminus B_\delta} V(-\Delta\varphi)\,dx = \int_{B_{1-\varepsilon}\setminus B_\delta} (-\Delta V)\varphi\,dx - \int_{\partial B_\delta}\left(-V\frac{\partial\varphi}{\partial r} + \varphi\frac{\partial V}{\partial r}\right)d\sigma$$

$$\geq \lambda \int_{B_{1-\varepsilon}\setminus B_\delta} f(V)\varphi\,dx - \|V\|_\infty\|\nabla\varphi\|_\infty|\partial B_\delta|.$$

Letting $\delta \to 0$, we deduce that

$$\int_{B_{1-\varepsilon}} V(-\Delta\varphi)\,dx \geq \lambda \int_{B_{1-\varepsilon}} f(V)\varphi\,dx.$$

It follows that the function U defined by $U(x) = V((1-\varepsilon)x)$ for $x \in B$ is a bounded weak supersolution to

$$\begin{cases} -\Delta u = (1-\varepsilon)^2 \lambda f(u) & \text{in } B, \\ u = 0 & \text{on } \partial B. \end{cases}$$

By the method of sub- and supersolutions, we deduce that there exists a classical solution to the aforementioned equation, contradicting the definition of λ^*. □

Exercise 3.3.1 Consider the semilinear boundary value problem

$$\begin{cases} \Delta u = 0 & \text{in } \Omega \\ \dfrac{\partial u}{\partial v} = \lambda f(u) & \text{on } \Gamma_1 \\ u = 0 & \text{on } \Gamma_2 \end{cases} \tag{3.25}$$

where Γ_1, Γ_2 is a partition of $\partial\Omega$ into surfaces separated by a smooth interface. We say that u is a weak solution to (3.25) if $u \in W^{1,1}(\Omega)$, $f(u) \in L^1(\Gamma_1)$, and

$$\int_\Omega u(-\Delta\varphi)\, dx = \int_{\Gamma_1} \lambda f(u)\varphi\, d\sigma \quad \forall\varphi \in C^2(\bar\Omega) \text{ s.t. } \varphi\Big|_{\Gamma_2} \equiv 0 \text{ and } \frac{\partial\varphi}{\partial\nu}\Big|_{\Gamma_1} \equiv 0.$$

Assume that $f \in C^1(\mathbb{R})$, f convex, and $f \geq 0$. Prove that there exists $\lambda^* \in (0,\infty]$ such that

- (3.25) has a smooth solution for $0 \leq \lambda < \lambda^*$, and

- (3.25) has no solution for $\lambda > \lambda^*$ (even in the weak sense).

Moreover, prove that for $0 \leq \lambda < \lambda^*$, there exists a minimal solution u_λ which is bounded, positive, and stable, in the sense that

$$\inf_{\varphi \in C^1(\bar\Omega), \varphi=0 \text{ on } \Gamma_2} \int_\Omega |\nabla\varphi|^2\, dx - \lambda \int_{\Gamma_1} f'(u_\lambda)\varphi^2\, d\sigma \geq 0.$$

Proposition 3.3.1 gives rise to a number of natural questions, which we address now.

3.3.1 When is λ^* finite?

Proposition 3.3.1 *Assume $N \geq 1$. Let $\Omega \subset \mathbb{R}^N$ denote a smoothly bounded domain. Assume that $f \in C^1(\mathbb{R})$, $f \geq 0$, $f(0) \neq 0$.*

- *If $\inf_{t\in\mathbb{R}^+} \frac{f(t)}{t} > 0$, then $\lambda^* < +\infty$.*

- *If $\inf_{t\in\mathbb{R}^+} \frac{\|f\|_{L^\infty(0,t)}}{t} = 0$, then $\lambda^* = +\infty$.*

Remark 3.3.3 *If $f(0) = 0$, then 0 is a trivial solution to (3.1) for all $\lambda > 0$ and so $\lambda^* = +\infty$.*

Remark 3.3.4 *Under the additional assumption that f is nondecreasing, the proposition is sharp: $\lambda^* < +\infty$ if and only if $\inf_{t\in\mathbb{R}^+} \frac{f(t)}{t} > 0$.*

Proof. Assume first that $a := \inf_{t\in\mathbb{R}^+} f(t)/t > 0$. Let $\varphi_1 > 0$ denote an eigenvector associated to the principal eigenvalue $\lambda_1 = \lambda_1(-\Delta;\Omega)$. Assume that $u \in C^2(\bar\Omega)$ is a solution to (3.1). Multiply (3.1) by φ_1 and integrate by parts:

$$\lambda \int_\Omega f(u)\varphi_1\, dx = \int_\Omega -\Delta u\, \varphi_1\, dx = \int_\Omega -\Delta\varphi_1\, u\, dx = \lambda_1 \int_\Omega u\varphi_1\, dx.$$

Since $f(t) \geq at$ for $t \geq 0$,

$$(a\lambda - \lambda_1) \int_\Omega u\varphi_1 \, dx \leq 0.$$

By the strong maximum principle (Proposition A.2.2) applied to (3.1), $u > 0$ and we deduce that $a\lambda \leq \lambda_1$. In particular, no classical solution to (3.1) exists for large values of λ.

Assume now that $\inf_{t \in \mathbb{R}^+} \frac{\|f\|_{L^\infty(0,t)}}{t} = 0$. Let ζ_0 denote the solution to (3.8). Also let $M = \|\zeta_0\|_{L^\infty(\Omega)}$. Fix $\lambda > 0$ and take $t_0 > 0$ such that $\frac{\|f\|_{L^\infty(0,t_0)}}{t_0} \leq \frac{1}{\lambda M}$. Set $\bar{u} = t_0 \frac{\zeta_0}{M}$. Then,

$$-\Delta \bar{u} = \frac{t_0}{M} \geq \lambda \|f\|_{L^\infty(0,t_0)} \geq \lambda f(\bar{u}).$$

Using the method of sub- and supersolutions, we obtain a solution to (3.1) for any $\lambda > 0$. □

3.3.2 What happens at $\lambda = \lambda^*$?

We now remain with the case where λ^* is finite. In particular, we must assume $f > 0$ (otherwise, if $t_0 \geq 0$ is such that $f(t_0) = 0$, then $\bar{u} = t_0$ is a supersolution to (3.1) and so the equation is solvable for all $\lambda > 0$). Results vary according to the behavior of f at infinity. We describe next the case of a superlinear nonlinearity.

The superlinear case

Proposition 3.3.2 ([63, 136, 137, 149]) *Assume $N \geq 1$. Let $\Omega \subset \mathbb{R}^N$ denote a smoothly bounded domain. Assume that $f \in C^1(\mathbb{R})$, $f > 0$. In addition, assume that f is nonndecreasing and that f is superlinear in the following sense:*

$$\lim_{t \to +\infty} \frac{f(t)}{t} = +\infty. \tag{3.26}$$

Then, the family $(u_\lambda)_{0 < \lambda < \lambda^*}$ of minimal solutions of (3.1) converges to a weak solution $u^* \in L^1(\Omega)$ of (3.1) for $\lambda = \lambda^*$. If in addition f is convex, then u^* is the unique weak solution to (3.1) for $\lambda = \lambda^*$.

Remark 3.3.5 *Since $f > 0$ and f satisfies (3.26), $\inf_{\mathbb{R}^+} f(t)/t > 0$. By Proposition 3.3.1, we are in the situation where $\lambda^* < +\infty$.*

Definition 3.3.1 *The solution u^* is called the extremal solution to (3.1).*

Before proving Proposition 3.3.2, we give a convenient criterion for identifying the extremal solution.

Corollary 3.3.1 ([32]) *Assume $N \geq 1$. Let $\Omega \subset \mathbb{R}^N$ denote a smoothly bounded domain. Assume that $f \in C^1(\mathbb{R})$, $f > 0$, and that f is nondecreasing and convex. If $u \in H_0^1(\Omega)$ is an unbounded stable solution to (3.1) associated to a parameter $\lambda > 0$, then necessarily $u = u^*$ and $\lambda = \lambda^*$.*

Proof of Corollary 3.3.1. Assume by contradiction that $\lambda < \lambda^*$. By Proposition 3.2.1, u must coincide with the minimal classical solution u_λ, a contradiction. So, $\lambda = \lambda^*$ and by minimality of $u_{\lambda'}$, $\lambda' < \lambda^*$, we must have $u_{\lambda'} \leq u$, hence $u^* \leq u$. Also,

$$\int_\Omega |\nabla u_{\lambda'}|^2 \, dx = \lambda' \int_\Omega f(u_{\lambda'}) u_{\lambda'} \, dx \leq \lambda^* \int_\Omega f(u) u \, dx = \int_\Omega |\nabla u|^2 \, dx < +\infty.$$

So, $u^* \in H_0^1(\Omega)$ and by Proposition 3.2.1, $u = u^*$. □

Proof of Proposition 3.3.2. We claim that there exists a constant $C > 0$ independent of λ, such that

$$\|u_\lambda\|_{L^1(\Omega)} \leq C \qquad \text{for all } \lambda \in (0, \lambda^*). \tag{3.27}$$

Fix $\lambda \in (0, \lambda^*)$ and multiply (3.1) by $\varphi_1 > 0$, an eigenvector associated to $\lambda_1 = \lambda_1(-\Delta; \Omega)$.

$$\lambda_1 \int_\Omega u_\lambda \varphi_1 \, dx = \lambda \int_\Omega f(u_\lambda) \varphi_1 \, dx. \tag{3.28}$$

f is superlinear, so for all $\varepsilon > 0$, there exists $C_\varepsilon > 0$ such that for all $t \geq 0$, $f(t) \geq \frac{1}{\varepsilon} t - C_\varepsilon$. Hence,

$$C \geq \left(\frac{\lambda}{\varepsilon} - \lambda_1 \right) \int_\Omega u_\lambda \varphi_1 \, dx.$$

Choosing $\varepsilon = \frac{\lambda}{2\lambda_1}$, we obtain that $(u_\lambda d_\Omega(x))$ is bounded in $L^1(\Omega)$. By (3.28), so is $(f(u_\lambda) d_\Omega)$. Test again (3.1), this time with ζ_0 solving (3.8).

$$\int_\Omega u_\lambda \cdot 1 \, dx = \int_\Omega (-\Delta u_\lambda) \zeta_0 \, dx = \lambda \int_\Omega f(u_\lambda) \zeta_0 \, dx$$

and (3.27) follows. Since u_λ is minimal, the sequence (u_λ) is nondecreasing in λ. So, $u_\lambda \nearrow u^* \in L^1(\Omega)$. Since f is nondecreasing, $f(u_\lambda) \nearrow f(u^*)$ in $L^1(\Omega, d_\Omega(x) \, dx)$. Passing to the limit in (3.5) as we may, we deduce that u^* is a weak solution to (3.1) for $\lambda = \lambda^*$. At last, u^* is unique when f is convex, thanks to Corollary 3.3.2 below. □

Theorem 3.3.1 ([149]) *Let $N \geq 1$, let $\Omega \subset \mathbb{R}^N$ denote a smoothly bounded domain and let $f \in C^1(\mathbb{R})$ denote a nondecreasing convex function such that $f > 0$. Let λ^* denote the associated extremal parameter. Assume that there exists $v \in L^1(\Omega)$ such that $v \geq 0$ a.e., $f(v)d_\Omega \in L^1(\Omega)$, where d_Ω is given by (3.4), and*

$$-\int_\Omega v\Delta\varphi \, dx \geq \lambda^* \int_\Omega f(v)\varphi \, dx \qquad \text{for all } \varphi \in C_0^2(\overline{\Omega}) \text{ such that } \varphi \geq 0.$$

Then, $v = u^*$ is the extremal solution to (3.1).

The following corollary is immediate.

Corollary 3.3.2 ([149]) *Under the assumptions of Theorem 3.3.1, there is at most one weak solution to (3.1) for $\lambda = \lambda^*$.*

We establish two intermediate lemmata in order to prove Theorem 3.3.1.

Lemma 3.3.1 ([149]) *Let $N \geq 1$, let $\Omega \subset \mathbb{R}^N$ denote a smoothly bounded domain and let $f \in C^1(\mathbb{R})$, $f > 0$, denote a nondecreasing convex function. Also fix $\epsilon > 0$. Assume that there exists a weak solution $w \in L^1(\Omega)$ to*

$$\begin{cases} -\Delta w = f(w) + \epsilon & \text{in } \Omega, \\ w = 0 & \text{on } \partial\Omega. \end{cases} \qquad (3.29)$$

Then, there exists a classical solution $u \in C^2(\overline{\Omega})$ to

$$\begin{cases} -\Delta u = (1 + \alpha)f(u) & \text{in } \Omega, \\ u = 0 & \text{on } \partial\Omega, \end{cases} \qquad (3.30)$$

for some $\alpha > 0$.

Proof.
Step 1. There exists a classical solution $v \in C^2(\overline{\Omega})$ of

$$\begin{cases} -\Delta v = f(v) + \dfrac{\epsilon}{2} & \text{in } \Omega, \\ v = 0 & \text{on } \partial\Omega. \end{cases}$$

Exercise 3.3.2 Prove Step 1, using the concave truncation technique of Theorem 3.2.1.

Step 2. Now consider the function $\zeta_0 \in C^2(\overline{\Omega})$ solving (3.8). Applying the mean-value theorem to v on the one hand and the boundary point lemma

64

(Lemma A.4.1) to ζ_0 on the other hand, it follows that there exists $\alpha > 0$ such that $2\alpha v \leq \epsilon \zeta_0$. Set

$$\bar{u} = v + \alpha v - \frac{\epsilon}{2}\zeta_0.$$

Clearly, $0 < \bar{u} \leq v$ in Ω. Furthermore, since f is nondecreasing, $\bar{u}$ satisfies

$$\begin{cases} -\Delta \bar{u} = f(v) + \alpha f(v) + \dfrac{\alpha \epsilon}{2} \geq (1+\alpha)f(\bar{u}) & \text{in } \Omega, \\ \bar{u} = 0 & \text{on } \partial\Omega. \end{cases}$$

In particular, $\bar{u}$ is a bounded supersolution to (3.30), while $\underline{u} = 0$ is a subsolution and $\underline{u} < \bar{u}$. The lemma follows. □

Taking advantage of this lemma, we establish that if (1.3) has a stable singular weak solution, then no strict supersolution to (1.3) exists.

Lemma 3.3.2 ([149]) *Let $N \geq 1$, let $\Omega \subset \mathbb{R}^N$ denote a smoothly bounded domain and let $f \in C^1(\mathbb{R})$, $f > 0$, denote a nondecreasing convex function. Let λ^* denote the associated extremal parameter. Assume $v \in L^1(\Omega)$, $v \geq 0$ a.e., verifies*

$$\int_\Omega f(v) d_\Omega \, dx < +\infty, \tag{3.31}$$

where d_Ω is given by (3.4). Assume further that for all $\varphi \in C_0^2(\overline{\Omega})$ such that $\varphi \geq 0$,

$$-\int_\Omega v \Delta \varphi \, dx \geq \lambda^* \int_\Omega f(v)\varphi \, dx. \tag{3.32}$$

Then, in fact

$$-\int_\Omega v \Delta \varphi \, dx = \lambda^* \int_\Omega f(v)\varphi \, dx, \tag{3.33}$$

for all $\varphi \in C_0^2(\overline{\Omega})$ such that $\varphi \geq 0$.

Proof. We argue by contradiction and assume that there exists a nonnegative measure $\mu \not\equiv 0$, such that d_Ω is μ-integrable and

$$-\int_\Omega v \Delta \varphi \, dx = \lambda^* \int_\Omega f(v)\varphi \, dx + \int_\Omega \varphi \, d\mu, \tag{3.34}$$

for all $\varphi \in C_0^2(\overline{\Omega})$. Consider $\zeta_1 \in L^1(\Omega)$, solving

$$\begin{cases} -\Delta \zeta_1 = \mu & \text{in } \Omega, \\ \zeta_1 = 0 & \text{on } \partial\Omega. \end{cases} \tag{3.35}$$

Such a solution exists and is unique by Corollary A.9.1. Since $\mu \not\equiv 0$, it follows from the boundary point lemma (Corollary A.9.2) on the one hand, and the mean value theorem on the other hand, that

$$\epsilon \zeta_0 \leq \zeta_1,$$

for some $\epsilon > 0$, and where ζ_0 is the solution to (3.8). Set $\bar{u} = v + \epsilon \zeta_0 - \zeta_1$. Clearly, $0 < \bar{u} \leq v$. In addition,

$$-\int_\Omega \bar{u} \Delta \varphi \, dx = \int_\Omega (\lambda^* f(v) + \epsilon) \varphi \, dx \geq \int_\Omega (\lambda^* f(\bar{u}) + \epsilon) \varphi \, dx$$

for all $\varphi \in C_0^2(\overline{\Omega})$, $\varphi \geq 0$. By the method of sub- and supersolutions, there exists a weak solution $0 \leq w \leq \bar{u}$ to

$$\begin{cases} -\Delta w = \lambda^* f(w) + \epsilon & \text{in } \Omega, \\ \quad\quad w = 0 & \text{on } \partial\Omega. \end{cases}$$

Lemma 3.3.1 now contradicts the definition of λ^*. $\qquad\square$

Proof of Theorem 3.3.1. Let v denote a weak supersolution to the equation, as defined in Theorem 3.3.1. By the method of sub- and supersolutions, there exists a minimal weak solution $\tilde{u}$ such that $0 \leq \tilde{u} \leq v$. Since $\tilde{u}$ is minimal, it follows that $\tilde{u} = u^*$ is the extremal solution. Assume now by contradiction that $v \neq u^*$.

Step 1. There exists $A \subset u^*(\Omega)$, $|A| \neq 0$, such that

$$f''(s) > 0 \quad \text{for all } s \in A.$$

If not, we have $f(u^*) = f(0) + f'(0)u^*$ a.e. in Ω, and everything happens for u^* as though f was linear. This contradicts Proposition 3.3.3. Therefore, there exists $\eta > 0$ and $0 < K_1 < K_2 \leq \|u^*\|_{L^\infty(\Omega)}$ such that

$$f''(s) \geq \eta \quad \text{for all } s \in [K_1, K_2].$$

Step 2. We show the existence of a weak strict supersolution to the equation. First note that by Lemma 3.3.2, the function v satisfies

$$-\int_\Omega v \Delta \varphi \, dx = \lambda^* \int_\Omega f(v) \varphi \, dx,$$

for all $\varphi \in C_0^2(\overline{\Omega})$, such that $\varphi \geq 0$. Take $\psi = \lambda^*(f(v) - f(u^*)) \geq 0$ and consider $w \in L^1(\Omega)$ the solution to

$$\begin{cases} -\Delta w = \psi & \text{in } \Omega, \\ \quad\quad w = 0 & \text{on } \partial\Omega. \end{cases}$$

By assumption, $\psi d_\Omega \in L^1(\Omega)$. In addition, $\psi \neq 0$, otherwise $f(v) = f(u^*)$ a.e. in Ω, and, by Lemma A.9.1, $v = u^*$ a.e. in Ω. Using the boundary point lemma (Lemma A.9.2), we deduce that $w \geq c\, d_\Omega$, for some constant $c > 0$. Since

$$-\int_\Omega (v - u^* - w)\Delta\varphi \, dx = 0,$$

for all $\varphi \in C_0^2(\overline{\Omega})$ such that $\varphi \geq 0$, we deduce that

$$v - u^* = w \geq c d_\Omega.$$

Step 3. Now set $\overline{u} = \frac{v+u^*}{2}$. Then,

$$-\int_\Omega \overline{u}\Delta\varphi \, dx = \frac{\lambda^*}{2} \int_\Omega (f(v) + f(u^*))\varphi \, dx = \int_\Omega (f(\overline{u}) + h)\varphi \, dx,$$

for all $\varphi \in C_0^2(\overline{\Omega})$ such that $\varphi \geq 0$ and where h is given by

$$h = \frac{1}{2}(f(v) + f(u^*)) - f\left(\frac{v+u^*}{2}\right) = \frac{1}{2}\int_{u^*}^{v} dt \int_{\frac{u^*+t}{2}}^{t} f''(s)\,ds.$$

Clearly, $h\, d_\Omega \in L^1(\Omega)$ and $h \geq 0$ in Ω. In addition, by Steps 1 and 2, we have $h \not\equiv 0$. It follows that $\overline{u}$ is a strict supersolution to the equation, contradicting Lemma 3.3.2. $\qquad\square$

Exercise 3.3.3 Consider the following semilinear problem involving the fractional Laplacian

$$\begin{cases} (-\Delta)^s u = \lambda f(u) & \text{in } B_1, \\ u = 0 & \text{on } \partial B_1. \end{cases} \tag{3.36}$$

Here, B_1 denotes the unit-ball in $\mathbb{R}^N$, $N \geq 2$, and $s \in (0,1)$. The operator $(-\Delta)^s$ is defined as follows. Let $\{\varphi_k\}_{k=1}^\infty$ denote an orthonormal basis of $L^2(B_1)$ consisting of eigenfunctions of $-\Delta$ in B_1 with homogeneous Dirichlet boundary conditions, associated to the eigenvalues $\{\lambda_k\}_{k=1}^\infty$. The operator $(-\Delta)^s$ is defined for any u in the Hilbert space

$$H = \{u \in L^2(B_1) : \|u\|_H^2 = \sum_{k=1}^\infty \lambda_k^s |u_k|^2 < +\infty\},$$

by

$$(-\Delta)^s u = \sum_{k=1}^\infty \lambda_k^s u_k \varphi_k,$$

where

$$u = \sum_{k=1}^{\infty} u_k \varphi_k, \quad \text{and } u_k = \int_{B_1} u \varphi_k \, dx.$$

- Let $\psi \in C_c^{\infty}(B_1)$, and let $\varphi := (-\Delta)^{-s}\psi$ denote the unique solution in H to $(-\Delta)^s \varphi = \psi$. Prove that there exists a constant $C > 0$ such that $|(-\Delta)^{-s}\psi| \leq C\varphi_1$.

We assume that the nonlinearity f is smooth, nondecreasing, positive, and superlinear. A measurable function u in B_1 such that $\int_{B_1} |u|\varphi_1 \, dx < +\infty$ and $\int_{B_1} f(u)\varphi_1 \, dx < +\infty$, is a weak solution to (3.36) if

$$\int_{B_1} u\psi \, dx = \lambda \int_{B_1} f(u)(-\Delta)^{-s}\psi \, dx, \quad \text{for all } \psi \in C_c^{\infty}(B_1).$$

In addition, we say that u is stable if for all $\psi \in C_c^{\infty}(B_1)$ we have

$$\int_{B_1} |(-\Delta)^{\frac{s}{2}}\psi|^2 \, dx \geq \int_{B_1} f'(u)\psi^2 \, dx.$$

Let $s \in (0, 1)$. Prove that there exists $\lambda^* > 0$ such that

- For $0 < \lambda < \lambda^*$, there exists a minimal solution $u_\lambda \in H \cap L^{\infty}(B_1)$ of (3.36). In addition, u_λ is stable and increasing with λ.

- For $\lambda = \lambda^*$, the function $u^* = \lim_{\lambda \nearrow \lambda^*} u_\lambda$ is a weak solution to (3.36).

- For $\lambda > \lambda^*$, (3.36) has no solution $u \in H \cap L^{\infty}(B_1)$.

The affine case

When the nonlinearity is not superlinear, the extremal solution need not exist, as the following simple example shows.

Proposition 3.3.3 ([152]) *Let $N \geq 1$, let Ω denote a smoothly bounded domain of $\mathbb{R}^N$, and let $\lambda_1 = \lambda_1(-\Delta; \Omega)$. If $f(t) = at + b$ when $t \geq 0$, with $a, b > 0$, then the extremal parameter associated to (3.1) verifies*

(i) $\lambda^* = \lambda_1/a$, and

(ii) (3.1) has no solution for $\lambda = \lambda^*$.

68

Proof. If $\lambda \in (0, \lambda_1/a)$ then the problem

$$\begin{cases} -\Delta u - \lambda a u = \lambda b & \text{in } \Omega, \\ \qquad\qquad u = 0 & \text{on } \partial\Omega, \end{cases} \tag{3.37}$$

has a unique solution $u \in H_0^1(\Omega)$. By elliptic regularity, $u \in C^2(\overline{\Omega})$. By the maximum principle, $u > 0$ in Ω. We claim that (3.37) has no solution for $\lambda^* = \lambda_1/a$. If u were such a solution, multiply (3.37) by $\varphi_1 > 0$, an eigenfunction associated to $\lambda_1 = \lambda_1(-\Delta, \Omega)$, to get $\int_\Omega \varphi_1 \, dx = 0$, contradicting $\varphi_1 > 0$. $\qquad \square$

For a thorough investigation of asymptotically linear nonlinearities, we refer the reader to [152].

3.3.3 Is the stable branch a (smooth) curve?

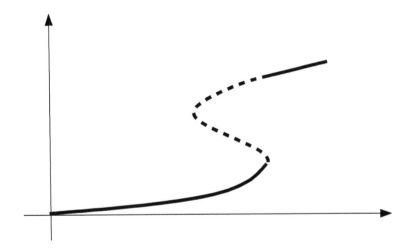

Figure 3.1: A possible piece of the solution curve. Stable solutions are represented by the solid line.

Let $(u_\lambda)_{\lambda \in (0, \lambda^*)}$ denote the branch of minimal solutions constructed in Proposition 3.3.1. The minimality property of u_λ implies that the mapping $\lambda \mapsto u_\lambda$ is nondecreasing. We claim that this mapping is also left-continuous. Take, for example, $0 < \mu \leq \lambda < \lambda^*$. Using elliptic regularity, $u_\mu \to v$, as $\mu \to \lambda^-$ where $v \in C^2(\overline{\Omega})$ is a solution to (3.1). Since $u_\mu \leq u_\lambda$, $v \leq u_\lambda$. Since u_λ is the minimal solution to (3.1), we also have $u_\lambda \leq v$. Hence, $u_\lambda = v$ and the claim follows.

But jump discontinuities can still occur. This is precisely what happens when the branch of solutions is S-shaped (see Figure 3.1).

In this section, we establish that if f is convex, then in fact the stable branch is a C^1 curve. For examples where the bifurcation diagram is S-shaped (so that the stable branch has a jump discontinuity), we refer the reader to [56].

Proposition 3.3.4 ([56]) *Let $N \geq 1$, let Ω denote a smoothly bounded domain of $\mathbb{R}^N$. Let $f \in C^2(\mathbb{R})$, $f \geq 0$, denote a nondecreasing function and let $(u_\lambda)_{\lambda \in (0, \lambda^*)}$ denote the branch of minimal solutions of (3.1). Given $\lambda \in (0, \lambda^*)$, the following properties are equivalent:*

(i) $\lambda_1(-\Delta - f'(u_\lambda); \Omega) > 0$.

(ii) *The map $\mu \mapsto u_\mu$ is C^1 from a neighborhood of λ to $L^\infty(\Omega)$.*

(iii) $\int_\Omega |u_\lambda - u_\mu|^2 d_\Omega(x)\, dx = o(|\lambda - \mu|)$, as $\mu \to \lambda$.

In addition, if f is convex (or if f is concave), these properties hold for all $\lambda \in (0, \lambda^*)$.

Proof.
Step 1. If f is convex (or if f is concave), then (i) holds for all $\lambda \in (0, \lambda^*)$.

Let $\lambda, \mu \in (0, \lambda^*)$, let u_λ, u_μ denote the corresponding minimal solutions to (3.1) and let $\varphi_1 > 0$ denote an eigenfunction associated to $\lambda_1 := \lambda_1(-\Delta - f'(u_\lambda); \Omega)$. Assume by contradiction that $\lambda_1 = 0$. Then,

$$-\Delta \varphi_1 = \lambda f'(u_\lambda)\varphi_1, \qquad \text{in } \Omega. \tag{3.38}$$

Multiply (3.38) by u_λ, (1.3) by φ_1, integrate and subtract these expressions. Then,

$$\int_\Omega (\lambda f(u_\lambda) - \lambda u_\lambda f'(u_\lambda))\varphi_1\, dx = 0. \tag{3.39}$$

Similarly,

$$\int_\Omega (\mu f(u_\mu) - \lambda u_\mu f'(u_\lambda))\varphi_1\, dx = 0. \tag{3.40}$$

Subtracting (3.39) from (3.40), we deduce that

$$\lambda \int_\Omega (f(u_\mu) - f(u_\lambda) - (u_\mu - u_\lambda)f'(u_\lambda))\varphi_1\, dx = (\lambda - \mu) \int_\Omega f(u_\mu)\varphi_1\, dx. \tag{3.41}$$

If f is convex, the left-hand side of (3.41) is nonnegative and we get a contradiction by choosing $\mu > \lambda$. If f is concave, then the left-hand side of (3.41) is nonpositive and we get a contradiction by choosing $\mu < \lambda$.

Step 2. Property (ii) implies (iii). This is immediate.

Step 3. Property (iii) implies (i). Assume by contradiction that $\lambda_1 := \lambda_1(-\Delta - f'(u_\lambda); \Omega) = 0$. Fix $\overline{\lambda} \in (\lambda, \lambda^*)$ and set $M = \|u_\lambda\|_{L^\infty(\Omega)}$. Since f is C^2, there exists $C > 0$ such that

$$|f(t) - f(s) - (t - s)f'(s)| \leq C|t - s|^2, \qquad \text{whenever } 0 \leq s, t \leq M.$$

So, we deduce from (3.41) that for $\lambda < \mu < \overline{\lambda}$,

$$|\lambda - \mu| \int_\Omega f(u_\mu) \varphi_1 \, dx \leq C\lambda \int_\Omega |u_\lambda - u_\mu|^2 \varphi_1 \, dx, \qquad (3.42)$$

where $\varphi_1 > 0$ is an eigenfunction associated to λ_1. Since $\varphi_1 \leq C d_\Omega$, we deduce from (iii) and (3.42) that $|\lambda - \mu| \int_\Omega f(u_\mu)\varphi_1 \, dx = o(|\lambda - \mu|)$. Recalling that $u_\mu(x) \to u_\lambda(x)$ as $\mu \to \lambda^-$, we deduce that $\int_\Omega f(u_\lambda)\varphi_1 \, dx = 0$. So, $f \equiv 0$ on the range of u_λ and $\lambda_1(-\Delta - f'(u_\lambda); \Omega) = \lambda_1(-\Delta; \Omega) > 0$, a contradiction.

Step 4. Property (i) implies (ii).

We first show that

$$\|u_\mu - u_\lambda\|_{L^\infty(\Omega)} \to 0, \qquad \text{as } \mu \to \lambda. \qquad (3.43)$$

Assume that f vanishes on the range of u_λ. Then, $u_\lambda \equiv 0$, and the same must be true of u_μ, for all $\mu \geq 0$.

So, we may assume that f does not vanish identically on the ranges of the functions u_μ, $\mu \in (0, \lambda^*)$. By the boundary point lemma (Proposition A.8.1 and Lemma A.5.1), given $\mu > \nu$, there exists $\varepsilon > 0$ such that $u_\mu \geq u_\nu + \varepsilon d_\Omega$. Set

$$\underline{u} = \lim_{\mu \uparrow \lambda} u_\mu \quad \text{and} \quad \overline{u} = \lim_{\mu \downarrow \lambda} u_\mu.$$

It is clear that $\underline{u} \leq u_\lambda$ and that $\underline{u}$ is a solution to (3.1). So, $\underline{u} = u_\lambda$. We claim that $\overline{u} = u_\lambda$. Indeed, since (i) holds, there exists the unique solution $\zeta_1 \in C_0^2(\overline{\Omega})$ to

$$\begin{cases} -\Delta \zeta_1 - \lambda f'(u_\lambda)\zeta_1 = 1 & \text{in } \Omega, \\ \qquad\qquad\quad \zeta_1 = 0 & \text{on } \partial\Omega. \end{cases}$$

We set $v = u_\lambda + \delta\zeta_1$ for $\delta > 0$, so that

$$-\Delta v - (\lambda + \theta)f(v) = (\delta - \theta f(v)) - \lambda \left(f(v) - f(u_\lambda) - (v - u_\lambda)f'(u_\lambda) \right).$$

Since $f(v) \leq \sup_{[0, \|u_\lambda\|_\infty + \delta\|\zeta_1\|_\infty]} f$ and $f(v) - f(u_\lambda) - (v - u_\lambda)f'(u_\lambda) = o(\delta)$, we deduce that for $\delta > 0$ sufficiently small, there exists $\theta = \theta(\delta) > 0$ such that $-\Delta v - (\lambda + \theta)f(v) \geq 0$ in Ω. In particular, $u_{\lambda+\theta} \leq v$ and so $\overline{u} \leq v$. Letting

$\delta \downarrow 0$, we obtain $\bar{u} \leq u_\lambda$, thus $\bar{u} = u_\lambda$. So, $u_\mu(x) \to u_\lambda(x)$ as $\mu \to \lambda$, for all $x \in \Omega$. Since u_μ is nondecreasing in μ and $u_\mu \in C(\overline{\Omega})$ for all $\mu < \lambda^*$, the convergence is uniform and (3.43) holds. It then follows easily from (3.43) that $\lambda_1(-\Delta - \mu f'(u_\mu); \Omega) \to \lambda_1(-\Delta - \lambda f'(u_\lambda); \Omega)$, as $\mu \to \lambda$. In particular, we deduce from (i) that there exists $\delta, \eta > 0$ such that

$$\lambda_1(-\Delta - \mu f'(u_\mu); \Omega) > \eta, \tag{3.44}$$

for $|\mu - \lambda| < \delta$. Hence, (i) holds with λ replaced by μ such that $|\mu - \lambda| < \delta$ and so we deduce from (3.43) that

the mapping $\mu \mapsto u_\mu$ is continuous from $(\lambda - \delta, \lambda + \delta)$ to $L^\infty(\Omega)$. (3.45)

We show next that there exists a constant $C > 0$ such that

$$\|u_\mu - u_\nu\|_\infty \leq C|\mu - \nu|, \tag{3.46}$$

for $|\mu - \lambda|, |\nu - \lambda| < \delta$. Indeed, it follows from (3.44) that

$$\eta\|u_\mu - u_\nu\|_{L^2(\Omega)}^2 \leq \int_\Omega \left|\nabla(u_\mu - u_\nu)\right|^2 dx - \mu \int_\Omega f'(u_\mu)(u_\mu - u_\nu)^2 \, dx$$

$$= \int_\Omega (u_\mu - u_\nu)[-\Delta(u_\mu - u_\nu) - \mu f'(u_\mu)(u_\mu - u_\nu)] \, dx$$

$$= \mu \int_\Omega (u_\mu - u_\nu)[f(u_\mu) - f(u_\nu) - f'(u_\mu)(u_\mu - u_\nu)] \, dx +$$

$$(\mu - \nu) \int_\Omega f(u_\nu)(u_\mu - u_\nu) \, dx.$$

Since $\mu|f(u_\mu) - f(u_\nu) - f'(u_\mu)(u_\mu - u_\nu)| \leq \varepsilon(|\mu - \nu|)|u_\mu - u_\nu|$, with $\varepsilon(t) \to 0$ as $t \to 0$, we obtain

$$\eta\|u_\mu - u_\nu\|_{L^2(\Omega)}^2 \leq \varepsilon(|\mu - \nu|)\|u_\mu - u_\nu\|_{L^2(\Omega)} + C|\mu - \nu|\|u_\mu - u_\nu\|_{L^2(\Omega)},$$

so that $\|u_\mu - u_\nu\|_{L^2(\Omega)} \leq C|\mu - \nu|$. Since

$$-\Delta(u_\mu - u_\nu) = \mu(f(u_\mu) - f(u_\nu)) + (\mu - \nu)f(u_\nu)$$

and $|\mu(f(u_\mu) - f(u_\nu)) + (\mu - \nu)f(u_\nu)| \leq C|u_\mu - u_\nu| + C|\mu - \nu|$, (3.46) follows from the L^2 estimate and a direct bootstrap argument. Suppose now that $|\mu - \lambda| < \delta$. By (3.44), there exists the unique solution $w_\mu \in C_0^2(\overline{\Omega})$ of

$$\begin{cases} -\Delta w_\mu - \mu f'(u_\mu)w_\mu = f(u_\mu) & \text{in } \Omega, \\ \qquad\qquad\qquad\quad w_\mu = 0 & \text{on } \partial\Omega. \end{cases}$$

By (3.44), (w_μ) is bounded in $H_0^1(\Omega)$ and so by standard elliptic regularity (w_μ) is bounded in $C^1(\overline{\Omega})$. Using (3.46), we deduce that w_μ is continuous from $(\lambda - \delta, \lambda + \delta)$ to $L^\infty(\Omega)$. Property (ii) follows if we show that $w_\mu = \frac{d}{d\mu}u_\mu$, that is,

$$\psi := \frac{u_\sigma - u_\mu - (\sigma - \mu)w_\mu}{\sigma - \mu} \to 0 \quad \text{in } L^\infty(\Omega), \text{ as } \sigma \to \mu.$$

We have

$$-\Delta\psi - \mu f'(u_\mu)\psi = (u_\sigma - u_\mu)f'(u_\mu) + \sigma\frac{u_\sigma - u_\mu}{\sigma - \mu}\frac{f(u_\sigma) - f(u_\mu) - (u_\sigma - u_\mu)f'(u_\mu)}{u_\sigma - u_\mu},$$

and it follows from (3.46) that the right-hand side in the above equality converges to 0 in $L^\infty(\Omega)$ as $\sigma \to \mu$. Using (3.44), we conclude that $\|\psi\|_\infty \to 0$ as $\sigma \to \mu$. $\qquad\square$

Exercise 3.3.4 Assume that $f \in C^2(\mathbb{R})$, $f \geq 0$, is a convex nondecreasing function. Prove that the map $\lambda \in (0, \lambda^*) \to u_\lambda \in L^\infty(\Omega)$ is convex.

3.3.4   Is the extremal solution bounded?

As follows from Chapter 2, when Ω is the unit ball and $f(u) = \exp(u)$, the extremal solution is bounded (hence smooth) if and only if $N \leq 9$. Deciding whether u^* is bounded or not in the general setting of (3.1) is a delicate question, which we shall address in the two following chapters.

Chapter 4

Regularity theory of stable solutions

Recall that there exists singular solutions to PDEs of the form (1.3). Take, for example,

$$u_s(x) = -2\ln|x|, \tag{4.1}$$

solving (1.3) in the unit ball $B \subset \mathbb{R}^N$, $N \geq 3$, for the nonlinearity $f(u) = \lambda e^u$, with $\lambda = \lambda_s := 2(N - 2)$. Also recall that, when $N \geq 10$, u_s is the monotone limit of a curve of (regular) stable solutions u_λ, $\lambda < \lambda_s$. So, u_s is stable. In addition, even if our interest rested solely in studying the nicest possible solutions—say the smooth, radial, and stable ones—no *a priori* bound holds true in full generality, as examplified by the aforementioned curve u_λ.

This chapter is devoted to the regularity theory of stable solutions. In other words, we shall try to answer the following questions: when is any stable solution to (1.3) regular? When can we obtain *a priori* estimates on families of (regular) stable solutions?

4.1 The radial case

Example 3.2.1 shows that no regularity result can hold in full generality for stable solutions that do not belong to the energy space $H_0^1(\Omega)$. This being said, we have the following sharp result, when Ω is the unit ball.

Theorem 4.1.1 ([38]) *Let $N \geq 2$ and let $\Omega = B$ denote the unit ball of $\mathbb{R}^N$. Assume that f is a locally Lipschitz function and let u denote a stable radial weak solution to (1.3), such that*

$$u \in H_0^1(B).$$

75

Then, u is either constant, radially increasing, or radially decreasing in $B \setminus \{0\}$ and there exists a constant $C = C(N) > 0$ such that

$$\|u\|_{L^{\infty}(B)} \leq C \left(\|u\|_{L^1(B)} + \|f(u)d_B\|_{L^1(B)} \right), \quad \text{if } 1 \leq N \leq 9, \qquad (4.2)$$

Remark 4.1.1 *Recall that for $N \geq 3$, the function $u_s(x) = -2\ln|x|$ solves*

$$\begin{cases} -\Delta u = \lambda_s e^u & \text{in } B, \\ u = 0 & \text{on } \partial B, \end{cases}$$

with $\lambda_s = 2(N-2)$. By Hardy's inequality, u_s is stable if and only if $N \geq 10$. In particular, (4.2) fails for $N \geq 10$. Still, sharp estimates of the (possibly) singular behavior of solutions are available (see [38] and [217]).

Proof of Theorem 4.1.1. We establish two auxiliary lemmata.

Lemma 4.1.1 ([38]) *Let $N \geq 1$ and let $u \in H^1(B)$ denote a radial solution to (1.4) in $\Omega = B \setminus \{0\}$. Then, for every $\varphi \in H^1(B) \cap L^{\infty}(B)$ with compact support in $B \setminus \{0\}$, there holds*

$$Q_u(ru_r\varphi) = \int_B u_r^2 \left(|\nabla(r\varphi)|^2 - (N-1)\varphi^2 \right) dx, \qquad (4.3)$$

where, as usual,

$$Q_u(\varphi) = \int_B |\nabla\varphi|^2 \, dx - \int_B f'(u)\varphi^2 \, dx.$$

Proof of Lemma 4.1.1. Take $\varphi \in H^1(B) \cap L^{\infty}(B)$ with compact support in $B \setminus \{0\}$. Also let c be any function in $H^2_{loc}(B \setminus \{0\}) \cap L^{\infty}(B \setminus \{0\})$. Note that $\phi = r\varphi c \in H^1(B) \cap L^{\infty}(B)$ has compact support in $B \setminus \{0\}$. Use ϕ as a test function in the definition of Q_u:

$$Q_u(r\varphi c) = \int_B \left(r^2\varphi^2 |\nabla c|^2 + c^2 |\nabla(r\varphi)|^2 + c\nabla c \cdot \nabla(r^2\varphi^2) - f'(u)r^2\varphi^2 c^2 \right) dx$$

$$= \int_B \left(r^2\varphi^2 |\nabla c|^2 + c^2 |\nabla(r\varphi)|^2 - r^2\varphi^2 \nabla \cdot (c\nabla c) - f'(u)r^2\varphi^2 c^2 \right) dx$$

$$= \int_B \left(c^2 |\nabla(r\varphi)|^2 - r^2\varphi^2(c\Delta c + f'(u)c^2) \right) dx.$$

Differentiate (1.3) with respect to r to get

$$-\Delta u_r + \frac{N-1}{r^2}u_r = f'(u)u_r \qquad \text{in } B \setminus \{0\}. \qquad (4.4)$$

Using local elliptic regularity and the fact that u is radial, we have that $u_r \in H^2_{loc}(B \setminus \{0\}) \cap L^\infty(B \setminus \{0\})$. So, we can take $c := u_r$ in the previous computations. Using (4.4), we deduce (4.3). $\qquad\square$

Lemma 4.1.2 ([38]) *Let $N \geq 2$ and let B denote the unit ball in $\mathbb{R}^N$. Assume that f is a locally Lipschitz function and let u denote a radial weak solution to (1.3). Let α satisfy*

$$1 \leq \alpha < 1 + \sqrt{N-1}. \tag{4.5}$$

Then,

$$\int_{B_{1/2}} u_r^2 r^{-2\alpha} \, dx \;\leq\; \frac{C_N}{(N-1)-(\alpha-1)^2} \left(\|u\|^2_{L^1(B)} + \|f(u)d_B\|^2_{L^1(B)} \right), \tag{4.6}$$

where C_N is a constant depending on N only and where d_B is the distance to the boundary of B, defined by (3.4).

Proof of Lemma 4.1.2. By approximation, since u is stable, $Q_u(\phi) \geq 0$ for every $\phi \in H^1(B)$ with compact support in $B \setminus \{0\}$. By Lemma 4.1.1,

$$(N-1) \int_B u_r^2 \varphi^2 \, dx \leq \int_B u_r^2 \left| \nabla(r\varphi) \right|^2 \, dx, \tag{4.7}$$

for every $\varphi \in H^1(B) \cap L^\infty(B)$ with compact support in $B \setminus \{0\}$. In fact, (4.7) also holds for $\varphi \in H^1(B) \cap L^\infty(B)$ with compact support in B such that $\left| \nabla(r\varphi) \right| \in L^\infty(B)$. To see this, take $\zeta \in C^1(\mathbb{R}^N)$ such that $0 \leq \zeta \leq 1$, $\zeta \equiv 0$ in B and $\zeta \equiv 1$ in $\mathbb{R}^N \setminus B_2$. Let $\zeta_\delta(x) = \zeta(x/\delta)$ for $\delta > 0$, $x \in \mathbb{R}^N$. Applying (4.7) with test function $\zeta_\delta \varphi$, we obtain

$$(N-1) \int_B u_r^2 \varphi^2 \zeta_\delta^2 \, dx \leq \int_B u_r^2 \left| \nabla(r\varphi\zeta_\delta) \right|^2 \, dx.$$

Now,

$$\int_B u_r^2 \left| \nabla(r\varphi\zeta_\delta) \right|^2 \, dx =$$

$$= \int_B u_r^2 \left(\left| \nabla(r\varphi) \right|^2 \zeta_\delta^2 + r^2\varphi^2 \left| \nabla\zeta_\delta \right|^2 + \zeta_\delta \nabla\zeta_\delta \cdot \nabla(r^2\varphi^2) \right) dx \leq$$

$$\leq \int_B u_r^2 \left| \nabla(r\varphi) \right|^2 \zeta_\delta^2 \, dx + C \int_{B_{2\delta} \setminus B_\delta} u_r^2 |\varphi| \left(\frac{r^2}{\delta^2}|\varphi| + \frac{r}{\delta}\zeta_\delta \left| \nabla(r\varphi) \right| \right) dx \leq$$

$$\leq \int_B u_r^2 \left| \nabla(r\varphi) \right|^2 \zeta_\delta^2 \, dx + C \int_{B_{2\delta} \setminus B_\delta} u_r^2 \, dx.$$

In the last inequality, we used that φ and $\left|\nabla(r\varphi)\right|$ are bounded. Since $u \in H^1_0(B)$, the last term tends to zero as $\delta \to 0$. By monotone convergence, we deduce that (4.7) holds for every $\varphi \in H^1(B) \cap L^\infty(B)$ with compact support in B and such that $\left|\nabla(r\varphi)\right| \in L^\infty(B)$.

Let $\varepsilon \in (0, 1/2)$. For $\alpha \geq 1$ in the range (4.5), apply (4.7) with $\varphi = \varphi_\varepsilon$ given by

$$\varphi_\varepsilon(r) = \begin{cases} \varepsilon^{-\alpha} - (1/2)^{-\alpha} & \text{if } 0 \leq r \leq \varepsilon, \\ r^{-\alpha} - (1/2)^{-\alpha} & \text{if } \varepsilon < r \leq 1/2, \\ 0 & \text{if } 1/2 < r. \end{cases}$$

Note that φ_ε and $\left|\nabla(r\varphi_\varepsilon)\right|$ are bounded. We obtain

$$(N-1)\int_{B_{1/2}\backslash B_\varepsilon} u_r^2(r^{-\alpha} - (1/2)^{-\alpha})^2\, dx + (N-1)(\varepsilon^{-\alpha} - (1/2)^{-\alpha})^2 \int_{B_\varepsilon} u_r^2\, dx \leq$$

$$\leq \int_{B_{1/2}\backslash B_\varepsilon} u_r^2((1-\alpha)r^{-\alpha} - (1/2)^{-\alpha})^2\, dx + (\varepsilon^{-\alpha} - (1/2)^{-\alpha})^2 \int_{B_\varepsilon} u_r^2\, dx.$$

Since $N \geq 2$, it follows that

$$(N-1)\int_{B_{1/2}\backslash B_\varepsilon} u_r^2(r^{-\alpha} - (1/2)^{-\alpha})^2\, dx \leq \int_{B_{1/2}\backslash B_\varepsilon} u_r^2((1-\alpha)r^{-\alpha} - (1/2)^{-\alpha})^2\, dx.$$

Expanding squares, using $N \geq 2$ and (4.5), we find the estimate

$$\int_{B_{1/2}\backslash B_\varepsilon} u_r^2 r^{-2\alpha}\, dx \leq \frac{C_N}{(N-1)-(\alpha-1)^2} \int_{B_{1/2}\backslash B_\varepsilon} u_r^2 r^{-\alpha}\, dx.$$

Now, choose a positive constant $C_{\alpha,N}$ such that

$$\frac{C_N}{(N-1)-(\alpha-1)^2} r^{-\alpha} \leq \frac{1}{2}r^{-2\alpha} + C_{\alpha,N} r^{N-1}, \qquad \text{for all } r \in (0,1).$$

It follows that

$$\int_{B_{1/2}\backslash B_\varepsilon} u_r^2 r^{-2\alpha}\, dx \leq C_{\alpha,N}\int_{B_{1/2}\backslash B_\varepsilon} u_r^2 r^{N-1}\, dx. \tag{4.8}$$

Next, we claim that

$$\int_{B_{1/2}} u_r^2 r^{N-1}\, dx = |\partial B| \int_0^{1/2} u_r^2 r^{2N-2}\, dr \leq C_N \left(\|u\|^2_{L^1(B)} + \|f(u)d_B\|^2_{L^1(B)} \right). \tag{4.9}$$

Assume this claim for the moment. Using (4.8) and (4.9) and letting $\varepsilon \to 0$, we obtain

$$\int_{B_{1/2}} u_r^2 r^{-2\alpha}\, dx \le C_{\alpha,N} \left(\|u\|_{L^1(B)}^2 + \|f(u)d_B\|_{L^1(B)}^2 \right).$$

In particular, taking $\alpha = (1 + \sqrt{N-1})/2$, we deduce

$$\int_{B_{1/2}} u_r^2 r^{-(1+\sqrt{N-1})}\, dx \le C_N \left(\|u\|_{L^1(B)}^2 + \|f(u)d_B\|_{L^1(B)}^2 \right). \tag{4.10}$$

Finally, since $r^{-\alpha} \le r^{-(1+\sqrt{N-1})}$ in B, the desired estimate follows after letting $\varepsilon \to 0$.

It remains to establish (4.9). First, since u is radially decreasing,

$$u(1/2) \le C_N \int_{1/4}^{1/2} u r^{N-1}\, dr \le C_N \|u\|_{L^1(B)}. \tag{4.11}$$

Let $\rho \in (1/2, 3/4)$ be chosen such that

$$-u_r(\rho) = -\frac{u(3/4) - u(1/2)}{1/4} = 4u(1/2) - 4u(3/4) \le 4u(1/2).$$

For $s \le 1/2$, integrate $(r^{N-1}u_r)_r = -f(u)r^{N-1}$ with respect to r, from s to ρ:

$$-u_r(s)s^{N-1} = -u_r(\rho)\rho^{N-1} - \int_s^\rho f(u)r^{N-1}\, dr \le$$

$$\le C_N \left(u(1/2) + \|f(u)d_B\|_{L^1(B)} \right).$$

Collecting the aforementioned estimates, it follows that

$$0 \le -u_r(s)s^{N-1} \le C_N \left(\|u\|_{L^1(B)}^2 + \|f(u)d_B\|_{L^1(B)}^2 \right),$$

for all $s \le 1/2$. Squaring this inequality and integrating it in s, from 0 to $1/2$, we conclude that (4.9) holds. $\qquad\qquad\qquad\qquad\qquad\qquad\qquad\qquad\square$

Proof of Theorem 4.1.1 Completed. Take α in the range (4.5). For $s \in (0, 1/2]$, we have

$$u(s) - u(1/2) = \int_s^{1/2} -u_r\, dr = \int_s^{1/2} -u_r r^{-\alpha + \frac{N-1}{2}} r^{\alpha - \frac{N-1}{2}}\, dr \le$$

$$\le C_N \left(\int_{B_1} u_r^2 r^{-2\alpha}\, dx \right)^{1/2} \left(\int_s^{1/2} r^{2\alpha+1-N}\, dr \right)^{1/2}. \tag{4.12}$$

Using Lemma 4.1.1, we deduce that for all $s \in (0, 1/2]$,

$$u(s) \leq u(1/2) +$$

$$+ \frac{C_N}{\sqrt{(N-1)-(\alpha-1)^2}} \left(\int_s^{1/2} r^{2\alpha+1-N} \, dr \right)^{1/2} \left(\|u\|_{L^1(B)} + \|f(u)d_B\|_{L^1(B)} \right).$$

$$(4.13)$$

Assume $N \leq 9$. The integral in the right-hand side of the above inequality is finite with $s = 0$ if we take $2\alpha + 1 - N > -1$. Such a choice of α in the range (4.5) is possible since $N \leq 9$.

□

Exercise 4.1.1 Under the assumptions of Theorem 4.1.1, prove that if $N = 10$, then any stable radial weak solution u of (1.3) such that $u \in H_0^1(B)$ satisfies

$$|u(r)| \leq C \left(\|u\|_{L^1(B)} + \|f(u)d_B\|_{L^1(B)} \right) |\ln r|.$$

Is the estimate sharp?

4.2 Back to the Gelfand problem

We will now turn to nonradial solutions. To this end, we begin by returning to the study of the extremal solution for the Gelfand problem, posed this time in an arbitrary bounded domain:

$$\begin{cases} -\Delta u = \lambda e^u & \text{in } \Omega, \\ \quad\; u = 0 & \text{on } \partial\Omega. \end{cases} \qquad (4.14)$$

Theorem 4.2.1 ([154], [63]) *Let $1 \leq N \leq 9$ and let $\Omega \subset \mathbb{R}^N$ denote a bounded domain with $C^{2,\alpha}$ boundary. Then, the extremal solution to (4.14) is classical. Furthermore, there exists a constant $C = C(N, \Omega) > 0$ such that for all $0 \leq \lambda \leq \lambda^*$,*

$$\|u_\lambda\|_{L^\infty(\Omega)} \leq C, \qquad (4.15)$$

where u_λ denotes the unique classical stable solution to (4.14).

Remark 4.2.1 *It follows from Chapter 2 that the restriction $N \leq 9$ is sharp since the extremal solution is given by $u^*(x) = -2\ln|x|$, when $\Omega = B$ is the unit ball and $N \geq 10$.*

Proof. By elliptic regularity, it suffices to establish (4.15). We may also restrict to the case $0 < \lambda < \lambda^*$ since u^* is obtained as the monotone limit of u_λ as $\lambda \nearrow \lambda^*$. Fix such λ and let $u = u_\lambda$. For $\alpha \in (0,2)$, introduce the test functions $\psi = e^{2\alpha u} - 1$ and $\varphi = e^{\alpha u} - 1$. We are going to multiply (4.14) by ψ on the one hand, and to use the stability inequality (1.5) with test function φ on the other hand. The former calculation reads as follows.

$$\int_\Omega \nabla u \nabla \psi \, dx = \lambda \int_\Omega e^u \psi \, dx =$$

$$2\alpha \int_\Omega |\nabla u|^2 e^{2\alpha u} \, dx = \lambda \int_\Omega e^u \left(e^{2\alpha u} - 1 \right) dx =$$

$$\frac{2}{\alpha} \int_\Omega |\nabla (e^{\alpha u} - 1)|^2 \, dx = \lambda \int_\Omega \left(e^{(2\alpha+1)u} - e^u \right) dx.$$

Using $\varphi = e^{\alpha u} - 1$ in (1.5), we obtain

$$\int_\Omega |\nabla (e^{\alpha u} - 1)|^2 \, dx \geq \lambda \int_\Omega e^u (e^{\alpha u} - 1)^2 \, dx$$

$$= \lambda \int_\Omega \left(e^{(2\alpha+1)u} + e^u - 2e^{(\alpha+1)u} \right) dx.$$

Combining the two equations, it follows that

$$\int_\Omega \left(e^{(2\alpha+1)u} - e^u \right) dx \geq \frac{2}{\alpha} \int_\Omega \left(e^{(2\alpha+1)u} + e^u - 2e^{(\alpha+1)u} \right) dx,$$

which simplifies to

$$\frac{1}{\alpha} \int_\Omega \left(4e^{(\alpha+1)u} - (\alpha+2)e^u \right) dx \geq \frac{2-\alpha}{\alpha} \int_\Omega e^{(2\alpha+1)u} \, dx.$$

In particular,

$$\int_\Omega e^{(\alpha+1)u} \, dx \geq \frac{2-\alpha}{4} \int_\Omega e^{(2\alpha+1)u} \, dx.$$

Applying Hölder's inequality, we obtain

$$|\Omega|^{\frac{\alpha}{2\alpha+1}} \left(\int_\Omega e^{(2\alpha+1)u} \, dx \right)^{\frac{\alpha+1}{2\alpha+1}} \geq \frac{2-\alpha}{4} \int_\Omega e^{(2\alpha+1)u} \, dx.$$

This implies that e^u is bounded in $L^p(\Omega)$ for any $p = 2\alpha + 1$ with $\alpha \in (0,2)$. Since $N \leq 9$, there exists $\alpha \in (0,2)$ such that $p = 2\alpha + 1 > N/2$. Since u solves (4.14), (4.15) follows by elliptic regularity. $\qquad\square$

Exercise 4.2.1 Given $p > 1$, consider the problem

$$\begin{cases} -\Delta u = \lambda(1+u)^p & \text{in } \Omega, \\ \quad u = 0 & \text{on } \partial\Omega. \end{cases}$$

Prove that the extremal solution is bounded if $p < p_c(N)$, where

$$p_c(N) = \begin{cases} +\infty & \text{if } N \leq 10, \\ \dfrac{(N-2)^2 - 4N + 8\sqrt{N-1}}{(N-2)(N-10)} & \text{if } N \geq 11. \end{cases} \qquad (4.16)$$

4.3 Dimensions $N = 1, 2, 3$

We return to the question of regularity for (1.3), with general nonlinearity f: if u is a stable weak solution to (1.3), is u in fact a classical solution? In dimensions $1 \leq N \leq 3$, a first partial answer is given by the following theorem.

Theorem 4.3.1 ([166]) *Let $1 \leq N \leq 3$ and let $\Omega \subset \mathbb{R}^N$ denote a smoothly bounded domain. Assume that $f \in C^1(\mathbb{R})$ is a nondecreasing convex function such that $f(0) > 0$. In addition, assume that f is superlinear, that is,*

$$\lim_{t \to +\infty} \frac{f(t)}{t} = +\infty. \qquad (4.17)$$

Then, any stable weak solution u to (1.3) such that $u \in H_0^1(\Omega)$, is in fact classical and there exists a constant $C = C(\Omega, N, f)$ such that

$$\|u\|_{L^\infty(\Omega)} \leq C. \qquad (4.18)$$

Proof. By Corollary 3.2.1, we need only prove (4.18) for classical solutions of (1.3). Following Theorem 4.2.1 in spirit, we introduce two related test functions $\varphi = \tilde{f}(u) = f(u) - f(0)$ and $\psi = g(u) = \int_0^u f'(t)^2 \, dt$. We multiply (1.3) by ψ on the one hand, and use the stability inequality (1.5) with test function φ on the other hand. The former calculation reads

$$\int_\Omega |\nabla u|^2 f'(u)^2 \, dx = \int_\Omega f(u)g(u) \, dx,$$

while testing the stability inequality (1.5) with $\varphi = \tilde{f}(u)$ leads to

$$\int_\Omega f'(u)\tilde{f}(u)^2\, dx \leq \int_\Omega |\nabla u|^2 f'(u)^2\, dx.$$

So,

$$\int_\Omega f'(u)\tilde{f}(u)^2\, dx \leq \int_\Omega \tilde{f}(u)g(u)\, dx + f(0)\int_\Omega g(u)\, dx. \tag{4.19}$$

Now we estimate the difference $\tilde{f}(t)^2 f'(t) - \tilde{f}(t)g(t)$ for $t > 0$. By definition of g, we have

$$\tilde{f}(t)^2 f'(t) - \tilde{f}(t)g(t) = \tilde{f}(t)f'(t)\int_0^t f'(s)\, ds - \tilde{f}(t)\int_0^t f'(s)^2\, ds$$

$$= \tilde{f}(t)\int_0^t f'(s)\left(f'(t) - f'(s)\right)\, ds.$$

So, denoting

$$h(t) = \int_0^t f'(s)\left(f'(t) - f'(s)\right)\, ds,$$

we obtain from (4.19)

$$\int_\Omega \tilde{f}(u)h(u)\, dx \leq f(0)\int_\Omega g(u)\, dx. \tag{4.20}$$

We claim that

$$\lim_{t\to+\infty} \frac{h(t)}{f'(t)} = +\infty. \tag{4.21}$$

Indeed, let $C > 0$ and choose $s_0 > 1$ such that $f'(s_0) > 2C$. By assumption, f is superlinear, that is, (4.17) holds. So, for t sufficiently large, there holds $f'(t) > 2f'(s_0 + 1)$. And, for such t,

$$h(t) = \int_0^t f'(s)\left(f'(t) - f'(s)\right)\, ds \geq \int_{s_0}^{s_0+1} f'(s)\left(f'(t) - f'(s)\right)\, ds$$

$$\geq \int_{s_0}^{s_0+1} f'(s)\left(f'(t) - f'(s_0 + 1)\right)\, ds \geq Cf'(t).$$

Claim (4.21) is proved. In addition,

$$g(t) = \int_0^t f'(s)^2\, ds \leq \int_0^t f'(s)f'(t)\, ds = f'(t)\tilde{f}(t).$$

By the above equation and (4.21), we deduce that $\lim_{t \to +\infty} \tilde{f}(t)h(t)/g(t) = +\infty$, which combined with (4.20) leads to

$$\int_\Omega \tilde{f}(u)h(u)\, dx \leq C.$$

Using (4.21) again, we deduce that

$$\int_\Omega \tilde{f}(u)f'(u)\, dx \leq C. \tag{4.22}$$

By Kato's inequality (Lemma 3.2.1),

$$-\Delta \tilde{f}(u) \leq f'(u)f(u), \qquad \text{in } \Omega.$$

Using the maximum principle, it follows that $0 \leq \tilde{f}(u) \leq v$, where v solves

$$\begin{cases} -\Delta v = f'(u)f(u) & \text{in } \Omega, \\ \quad v = 0 & \text{on } \partial\Omega. \end{cases}$$

By (4.22) and elliptic regularity (see Exercise A.9.1), it follows that v is uniformly bounded in $L^p(\Omega)$, for all $1 \leq p < \frac{N}{N-2}$. Hence, $f(u)$ is uniformly bounded in $L^p(\Omega)$. Applying elliptic regularity (Theorem B.3.1) again in (1.3), we deduce that u is uniformly bounded, provided $N \leq 3$. $\qquad\square$

Exercise 4.3.1 Under the assumptions of Theorem 4.3.1, prove that for any $N \geq 4$, u is bounded in $L^p(\Omega)$ for all $p < \frac{N}{N-4}$ ($p < +\infty$ if $N = 4$).

Theorem 4.3.1 applies to solutions belonging to $H_0^1(\Omega)$. In fact, using the same proof, the theorem remains valid for extremal solutions. It is interesting to note that in any dimension, extremal solutions always belong to $H_0^1(\Omega)$, at least if the domain is convex.

Theorem 4.3.2 ([167]) *Let $1 \leq N$ and let $\Omega \subset \mathbb{R}^N$ denote a smoothly bounded convex domain. Assume that $f \in C^1(\mathbb{R})$ is a nondecreasing function such that $f(0) > 0$. In addition, assume that f is superlinear, that is, (4.17) holds. Then, the extremal solution u^* to (3.1) belongs to $H_0^1(\Omega)$.*

Proof. Take $\lambda < \lambda^*$ and u_λ the associated minimal solution. Then, by Lemma 1.1.1, u_λ minimizes the energy

$$\mathcal{E}_\Omega(u) = \frac{1}{2}\int_\Omega |\nabla u|^2\, dx - \lambda \int_\Omega F(u)\, dx,$$

among all functions lying between 0 and u_λ. In particular,

$$\mathscr{E}_\Omega(u) \le \mathscr{E}_\Omega(0) = 0. \qquad (4.23)$$

Pohozaev's identity (see (8.9)) asserts

$$\frac{2N}{N-2}\lambda \int_\Omega F(u_\lambda)\,dx - \int_\Omega |\nabla u_\lambda|^2\,dx = \frac{1}{N-2}\int_{\partial\Omega} |\nabla u_\lambda|^2 x \cdot n\,d\sigma. \qquad (4.24)$$

We may suppose that $N > 2$ (since for $N = 2, 3$, we already know that u^* is regular by Theorem 4.3.1). Combining (4.23) and (4.24), we obtain

$$\int_\Omega |\nabla u_\lambda|^2\,dx \le \frac{1}{2}\int_{\partial\Omega} |\nabla u_\lambda|^2 x \cdot n\,d\sigma.$$

Finally, using the boundary estimates of Theorem 4.5.3 below, we deduce that (u_λ) is bounded in $H_0^1(\Omega)$. □

4.4 A geometric Poincaré formula

In the light of Theorem 4.3.1, two natural questions arise: are stable solutions classical for $N = 4, \ldots, 9$ and functions f that are nondecreasing, convex, and superlinear? Are the assumptions on f needed for the result to hold? In order to go any further, we pause to establish the following geometric restatement of stability.

Theorem 4.4.1 ([207, 208]) *Let $N \ge 1$ and let $\Omega \subset \mathbb{R}^N$ denote an open set. Let $u \in C^2(\Omega)$ denote a stable solution to (1.4). Then, for any $\varphi \in C_c^1(\Omega)$, there holds*

$$\int_{[\nabla u \ne 0]} \left(\left| \nabla_T |\nabla u| \right|^2 + |B|^2 |\nabla u|^2 \right) \varphi^2\,dx \le \int_\Omega |\nabla u|^2 |\nabla \varphi|^2\,dx, \qquad (4.25)$$

where ∇_T denotes the tangential gradient along a given level set of u and where $|B|^2$ denotes the sum of the squares of the principal curvatures of such a level set.

Remark 4.4.1 *By the implicit function theorem, each level set $M = [u = t]$ of u is an $N - 1$ dimensional submanifold on $[\nabla u \ne 0]$. In particular, the tangential gradient (defined as the orthogonal projection of the gradient on the tangent space to M), as well as the principal curvatures of M are well defined.*

Remark 4.4.2 *In the case where Ω is a ball and u is a radial monotone function, any given level set of u is a hypersphere. It follows that its principal curvatures are equal and constant, so $|B|^2 = (N-1)/r^2$ with $r = |x|$, $x \in B$. The gradient of u is also radial so that $\nabla_T |\nabla u| = 0$. Gathering these facts, we see that (4.25) reduces to*

$$0 \le \int_B u_r^2 \left(|\nabla \varphi|^2 - \frac{N-1}{r^2} \varphi^2 \right) dx.$$

In particular, observe that (4.3) is simply a restatement of the geometric Poincaré formula in the radial setting.

Proof of Theorem 4.4.1.

Differentiate (1.3) with respect to x_i, $i = 1, \ldots, N$. Then, $u_i = \frac{\partial u}{\partial x_i}$ solves the linearized equation

$$-\Delta u_i = f'(u) u_i, \quad \text{in } \Omega.$$

Take a test function $\varphi \in C_c^1(\Omega)$, multiply the previous equation by $u_i \varphi^2$, integrate, and sum over i. Then,

$$\int_\Omega f'(u) |\nabla u|^2 \varphi^2 \, dx = \sum_{i=1}^N \int_\Omega (-\Delta u_i) u_i \varphi^2 \, dx$$

$$= \sum_{i=1}^N \left(\int_\Omega |\nabla u_i|^2 \varphi^2 \, dx + \int_\Omega u_i \nabla u_i \cdot \nabla \varphi^2 \, dx \right)$$

$$= \int_\Omega |D^2 u|^2 \, dx + \frac{1}{2} \int_\Omega \nabla |\nabla u|^2 \cdot \nabla \varphi^2 \, dx, \qquad (4.26)$$

where $|D^2 u|^2 = \sum_{i,j=1}^N u_{ij}^2$.

Let $\psi = |\nabla u| \varphi$. Then, $\psi \in H^1(\Omega)$, ψ has compact support, and

$$\nabla \psi = \varphi \nabla |\nabla u| + |\nabla u| \nabla \varphi.$$

Hence,

$$|\nabla \psi|^2 = |\nabla |\nabla u||^2 \varphi^2 + |\nabla u|^2 |\nabla \varphi|^2 + 2\varphi |\nabla u| \nabla \varphi \cdot \nabla |\nabla u|$$

$$= |\nabla |\nabla u||^2 \varphi^2 + |\nabla u|^2 |\nabla \varphi|^2 + \frac{1}{2} \nabla \varphi^2 \cdot \nabla |\nabla u|^2.$$

Apply the stability inequality (1.5) with test function ψ. Then,

$$\int_\Omega f'(u) |\nabla u|^2 \varphi^2 \, dx \le \int_\Omega \left(|\nabla |\nabla u||^2 \varphi^2 + |\nabla u|^2 |\nabla \varphi|^2 + \frac{1}{2} \nabla \varphi^2 \cdot \nabla |\nabla u|^2 \right) dx.$$

Using (4.26), we obtain

$$\int_\Omega \left(|D^2 u|^2 - |\nabla|\nabla u||^2 \right) \varphi^2 \, dx \leq \int_\Omega |\nabla u|^2 |\nabla \varphi|^2 \, dx.$$

Equation (4.25) then follows from the following geometric identity.

$$|D^2 u|^2 - |\nabla|\nabla u||^2 = \left| \nabla_T |\nabla u| \right|^2 + |B|^2 |\nabla u|^2 \quad \text{on } \{x \in \Omega \ : \ \nabla u(x) \neq 0\}. \quad (4.27)$$

We prove at last Identity (4.27). Take a point $x \in \Omega$ such that $\nabla u(x) \neq 0$ and let $t = u(x)$. Without loss of generality, we may assume that $x = 0$, that $\nabla u(0) = |\nabla u(0)| e_N$, and that the level set $L_t := \{y \ : \ y \in \Omega, \nabla u(y) \neq 0, u(y) = t\}$ takes the form

$$L_t = \{y \ : \ y = (y', y_N) \in \Omega, \nabla u(y) \neq 0, y_N = \Phi(y')\},$$

for some C^2 function Φ, such that $D^2 \Phi(0)$ is a diagonal matrix which eigenvalues $\lambda_1, \ldots, \lambda_{N-1}$ are the principal curvatures of the level set L_t.

Let us compute $|\nabla|\nabla u||^2$ at $x = 0$.

$$|\nabla|\nabla u||^2 = \sum_{i=1}^N (\partial_i |\nabla u|)^2 = \sum_{i=1}^N \left(\frac{\nabla u \cdot \nabla u_i}{|\nabla u|} \right)^2 = \sum_{i=1}^N u_{iN}^2. \quad (4.28)$$

Similarly, at $x = 0$,

$$\left| \nabla_T |\nabla u| \right|^2 = \sum_{i=1}^{N-1} (\partial_i |\nabla u|)^2 = \sum_{i=1}^{N-1} u_{iN}^2. \quad (4.29)$$

By definition of Φ,

$$u(x', \Phi(x')) = t, \quad \text{for all } x = (x', \Phi(x')) \in L_t.$$

Differentiating with respect to x_i, $i = 1, \ldots, N-1$, it follows that

$$\partial_i u(x', \Phi(x')) + \partial_N u(x', \Phi(x')) \partial_i \Phi(x') = 0.$$

Differentiating again with respect to x_j, $j = 1, \ldots, N-1$, we get

$$\partial_{ij} u + \partial_{iN} u \partial_j \phi + \partial_{jN} u \partial_i \Phi + \partial_{NN} u \partial_i \Phi \partial_j \Phi + \partial_N u \partial_{ij} \Phi = 0$$

at $x = (x', \Phi(x'))$. Evaluated at $x = 0$, the previous expression simplifies:

$$\partial_{ij} u + |\nabla u| \partial_{ij} \Phi = 0.$$

So, at $x = 0$,

$$D^2 u = \begin{bmatrix} -|\nabla u|\lambda_1 & & & u_{1N} \\ & \ddots & & \vdots \\ & & -|\nabla u|\lambda_{N-1} & u_{N-1,N} \\ \hline u_{1N} & \cdots & u_{N-1,N} & u_{NN} \end{bmatrix}.$$

Hence, at $x = 0$,

$$|D^2 u|^2 = |B|^2 |\nabla u|^2 + 2\sum_{i=1}^{N-1} u_{iN}^2 + u_{NN}^2. \tag{4.30}$$

Using (4.28) and (4.29), Identity (4.27) follows. $\qquad\square$

4.5   Dimension $N = 4$

In this section, we exploit the geometric Poincaré formula to obtain *a priori* bounds on positive classical stable solutions of (1.3) posed in a convex domain $\Omega \subset \mathbb{R}^N$, $N \leq 4$.

Theorem 4.5.1 ([34]) *Let $f \in C^\infty(\mathbb{R})$ and $\Omega \subset \mathbb{R}^N$ denote a smoothly bounded and convex domain. Assume that $f \geq 0$. Then, there exists a constant c_Ω, depending on Ω only, such that given any stable solution $u \in C^2(\overline{\Omega})$ to (1.3), we have*

$$\|u\|_{L^\infty(\Omega)} \leq C(\Omega, \|u\|_{L^1(\Omega)}, \|f\|_{L^\infty(c_\Omega\|u\|_{L^1(\Omega)})}), \tag{4.31}$$

where $C(\cdot)$ depends only on the quantities within the parentheses.

As a consequence, we obtain the following regularity result for extremal solutions.

Corollary 4.5.1 ([34]) *Let $1 \leq N \leq 4$ and let $\Omega \subset \mathbb{R}^N$ be a smoothly bounded and convex domain of $\mathbb{R}^N$. Assume that $f \in C^1(\mathbb{R})$ is nondecreasing, $f(0) > 0$, and f is superlinear in the sense of (4.17). Then, the extremal solution to (3.1) is classical.*

4.5.1   Interior estimates

Theorem 4.5.2 ([34]) *Let $f \in C^\infty(\mathbb{R})$ and let $\Omega \subset \mathbb{R}^N$ a smoothly bounded domain. Assume $2 \leq N \leq 4$. Let u denote a classical stable solution to (1.3). Assume*

$$u > 0 \qquad in \ \Omega.$$

88

Then, for every $t > 0$,

$$\|u\|_{L^\infty(\Omega)} \le t + \frac{C}{t}|\Omega|^{\frac{4-N}{2N}}\left(\int_{[u<t]} |\nabla u|^4 \, dx\right)^2, \qquad (4.32)$$

where C is a universal constant, independent of f, Ω, or u.

Proof. Let $T = \|u\|_{L^\infty(\Omega)}$ and note that by Sard's theorem (Theorem C.3.1), almost every $s \in (0, T)$ is a regular value of u, that is, $\nabla u(x) \neq 0$ for $x \in [u = s]$. Now apply the geometric Poincaré formula (Theorem 4.4.1) with $\varphi = \varphi(u(x))$, where φ is Lipschitz in $[0, T]$ and $\varphi(0) = 0$. Using the coarea formula (C.23), the right-hand side of (4.25) becomes

$$\int_\Omega |\nabla u|^2 |\nabla \varphi|^2 \, dx = \int_\Omega |\nabla u|^4 \varphi'(u)^2 \, dx$$

$$= \int_0^T \left(\int_{[u=s]} |\nabla u|^3 d\sigma\right) \varphi'(s)^2 \, ds.$$

Given $\delta > 0$, the left-hand side of (4.25) is bounded below by

$$\int_{[|\nabla u|>\delta]} \left(\left|\nabla_T |\nabla u|\right|^2 + |B|^2 |\nabla u|^2\right) \varphi(u)^2 \, dx =$$

$$\int_0^T \left(\int_{[u=s]\cap[|\nabla u|>\delta]} \frac{1}{|\nabla u|} \left(\left|\nabla_T |\nabla u|\right|^2 + |B|^2 |\nabla u|^2\right) d\sigma\right) \varphi(s)^2 \, ds$$

$$= \int_0^T \int_{[u=s]\cap[|\nabla u|>\delta]} \left(4\left|\nabla_T |\nabla u|^{1/2}\right|^2 + \left(|B| \, |\nabla u|^{1/2}\right)^2\right) d\sigma \, \varphi(s)^2 \, ds.$$

Letting $\delta \searrow 0$ and using the monotone convergence theorem, we deduce that

$$\int_0^T h_1(s)\varphi(s)^2 \, ds \le \int_0^T h_2(s)\varphi'(s)^2 \, ds, \qquad (4.33)$$

where

$$h_1(s) = \int_{[u=s]} \left(4\left|\nabla_T |\nabla u|^{1/2}\right|^2 + \left(|B| \, |\nabla u|^{1/2}\right)^2\right) d\sigma \qquad (4.34)$$

and

$$h_2(s) = \int_{[u=s]} |\nabla u|^3 d\sigma, \qquad (4.35)$$

are defined at every regular value s of u. The rest of the proof differs in every dimension $N = 4, 3, 2$.

Case $N = 4$. Given a regular value s of u, apply the Sobolev inequality (C.13) with $M = [u = s]$, $p = 2$, and $v = |\nabla u|^{1/2}$. Noting that the mean curvature H of $[u = s]$ satisfies $|H| \le |B|$, we obtain

$$\left(\int_{[u=s]} |\nabla u|^{\frac{N-1}{N-3}} d\sigma \right)^{\frac{N-3}{N-1}} \le C(N) h_1(s). \tag{4.36}$$

For $N = 4$, (4.36) reduces to

$$h_2^{1/3} \le C h_1 \qquad \text{a.e. in } (0, T). \tag{4.37}$$

For every regular value s of u, we have $h_2(s) > 0$ and $h_1(s) < \infty$. This together with (4.37) gives $h_1/h_2 \in (0, +\infty)$ a.e. in $(0, T)$. So, for any regular value s and any integer $k \ge 1$,

$$g_k(s) = \min \left(k, \frac{h_1(s)}{h_2(s)} \right)$$

is well defined and $g_k \in L^\infty(0, T)$. In addition,

$$g_k(s) \nearrow \frac{h_1(s)}{h_2(s)} \qquad \text{for a.e. } s \in (0, T), \text{ as } k \to +\infty. \tag{4.38}$$

Since g_k is bounded, the function

$$\varphi_k(t) = \begin{cases} s/t & \text{if } s \le t \\ \exp \left(\frac{1}{\sqrt{2}} \int_t^s \sqrt{g_k(\tau)} d\tau \right) & \text{if } t < s \le T \end{cases} \tag{4.39}$$

is well defined, Lipschitz in $[0, T]$, and satisfies $\varphi_k(0) = 0$. Since

$$h_2 \left(\varphi_k' \right)^2 = h_2 \frac{1}{2} g_k \varphi_k^2 \le \frac{1}{2} h_1 \varphi_k^2 \qquad \text{in } (t, T),$$

(4.33) used with $\varphi = \varphi_k$ leads to

$$\int_t^T h_1 \varphi_k^2 \, ds \le \frac{2}{t^2} \int_0^t h_2 \, ds = \frac{2}{t^2} \int_{[u<t]} |\nabla u|^4 \, dx. \tag{4.40}$$

Let

$$B_t = \frac{1}{t^2} \int_{[u<t]} |\nabla u|^4 \, dx = \frac{1}{t^2} h_2(s) \, ds. \tag{4.41}$$

We need to establish that $T - t \leq C B_t^{1/2}$. By (4.38), we have

$$T - t = \int_t^T ds = \sup_{k \geq 1} \int_t^T \left(\frac{h_2 g_k}{h_1} \right)^{1/4} ds. \tag{4.42}$$

Using (4.40) and the Cauchy-Schwarz inequality, we have that

$$\int_t^T \left(\frac{h_2 g_k}{h_1} \right)^{1/4} ds = \int_t^T \sqrt{h_1 \varphi_k} \left(\frac{1}{\varphi_k} \left(\frac{h_2 g_k}{h_1^3} \right)^{1/4} \right) ds \tag{4.43}$$

$$\leq (2B_t)^{1/2} \left(\int_t^T \sqrt{\frac{h_2 g_k}{h_1^3}} \frac{1}{\varphi_k^2} ds \right)^{1/2}$$

$$\leq (2B_t)^{1/2} \left(C \int_t^T \sqrt{g_k} \frac{1}{\varphi_k^2} ds \right)^{1/2}. \tag{4.44}$$

In the last inequality, we used the crucial estimate (4.37). Finally, using the definition (4.39) of φ_k, we bound the integral in (4.44) as follows.

$$\int_t^T \sqrt{g_k} \frac{1}{\varphi_k^2} ds = \int_t^T \sqrt{g_k} \frac{1}{\varphi_k^2} \frac{\varphi_k'}{\frac{1}{\sqrt{2}} \sqrt{g_k} \varphi_k} ds$$

$$= \sqrt{2} \int_t^T \frac{\varphi_k'}{\varphi_k^3} ds = \frac{\sqrt{2}}{2} [\varphi_k^{-2}(s)]_{s=T}^{s=t}$$

$$\leq \frac{\sqrt{2}}{2} \varphi_k^{-2}(t) = \frac{\sqrt{2}}{2}.$$

This bound, together with (4.42), (4.43), and (4.44), yields the desired proof for $N = 4$.

Case $N = 2, 3$. In this case, we use the simpler test function

$$\varphi(t) = \begin{cases} s/t & \text{if } s \leq t \\ 1 & \text{if } t < s \leq T. \end{cases}$$

By (4.34), $h_1(s) \geq \int_{[u=s]} |B|^2 |\nabla u| \, d\sigma$, hence, using the aforementioned test function and (4.33), we obtain

$$\int_t^T \int_{[u=s]} |B|^2 |\nabla u| \, d\sigma \, ds \leq \int_0^T h_1(s) \varphi^2(s) \, ds$$

$$\leq \int_0^t h_2(s) \frac{1}{t^2} ds = \frac{1}{t^2} \int_{[u<t]} |\nabla u|^4 \, dx = B_t. \tag{4.45}$$

Next, we use the following geometric inequality:

$$|[u=s]|^{\frac{N-2}{N-1}} \le C(N) \int_{[u=s]} |H| \, d\sigma, \qquad (4.46)$$

where H is the mean curvature of $[u=s]$, $C(N)$ is a constant depending only on N, and s is a regular value of u. In dimension $N=2$, this simply follows from the Gauss-Bonnet formula. For $N \ge 3$, (4.46) follows from the Sobolev inequality (C.13). Indeed, taking $v=1$ and $n=N-1>1=p$ in (C.13), we deduce (4.46). In addition, the classical isoperimetric inequality (C.1) gives

$$|[u>s]| \le C(N)|[u=s]|^{\frac{N}{N-1}}. \qquad (4.47)$$

Now, using (4.46) and (4.47), there exists a constant $C=C(N)$ such that

$$|[u>s]|^{\frac{N-2}{N}} \le C \int_{[u=s]} |H| \, d\sigma \le C \left(\int_{[u=s]} |B|^2 |\nabla u| \, d\sigma \right)^{\frac{1}{2}} \left(\int_{[u=s]} \frac{d\sigma}{|\nabla u|} \right)^{\frac{1}{2}}.$$

In the above, we assumed that s is a regular value and we used the Cauchy-Schwarz inequality as well as the inequality $|H| \le |B|$. From this, we deduce that

$$T - t = \int_t^T ds \le$$

$$C \int_t^T \left(\int_{[u=s]} |B|^2 |\nabla u| \, d\sigma \right)^{1/2} \left(|[u>s]|^{\frac{2(2-N)}{N}} \int_{[u=s]} \frac{d\sigma}{|\nabla u|} \right)^{1/2} ds$$

$$\le C \left(\int_t^T \int_{[u=s]} |B|^2 |\nabla u| \, d\sigma \, ds \right)^{1/2} \left(\int_t^T |[u>s]|^{\frac{2(2-N)}{N}} \int_{[u=s]} \frac{d\sigma}{|\nabla u|} \, ds \right)^{1/2}$$

$$\le C B_t^{1/2} \left(\int_t^T |[u>s]|^{\frac{2(2-N)}{N}} \int_{[u=s]} \frac{d\sigma}{|\nabla u|} \, ds \right)^{1/2}, \qquad (4.48)$$

where we used (4.45) in the last inequality. By the coarea formula, the mapping $s \to |[u>s]|$ is differentiable almost everywhere and

$$-\frac{d}{ds}|[u>s]| = \int_{[u=s]} \frac{d\sigma}{|\nabla u|} \quad \text{for a.e. } s \in (0, T).$$

In addition, for $N \leq 3$, $|[u > s]|^{\frac{4-N}{N}}$ is nonincreasing in s. Thus, its total variation satisfies

$$|\Omega|^{\frac{4-N}{N}} \geq |[u > t]|^{\frac{4-N}{N}} = \left[|[u > s]|^{\frac{4-N}{N}} \right]_{s=T}^{s=t}$$

$$\geq \int_t^T \frac{4-N}{N} |[u > s]|^{\frac{2(2-N)}{N}} \left(-\frac{d}{ds} |[u > s]| \right) ds$$

$$= \frac{4-N}{N} \int_t^T |[u > s]|^{\frac{2(2-N)}{N}} \int_{[u=s]} \frac{d\sigma}{|\nabla u|} ds.$$

From this and (4.48), we conclude that for $N \leq 3$,

$$T - t \leq C(N) B_t^{1/2} |\Omega|^{\frac{4-N}{2N}},$$

which is the desired inequality. $\qquad\square$

Using the interior estimates of Theorem 4.5.2, we easily obtain the following proposition.

Proposition 4.5.1 *Let $f \in C^\infty(\mathbb{R})$. Let Ω be a smoothly bounded domain of $\mathbb{R}^N$, with $2 \leq N \leq 4$. Let u be a classical stable solution to (1.3). In addition, assume that*

$$u(x) \geq c_1 \operatorname{dist}(x, \partial\Omega) \qquad \text{for all } x \in \Omega \tag{4.49}$$

and

$$\|u\|_{L^\infty(\Omega_\rho)} \leq c_2, \tag{4.50}$$

where $\Omega_\rho = \{x \in \Omega : \operatorname{dist}(x, \partial\Omega) < \rho\}$. Then,

$$\|u\|_{L^\infty(\Omega)} \leq C(\Omega, \rho, c_1, c_2, \|f\|_{L^\infty(0, c_2)}). \tag{4.51}$$

Proof. Taking ρ smaller if necessary, we may assume that Ω_δ is smooth for every $\delta \in (0, \rho)$. We apply Theorem 4.5.2 with $t = c_1 \frac{\rho}{2}$. So, by (4.49), $[u < t] \subset \Omega_{\rho/2}$. By (4.32), it suffices to estimate $\|u\|_{W^{1,4}(\Omega_{\rho/2})}$.

But u solves $-\Delta u = f(u)$ in Ω_ρ and $u = 0$ on $\partial\Omega$ (which is one part of $\partial\Omega_\rho$). In addition, $\partial\Omega \cup \Omega_{\rho/2}$ has compact closure contained in $\partial\Omega \cup \Omega_\rho$, and both sets are smooth. By (4.50), $\|u\|_{L^\infty(\Omega_\rho)} \leq c_2$ and thus $\|f(u)\|_{L^\infty(\Omega_\rho)} \leq \|f\|_{L^\infty(0, c_2)}$. By elliptic regularity, we deduce a bound on $\|u\|_{W^{1,4}(\Omega_{\rho/2})}$ depending on the quantities in the right-hand side of (4.51). $\qquad\square$

4.5.2 Boundary estimates

In order to apply Proposition 4.5.1, we need to estimate solutions near the boundary.

Theorem 4.5.3 ([57, 83, 123]) *Let f be a locally Lipschitz function and let Ω be a smoothly bounded and convex domain of $\mathbb{R}^N$, $N \geq 2$. Let u denote any positive classical solution to (1.3).*

Then, there exists constants $\rho, \gamma > 0$, depending only on Ω, such that

$$\|u\|_{L^\infty(\Omega_\rho)} \leq \frac{1}{\gamma} \|u\|_{L^1(\Omega)}, \tag{4.52}$$

where $\Omega_\rho = \{x \in \Omega : \text{dist}(x, \partial\Omega) < \rho\}$.

Proof. We follow [185]. Clearly, it suffices to prove that for every $x \in \Omega_\rho$, there exists a set I_x such that $|I_x| \geq \gamma$ and $u(x) \leq u(y)$, for all $y \in I_x$. To see this, we apply the moving planes device.

For $\lambda > 0$, $x_0 \in \partial\Omega$, let $n = n(x_0)$ denote the exterior unit normal to $\partial\Omega$ at x_0 and

$$\Sigma_\lambda := \{x \in \Omega : 0 < -(x - x_0) \cdot n < \lambda\}.$$

Step 1. There exists a value $\lambda_0 > 0$, depending on Ω only, such that $\partial_n u < 0$ in Σ_{λ_0}.

Indeed, since Ω is convex, there exists $\lambda_0 > 0$ such that the reflection of Σ_λ through the hyperplane $T_\lambda = \{x : -(x - x_0) \cdot n = \lambda\}$ remains inside Ω, for evey $\lambda \leq \lambda_0$. One may then apply the moving planes procedure, exactly as in the proof of Lemma 1.2.1.

Step 2. There exists a neighborhood Θ of the direction $n = n(x_0)$ in $\mathscr{S}^{N-1}$, depending on Ω only, such that for all $\theta \in \Theta$,

$$\partial_\theta u < 0 \quad \text{in } \Sigma := \left\{x \in \Omega : \frac{1}{8}\lambda_0 < -(x - x_0) \cdot n(x_0) < \frac{3}{8}\lambda_0\right\}.$$

Indeed, apply Step 1 at every point $\tilde{x}_0 \in \partial\Omega$ in a neighborhood of x_0. In particular, assuming for simplicity that all curvatures of $\partial\Omega$ are positive at x_0, we obtain a neighborhood Θ of $n(x_0)$ in $\mathscr{S}^{N-1}$, such that for every $\theta \in \Theta$,

$$\partial_\theta u < 0 \quad \text{in } \left\{x \in \Omega : 0 < -(x - x_0) \cdot \theta < \frac{\lambda_0}{2}\right\}.$$

By taking a smaller neighborhood Θ, we may assume that

$$|(x - x_0) \cdot (\theta - n(x_0))| < \frac{1}{8}\lambda_0, \quad \text{for all } x \in \Sigma_{\lambda_0} \text{ and } \theta \in \Theta.$$

Now, since $-(x - x_0) \cdot \theta = -(x - x_0) \cdot (\theta - n(x_0)) - (x - x_0) \cdot n(x_0)$, we have for any $x \in \Sigma$,

$$\frac{\lambda_0}{2} = \frac{1}{8}\lambda_0 + \frac{3}{8}\lambda_0 > -(x - x_0) \cdot \theta > \frac{1}{8}\lambda_0 - \frac{1}{8}\lambda_0 = 0$$

and so

$$\partial_\theta u(x) < 0.$$

Step 3. Now take $\rho = \lambda_0/8$, where λ_0 is given in Step 1. Fix a point $x \in \Omega_\rho = \{x \in \Omega : \text{dist}(x, \partial\Omega) < \rho\}$ and let x_0 denote its projection on $\partial\Omega$. By Step 1, $u(x) \leq u(x_1)$, where $x_1 = x_0 - \rho n(x_0)$. By Step 2, $u(x_1) \leq u(y)$, for all y in the cone $I_x \subset \Sigma$ having vertex at x_1, opening angle Θ, and height $\lambda_0/4$. $\qquad\square$

4.5.3 Proof of Theorem 4.5.1 and Corollary 4.5.1

Proof of Theorem 4.5.1. By the boundary point lemma (Proposition A.4.2), there exists a constant $c = c(\Omega)$ such that

$$u \geq c \left(\int_\Omega f(u) d_\Omega(x)\, dx \right) d_\Omega,$$

where $d_\Omega(x) = \text{dist}(x, \partial\Omega)$. Let ζ_0 be the solution to (3.8). Then, integrating (1.3) against ζ_0 yields

$$\|u\|_{L^1(\Omega)} = \int_\Omega f(u)\zeta_0(x)\, dx \leq C \int_\Omega f(u)d_\Omega(x)\, dx.$$

So, $u \geq c_1 d_\Omega$, for some c_1 depending on Ω and $\|u\|_{L^1(\Omega)}$ only.

In addition, by the boundary estimate of Theorem 4.5.3, $\|u\|_{L^\infty(\Omega_\rho)} \leq c_2$, for some constants ρ, c_2 depending on Ω and $\|u\|_{L^1(\Omega)}$ only. Applying Proposition 4.5.1, Theorem 4.5.1 follows. $\qquad\square$

Proof of Corollary 4.5.1. Since $f > 0$, all solutions of (3.1) are positive. In particular, up to modifying f for negative values of t, we may always assume that $f(t) \geq f(0)/2 > 0$ for all $t \in \mathbb{R}$. Recall that the extremal solution u^* is the increasing L^1-limit, as $\lambda \nearrow \lambda^*$, of the minimal solutions u_λ to (3.1). So, if $f \in C^\infty$, we may simply apply Theorem 4.5.1 with nonlinearity λf, $\lambda^*/2 < \lambda < \lambda^*$ and obtain estimates for $\|u_\lambda\|_{L^\infty(\Omega)}$, which are uniform in λ. Letting $\lambda \nearrow \lambda^*$, we conclude that $u^* \in L^\infty(\Omega)$.

If f is only C^1, let ρ_k be a C^∞ mollifier with support in $(0, 1/k)$, of the form $\rho_k(t) = k\rho_k(kt)$. We replace f by

$$f_k(s) = \int_{s-1/k}^s f(t)\rho_k(s-t)\,dt = \int_0^1 f(s-t/k)\rho(t)\,dt.$$

Given any $k \in \mathbb{N}$, note that $f_k \leq f_{k+1} \leq f$ in $\mathbb{R}$, $f_k \in C^\infty(\mathbb{R})$, and f_k is nondecreasing. In addition, $f_k(0) > 0$ and f_k is superlinear in the sense of (4.17). Since $f(u^*) \geq f_k(u^*)$, u^* is a weak supersolution to

$$\begin{cases} -\Delta u = \lambda f_k(u) & \text{in } \Omega, \\ \quad\;\, u = 0 & \text{on } \partial\Omega, \end{cases} \tag{4.53}$$

for $\lambda = \lambda^*$. By the method of sub- and supersolutions (Lemma 1.1.1), (4.53) is solvable for $\lambda = \lambda^*$, and so $\lambda_k^* \geq \lambda^*$, where λ_k^* is the extremal parameter associated to (4.53). Hence u_k, the minimal solution to (4.53) for $\lambda = \lambda^* - 1/k$ is classical. By Theorem 4.5.1, (u_k) is uniformly bounded in $L^\infty(\Omega)$. Note that $u_k \leq u_{k+1} \leq u^*$. Thus, u_k increases in $L^1(\Omega)$ toward a solution to (3.1) (with $\lambda = \lambda^*$) smaller than or equal to u^*, hence identical to u^*. Since u_k was uniformly bounded in $L^\infty(\Omega)$, $u^* \in L^\infty(\Omega)$. $\qquad\square$

4.6 Regularity of solutions of bounded Morse index

Consider again the bifurcation diagrams for the Gelfand problem in the unit ball, Figure P.1. From the analysis in Chapter 2, we see that no regularity theory can be developped for solutions of arbitrarily large Morse index. On the one hand, a singular solution exists in any dimension $N \geq 3$ (and its Morse index is infinite for $3 \leq N \leq 9$). On the other hand, there exists smooth solutions (thus having finite Morse index) having arbitrarily large L^∞ norm.

Still, one may ask whether solutions with bounded Morse index are bounded. This is indeed the case for the Gelfand problem and $1 \leq N \leq 9$.

Theorem 4.6.1 ([71]) *Assume that $3 \leq N \leq 9$ and let $\Omega \subset \mathbb{R}^N$ denote a bounded domain with $C^{2,\alpha}$ boundary. Fix $M \in \mathbb{N}$, and let (λ, u) be a solution to the Gelfand problem (4.14), with Morse index at most M. Then, there exists a constant $C = C(N, \Omega, M) > 0$, such that*

$$\|u\|_{L^\infty(\Omega)} \leq C. \tag{4.54}$$

Remark 4.6.1 *As follows from Chapter 2, in dimension $1 \leq N \leq 2$ (look at Figure P.1), (4.54) cannot hold with a constant independant of λ.*

Remark 4.6.2 *A similar result holds for the Lane-Emden nonlinearity $f(u) = |u|^{p-1}u$, see [97].*

Proof. By standard elliptic regularity, it suffices to show that $u \leq C$ in Ω. We assume to the contrary that there exists a sequence of solutions (λ_n, u_n), with index at most M, such that $M_n := \max_{\overline{\Omega}} u_n \to +\infty$, as $n \to +\infty$. Let x_n denote a corresponding point of maximum of u_n and $d_n = \mathrm{dist}(x_n, \partial\Omega)$. Passing to a subsequence if necessary, we may assume that there exists $\lambda_0 \in \mathbb{R}^+$, $x_0 \in \overline{\Omega}$, such that $\lambda_n \to \lambda_0$ and $x_n \to x_0$, as $n \to +\infty$. Furthermore, $\lambda_0 > 0$, by Theorems 4.2.1 and 8.3.4.

We use a rescaling (also called a blow-up) argument. Let $r_n = e^{-M_n/2}$ and $v_n(x) = u_n(x_n + r_n x) - M_n$, for $x \in \Omega_n := \frac{1}{r_n}(\Omega - x_n)$. Then, v_n solves

$$\begin{cases} -\Delta v_n = \lambda_n e^{v_n} & \text{in } \Omega_n, \\ v_n = -M_n & \text{on } \partial\Omega_n. \end{cases} \tag{4.55}$$

We distinguish two cases.
Case 1. Assume that (d_n/r_n) is unbounded. Taking a subsequence if necessary, we may assume that $\lim_{n \to +\infty} d_n/r_n = +\infty$. This implies that $\Omega_n \to \mathbb{R}^N$, as $n \to +\infty$. We claim that (v_n) is uniformly bounded on compact sets of $\mathbb{R}^N$. To see this, fix a ball B_R and n so large that $B_R \subset \Omega_n$. On the one hand, since $v_n \leq 0$, the solution w_n to

$$\begin{cases} -\Delta w_n = \lambda_n e^{v_n} & \text{in } B_R, \\ w_n = 0 & \text{on } \partial B_R, \end{cases}$$

is uniformly bounded in B_R. On the other hand, $z_n = v_n - w_n$ is harmonic in B_R, and $z_n(0) = -w_n(0)$ is bounded. By Harnack's inequality (Proposition A.3.1), z_n is uniformly bounded in $B_{R/2}$, and so must be v_n. So, we may pass to the limit in (4.55) and find a solution v of Morse index at most M to

$$-\Delta v = \lambda_0 e^v \quad \text{in } \mathbb{R}^N.$$

This is impossible, in virtue of Theorem 6.3.3.
Case 2. Assume that (d_n/r_n) is bounded. Taking a subsequence if necessary, we may assume that $\lim_{n \to +\infty} d_n/r_n = c \geq 0$. Let $\tilde{x}_n \in \partial\Omega$ be such that $d_n = |x_n - \tilde{x}_n|$. For any fixed n, take a coordinate chart $y = (y_1, \ldots, y_N)$ at $\tilde{x}_n$, mapping some neighborhood V of $\tilde{x}_n$ onto the cylinder $B(0,1) \times (-1,1)$, and such that $\Omega \cap V$ is mapped onto $B(0,1) \cap (0,1)$, $\partial\Omega \cap V$ onto $B(0,1) \times \{0\}$, $\tilde{x}_n$

to 0, and x_n to $y_n = (0, \ldots, d_n)$. We may always assume that the local charts are uniformly bounded in C^2, independently of n. Then, $v = v_n(y) = u_n(x)$ solves

$$\begin{cases} -Lv_n = \lambda_n e^{v_n} & \text{in } B(0,1) \times (0,1), \\ v_n = 0 & \text{on } B(0,1) \times \{0\}, \end{cases}$$

where $Lv = \sum_{i,j} a_{ij} \partial_{y_i y_j} v + \sum_i b_i \partial_{y_i} v$, $a_{ij} = \sum_k \frac{\partial y_i}{\partial x_k} \frac{\partial y_j}{\partial x_k}$, and $b_i = \Delta y_i$. The local charts being uniformly bounded in C^2 independently of n, the same holds true of a_{ij} in C^1 and b_i in L^∞. Also note that $a_{ij}(0) = \delta_{ij}$.

Now set

$$w_n(x) = \frac{1}{M_n} \left(M_n - v_n(y_n + r_n x) \right),$$

for $x \in \Omega_n := B(0, 1/r_n) \times (-d_n/r_n, (1 - d_n)/r_n)$. Then, w_n solves

$$\begin{cases} -L'w_n = \dfrac{\lambda_n}{M_n} e^{-M_n w_n} & \text{in } \Omega_n, \\ w_n = 1 & \text{on } B(0, 1/r_n) \times \{-d_n/r_n\}, \end{cases}$$

where $L'w = \sum_{i,j} a_{ij}(y_n + r_n x) \partial_{x_i x_j} w + r_n \sum_i b_i(y_n + r_n x) \partial_{x_i} w$.

Note that $0 \le w_n \le 1$ in Ω_n. By elliptic regularity, w_n converges to a function w such that $0 \le w \le 1$ and

$$\begin{cases} -\Delta w = 0 & \text{in } [x_N > -c], \\ w = 1 & \text{on } [x_N = -c]. \end{cases}$$

By the strong maximum principle, $w > 0$ in $[x_N \ge -c]$. But $w(0) = \lim_{n \to +\infty} w_n(0) = 0$, a contradiction. $\qquad\square$

Chapter 5

Singular stable solutions

In the previous chapter, we looked for optimal conditions under which stable solutions to (1.3) must be bounded. This chapter is concerned with the study of singular stable solutions. In the first section, we prove that in large dimensions, the extremal solution to the Gelfand problem (4.14) is singular for a large class of domains obtained as perturbations of the unit ball. It should be pointed out that these solutions are singular at only one point and that the construction of stable solutions having a bigger singular set is an open problem. In the second section, we prove that this result is optimal: in any dimension, one can find many (smoothly bounded convex) domains such that the extremal solution remains bounded. In the last section, in the context of the Lane-Emden nonlinearity, we prove that if a weak solution is stable, then the Hausdorff dimension of its singular set cannot be large.

5.1   The Gelfand problem in the perturbed ball

According to Theorem 4.2.1, singular stable solutions to the Gelfand problem (4.14) can only exist in dimension $N \geq 10$. But until now, we know of only one domain where the extremal solution is singular: the ball. In this section, we show that singular stable solutions persist in any domain obtained as a C^2-diffeomorphic perturbation of the ball B.

Theorem 5.1.1 ([73]) *Let* $N \geq 11$, *let* $\psi : \overline{B} \to \mathbb{R}^N$ *denote a* C^2 *map,* $t > 0$ *and define*

$$\Omega_t = \{x + t\psi(x) : x \in B\}.$$

Let $u^*(t)$ denote the extremal solution to

$$\begin{cases} -\Delta u = \lambda e^u & \text{in } \Omega_t, \\ \quad u = 0 & \text{on } \partial\Omega_t. \end{cases} \tag{5.1}$$

Then there exists $t_0 = t_0(N,\psi) > 0$ such that if $t < t_0$, $u^*(t)$ is singular. In addition, there exists $\xi(t) \in B$ such that $\|u^*(t) - \log\frac{1}{|x-\xi(t)|^2}\|_{L^\infty(\Omega_t)} \to 0$ as $t \to 0$.

The behavior of the singular solution at the origin is characterized as follows:

Corollary 5.1.1 ([73]) *Fix $t < t_0$ and let $(\lambda^*(t), u^*(t), \xi(t))$ denote the extremal solution to (5.1) given by Theorem 5.1.1. Then,*

$$u^*(t) = \ln\frac{1}{|x - \xi(t)|^2} + \ln\left(\frac{\lambda^*(0)}{\lambda^*(t)}\right) + \epsilon(|x - \xi(t)|), \tag{5.2}$$

where $\lim_{s\to 0} \epsilon(s) = 0$.

Remark 5.1.1 *It is not known whether Theorem 5.1.1 holds in dimension $N = 10$.*

Proof of Theorem 5.1.1. Recall that

$$\Omega_t = \{x + t\psi(x) : x \in B\},$$

where t is small and $\psi : \bar{B} \to \mathbb{R}^N$ a C^2 map. We change variables to replace (5.1) with a problem in the unit ball. The map $id + t\psi$ is invertible for t small and we write the inverse of $y = x + t\psi(x)$ as $x = y + t\tilde{\psi}(y)$. Define v by

$$u(y) = v(y + t\tilde{\psi}(y)).$$

Then,

$$\Delta_y u = \Delta_x v + L_t v,$$

where L_t is a second-order operator given by

$$L_t v = 2t \sum_{i,k} v_{x_i x_k} \frac{\partial \tilde{\psi}_k}{\partial y_i} + t \sum_{i,k} v_{x_k} \frac{\partial^2 \tilde{\psi}_k}{\partial y_i^2} + t^2 \sum_{i,j,k} v_{x_j x_k} \frac{\partial \tilde{\psi}_j}{\partial y_i} \frac{\partial \tilde{\psi}_k}{\partial y_i}.$$

We look for a solution to (5.1) of the form

$$v(x) = \log \frac{1}{|x - \xi|^2} + \phi, \quad \lambda = \lambda_s + \mu,$$

where $\lambda_s = 2(N - 2)$, where $\xi \in B$ is close to the origin, $\mu \in \mathbb{R}$ is small, and where ϕ is a small *bounded* perturbation. Then (5.1) is equivalent to

$$
\begin{cases}
-\Delta\phi - \dfrac{\lambda_s}{|x - \xi|^2}\phi = L_t\phi + \dfrac{\lambda_s}{|x - \xi|^2}(e^\phi - 1 - \phi) + \dfrac{\mu}{|x - \xi|^2}e^\phi \\
\qquad\qquad + L_t\left(\log \dfrac{1}{|x - \xi|^2} \right) & \text{in } B, \\
\phi = -\log \dfrac{1}{|x - \xi|^2} & \text{on } \partial B.
\end{cases}
\tag{5.3}
$$

To solve the above equation, we shall use the elliptic regularity theory developed for the inverse-square potential in Section B.5. Observe that for $N \geq 11$ and $\lambda_s = 2(N - 2)$, we have $0 < \lambda_s < (N - 2)^2/4$ and $\alpha_1^- > 0$, $\alpha_2^- < 0$, where α_k^- are the indicial roots defined in Section B.5.1. So, Theorem B.5.1 holds with $k_1 = 1$ and $\nu = 0$. For such ν, k_1, Theorem B.5.1 states that in order to find a *bounded* solution to (5.3) one must guarantee that the right-hand side and the boundary data of the equation satisfy the $N + 1$ orthogonality conditions (B.88). These conditions need not hold *a priori* for (5.3). So, we first modify the right-hand side of (5.3) so that the orthogonality conditions automatically hold: let $\epsilon_0 > 0$ and $\eta \in C^\infty(\mathbb{R})$ such that $0 \leq \eta \leq 1$, $\eta \not\equiv 0$ and $\text{supp}(\eta) \subset [\frac{1}{4}, \frac{1}{2}]$. For $\xi \in B_{1/2}$ we construct functions $V_{\ell,\xi}$ as

$$V_{l,\xi}(x) = \eta(|x - \xi|)W_{1,l,\xi}(x) \quad \ell = 1, \ldots, N,$$

where $W_{1,\ell,\xi}$ is constructed in Section B.5. Also let

$$\tilde{f}(x, t) = L_t\left(\log \frac{1}{|x - \xi|^2} \right)$$

and note that

$$\|\tilde{f}(x, t)|x - \xi|^2\|_{0,\alpha,-2,\xi} \leq C|t|. \tag{5.4}$$

101

Instead of (5.3), we first consider

$$
\begin{cases}
-\Delta\phi - \dfrac{\lambda_s}{|x-\xi|^2}\phi = L_t\phi + \dfrac{\lambda_s}{|x-\xi|^2}(e^\phi - 1 - \phi) + \mu_0\dfrac{1}{|x-\xi|^2}e^\phi + \\
\qquad\qquad\qquad + \tilde{f}(x,t) + \displaystyle\sum_{i=1}^{N}\mu_i V_{i,\xi} \qquad\qquad\qquad \text{in } B, \\[2ex]
\qquad\qquad \phi = -\log\dfrac{1}{|x-\xi|^2} \qquad\qquad\qquad\qquad \text{on } \partial B,
\end{cases}
$$

(5.5)

and prove the following lemma.

Lemma 5.1.1 *There exists $\epsilon_0 > 0$ such that if $|\xi| < \epsilon_0$, $|t| < \epsilon_0$, there exists unique $\phi \in C^{2,\alpha}_{0,\xi}(B)$ and $\mu_0, \dots, \mu_N \in \mathbb{R}$ solving (5.5).*

Proof. Let $\epsilon_0 > 0$ and consider the Banach space X of functions $\phi(x, \xi)$ defined for $x \in B$, $\xi \in B_{\epsilon_0}$, which are twice continuously differentiable with respect to x and continuous with respect to ξ for $x \neq \xi$ for which the following norm is finite:

$$
\|\phi\|_X = \sup_{\xi \in B_{\epsilon_0}} \|\phi(\cdot, \xi)\|_{2,\alpha,0,\xi;B}.
$$

Let $\mathcal{B}_R = \{\phi \in X \mid \|\phi\|_X \le R\}$. By Theorem B.5.1, given $\psi \in \mathcal{B}_R$, there exists a unique $\phi \in X$ solving

$$
\begin{cases}
-\Delta\phi - \dfrac{\lambda_s}{|x-\xi|^2}\phi = g & \text{in } B, \\[1.5ex]
\qquad\qquad\quad \phi = h & \text{on } \partial B,
\end{cases}
$$

(5.6)

with

$$
g = L_t\psi + \dfrac{\lambda_s}{|x-\xi|^2}(e^\psi - 1 - \psi) + \mu_0\dfrac{1}{|x-\xi|^2}e^\psi + \tilde{f}(x,t) + \sum_{i=1}^{N}\mu_i V_{i,\xi} \quad (5.7)
$$

and

$$
h = -\log\dfrac{1}{|x-\xi|^2}, \tag{5.8}
$$

if and only if the $N + 1$ orthogonality relations

$$\int_B g W_{k,l,\xi}\, dx = \int_{\partial B} h \frac{\partial W_{k,l,\xi}}{\partial n}\, d\sigma, \quad \text{for } (k,l) = (0,1) \text{ and } k = 1, l = 1 \ldots N$$
(5.9)

are satisfied. Equation (5.9) is a linear system in the unknown $\mu_0, \mu_1, \ldots, \mu_N$ and it is uniquely solvable provided the associated matrix of coefficients is nonsingular. This is indeed the case when $\xi = 0$ and $\psi = 0$, since by definition of $W_{k,l}$, the matrix is diagonal with nonzero diagonal entries. By continuity, the matrix remains nonsingular for small ξ and R.

So, we may define a nonlinear map $F : \mathcal{B}_R \to X$ by $F(\psi) = \phi$, where $\phi(\cdot, \xi)$ is the solution to (5.6) with g, h given by (5.7) and (5.8). Let us show that if t is small then one can choose R small so that $F : \mathcal{B}_R \to \mathcal{B}_R$. Indeed, let $\psi \in \mathcal{B}_R$ and $\phi = F(\psi)$. Then by Theorem B.5.1, we have

$$\|\phi\|_{2,\alpha,0,\xi;B} \leq C \left(t\|\psi\|_{0,\alpha,0,\xi;B} + \|\lambda_s(e^{\psi} - 1 - \psi) + |x - \xi|^2 \tilde{f}(x,t)\|_{0,\alpha,0,\xi;B} + \right.$$
$$\left. + \sum_{i=0}^N |\mu_i| \right). \quad (5.10)$$

From (5.9), we infer that for t, R and $|\xi|$ small,

$$\sum_{i=0}^N |\mu_i| \leq C \left(t\|\psi\|_{0,\alpha,0,\xi;B} + \|\lambda_s(e^{\psi} - 1 - \psi) + |x - \xi|^2 \tilde{f}(x,t)\|_{0,\alpha,0,\xi;B} + |\xi| \right)$$

and so

$$\|\phi\|_{2,\alpha,0,\xi;B} \leq$$
$$\leq C \left(t\|\psi\|_{0,\alpha,0,\xi;B} + \|\lambda_s(e^{\psi} - 1 - \psi) + |x - \xi|^2 \tilde{f}(x,t)\|_{0,\alpha,0,\xi;B} + |\xi| \right)$$
$$\leq C(|t|R + R^2 + |t| + |\xi|) < \frac{R}{2}, \quad (5.11)$$

provided R is first taken small enough and then $|t|$ and $|\xi| < \epsilon_0$ are chosen small.

Next we show that F is a contraction on $\mathcal{B}_R$. Let $\psi_1, \psi_2 \in \mathcal{B}_R$ and $\phi_\ell = F(\psi_\ell)$, $\ell = 1, 2$. Let $\mu_i^{(\ell)}$, $i = 0, \ldots, N$ be the constants in (5.6) associated with ψ_ℓ.

Let $\phi = \phi_1 - \phi_2$. Then ϕ satisfies

$$
\begin{cases}
-\Delta\phi - \dfrac{\lambda_s}{|x-\xi|^2}\phi = L_t(\psi_1 - \psi_2)+ \\
\qquad + \lambda_s\left(\dfrac{e^{\psi_1}-1-\psi_1}{|x-\xi|^2} - \dfrac{e^{\psi_2}-1-\psi_2}{|x-\xi|^2}\right) \\
\qquad + \mu_0^{(2)}\dfrac{e^{\psi_1}-e^{\psi_2}}{|x-\xi|^2} \\
\qquad + (\mu_0^{(1)}-\mu_0^{(2)})\dfrac{e^{\psi_1}}{|x-\xi|^2} \\
\qquad + \sum_{i=1}^{N}(\mu_i^{(1)}-\mu_i^{(2)})V_{i,\xi} \qquad\qquad \text{in } B, \\
\qquad\quad \phi = 0 \qquad\qquad\qquad\qquad\qquad\qquad \text{on } \partial B.
\end{cases}
\tag{5.12}
$$

Using Theorem B.5.1 and working as previously, we deduce that

$$
\|\phi\|_{2,\alpha,0,\xi} + \sum_{i=0}^{N}|\mu_i^{(1)} - \mu_i^{(2)}| \le C\|g\|_{0,\alpha,0,\xi},
\tag{5.13}
$$

where

$$
g := L_t(\psi_1-\psi_2)+\lambda_s\left(\frac{e^{\psi_1}-1-\psi_1}{|x-\xi|^2} - \frac{e^{\psi_2}-1-\psi_2}{|x-\xi|^2}\right)+\mu_0^{(2)}\frac{e^{\psi_1}-e^{\psi_2}}{|x-\xi|^2}.
\tag{5.14}
$$

Using (5.9), we have in particular that $|\mu_0^{(2)}| \le CR$ and it follows from (5.14) and (5.13) that

$$
\|\phi_1 - \phi_2\|_{2,\alpha,0,\xi} \le CR\|\psi_1 - \psi_2\|_{2,\alpha,0,\xi}.
\tag{5.15}
$$

and so

$$
\|F(\psi_1) - F(\psi_2)\|_X \le CR\|\psi_1 - \psi_2\|_X.
$$

So, F is a contraction if R is small enough, and the lemma follows. $\qquad\square$

Proof of Theorem 5.1.1 continued. Recall that we want to find a bounded solution to (5.3). Thanks to Lemma 5.1.1, it suffices to prove that the point ξ can be selected so that the parameters $\mu_1,\ldots,\mu_N$ in (5.5) all vanish. To do so, we define the map $(\xi, t) \mapsto \phi(\xi, t)$ as the small solution to (5.5) constructed in Lemma 5.1.1 for t, ξ small. We need to show that for t small enough there is a choice of ξ such that $\mu_i = 0$ for $i = 1, \ldots, N$. Let

$$\widehat{V}_{j,\xi} = W_{1,j}(x - \xi)\eta_1(|x - \xi|), \quad j = 1, \ldots, N, \tag{5.16}$$

where $\eta_1 \in C^\infty(\mathbb{R})$ is a cutoff function such that $0 \leq \eta_1 \leq 1$,

$$\begin{cases} \eta_1(r) = 0 \text{ for } r \leq \frac{1}{8}, \\ \eta_1(r) = 1 \text{ for } r \geq \frac{1}{4} \end{cases} \tag{5.17}$$

and where $W_{1,j}$ is defined in Section B.5.1. Multiplication of (5.5) by $\widehat{V}_{j,\xi}$ and integration in B gives

$$\int_B \left(-\Delta \widehat{V}_{j,\xi} - L_t \widehat{V}_{j,\xi} - \frac{\lambda_s}{|x - \xi|^2} \widehat{V}_{j,\xi} \right) \phi \, dx + \int_{\partial B} \log \frac{1}{|x - \xi|^2} \frac{\partial \widehat{V}_{j,\xi}}{\partial n} \, d\sigma$$

$$- \int_{\partial B} \frac{\partial \phi}{\partial n} \widehat{V}_{j,\xi} \, d\sigma$$

$$= \int_B \frac{\lambda_s}{|x - \xi|^2}(e^\phi - 1 - \phi)\widehat{V}_{j,\xi} \, dx + \mu_0 \int_B \frac{e^\phi}{|x - \xi|^2} \widehat{V}_{j,\xi} \, dx$$

$$+ \int_B \tilde{f}(x, t)\widehat{V}_{j,\xi} \, dx + \sum_{i=1}^{N} \mu_i \int_B V_{i,\xi}\widehat{V}_{j,\xi} \, dx.$$

When $\xi = 0$ the matrix $A = A(\xi)$ defined by

$$A_{i,j}(\xi) = \int_B V_{i,\xi}\widehat{V}_{j,\xi} \, dx \qquad \text{for } i, j = 1 \ldots N$$

is diagonal and invertible and by continuity it is still invertible for small ξ. Thus, we see that $\mu_i = 0$ for $i = 1, \ldots, N$ if and only if

$$H_j(\xi, t) = 0, \quad \forall j = 1, \ldots, N, \tag{5.18}$$

where, given $j = 1, \ldots, N$,

$$H_j(\xi, t) = \int_B \frac{\lambda_s}{|x - \xi|^2}(e^\phi - 1 - \phi)\widehat{V}_{j,\xi} \, dx + \mu_0 \int_B \frac{e^\phi}{|x - \xi|^2}\widehat{V}_{j,\xi} \, dx$$

$$+ \int_B \tilde{f}(x, t)\widehat{V}_{j,\xi} \, dx - \int_{\partial B} \log \frac{1}{|x - \xi|^2} \frac{\partial \widehat{V}_{j,\xi}}{\partial n} \, d\sigma + \int_{\partial B_1} \frac{\partial \phi}{\partial n}\widehat{V}_{j,\xi} \, d\sigma$$

$$- \int_B \left(-\Delta \widehat{V}_{j,\xi} - L_t \widehat{V}_{j,\xi} - \frac{\lambda_s}{|x - \xi|^2}\widehat{V}_{j,\xi} \right) \phi \, dx.$$

If this holds, then $\mu_1(\xi, t) = \cdots = \mu_N(\xi, t) = 0$ and $\phi(\xi, t)$ is the desired solution to (5.3) (with μ in (5.3) equal to $\mu_0(\xi, t)$). Observe that

$$\frac{\partial}{\partial \xi_k} \left[\int_{\partial B} \log \frac{1}{|x - \xi|^2} \frac{\partial \widehat{V}_{j,\xi}}{\partial n} \, d\sigma \right]_{\xi=0} = 2 \int_{\partial B} x_k \frac{\partial \widehat{V}_{j,0}}{\partial n} \, d\sigma$$

$$+ \int_{\partial B} \log \frac{1}{|x - \xi|^2} \frac{\partial}{\partial \xi_k} \frac{\partial \widehat{V}_{j,\xi}}{\partial n} \bigg|_{\xi=0} \, d\sigma$$

$$= 2 \int_{\partial B} x_k \frac{\partial \widehat{V}_{j,0}}{\partial n} \, d\sigma. \qquad (5.19)$$

We shall use the Brouwer fixed point theorem as follows. Define $H = (H_1, \ldots, H_N)$ and

$$B(\xi) = (B_1, \ldots, B_N) \quad \text{with} \quad B_j(\xi) = \int_{\partial B} \log \frac{1}{|x - \xi|^2} \frac{\partial W_{1,j}}{\partial n} (x - \xi) \, d\sigma.$$

For $j = 1, \ldots, N$ we have $W_{1,j}(x) = (|x|^{-\alpha_j^+} - |x|^{-\alpha_j^-}) \varphi_j(\frac{x}{|x|})$, and hence $\frac{\partial W_{1,j}}{\partial n}(x) = (\alpha_j^- - \alpha_j^+) \varphi_j(x) = \frac{\alpha_j^- - \alpha_j^+}{(\int_{S^{N-1}} x_j^2)^{1/2}} x_j$, for $x \in \partial B$. Using this and (5.19), we deduce that B is differentiable and $DB(0)$ is invertible. Equation (5.18) is then equivalent to

$$\xi = G(\xi),$$

where

$$G(\xi) = DB(0)^{-1} (DB(0)\xi - H(\xi, t)).$$

To apply the Brouwer fixed point theorem, it suffices to prove that for t, ρ small, G is a continuous function of ξ and $G : \overline{B}_\rho \to \overline{B}_\rho$. This is the object of the next two lemmata.

Lemma 5.1.2 *G is continuous for t, ξ small.*

Proof. Observe first that for t, ξ small such that $\|\phi\|_{L^\infty(B)} \leq R$ we have

$$\|\phi\|_{L^\infty(B)} \leq C(\|\lambda_s(e^\phi - 1 - \phi) + |x - \xi|^2 \tilde{f}(x, t)\|_{L^\infty(B)} + |\xi|)$$

$$\leq C(R\|\phi\|_{L^\infty(B)} + |t| + |\xi|),$$

106

and we deduce (taking R smaller if necessary)

$$\|\phi\|_{L^\infty(B)} \le C(|t|+|\xi|). \tag{5.20}$$

Similarly,

$$|\mu_i| \le C(|t|+|\xi|), \quad \forall i = 0,\dots,N. \tag{5.21}$$

Now, take a sequence $\xi_k \to \xi$ and let $\phi_k = \phi(\xi_k,t)$, $\mu_i^{(k)}$ be the solutions and parameters associated to (5.5). By (5.20) and (5.5) and using elliptic estimates we see that (ϕ_k) is bounded in $C^{1,\alpha}$ on compact sets of $\overline{B}\setminus\{\xi\}$. By passing to a subsequence we may assume that $\phi_k \to \phi$ uniformly on compact sets of $\overline{B}\setminus\{\xi\}$ and by (5.21) that $\mu_i^{(k)} \to \mu_i$. Then ϕ is a solution to (5.5) with $\|\phi\|_{L^\infty(B)} \le R$ and with parameters ξ and μ_i. This solution is unique by Lemma 5.1.1 and this shows that in fact, the complete sequence converges. Then all terms in the definition of $H(\xi,t)$ converge. In fact

$$\int_B \lambda_s \frac{e^{\phi_k}-1-\phi_k}{|x-\xi|^2}\widehat{V}_{j,\xi}\,dx \to \int_B \lambda_s \frac{e^{\phi}-1-\phi}{|x-\xi|^2}\widehat{V}_{j,\xi}\,dx \quad \text{as } k\to\infty,$$

by dominated convergence, because

$$\left|\frac{e^{\phi_k}-1-\phi_k}{|x-\xi|^2}\widehat{V}_{j,\xi}\right| \le \frac{C}{|x-\xi|^2}.$$

Similarly,

$$\mu_0^{(k)}\int_B \frac{e^{\phi_k}}{|x-\xi|^2}\widehat{V}_{j,\xi}\,dx \to \mu_0\int_B \frac{e^{\phi}}{|x-\xi|^2}\widehat{V}_{j,\xi}\,dx \quad \text{as } k\to\infty. \tag{5.22}$$

$\square$

Lemma 5.1.3 *If $\rho > 0$ and $|t|$ are small enough then $G : \overline{B_\rho} \to \overline{B_\rho}$.*

Proof. By (5.20)

$$\left|\int_B \lambda_s \frac{e^{\phi(\xi)}-1-\phi(\xi)}{|x-\xi|^2}\widehat{V}_{j,\xi}\,dx\right| \le C\|\phi\|_{L^\infty(B)}^2 \le C(|\xi|+|t|)^2.$$

Let $\sigma > 0$ to be fixed later. From (5.22) we have

$$\left|\int_B \frac{e^{\phi}}{|x-\xi|^2}\widehat{V}_{j,\xi}\,dx\right| \le \sigma$$

if t and ξ are small enough. Also,

$$|DB(0)\xi - B(\xi)| \le C|\xi|^2,$$

and

$$\left| \int_B \tilde{f}(x,t) \widehat{V}_{j,\xi} \, dx \right| \le C|t|$$

for some constant C. Thus if $|\xi| < \rho$ and ρ is small we have

$$|G(\xi)| \le C \left(\rho^2 + |t| + \sigma\rho \right).$$

First, fix σ such that $C\sigma < \frac{1}{4}$. We can then fix $\rho > 0$ so small that $C(\rho^2 + \sigma\rho) < \frac{\rho}{2}$. Then, for $|t|$ small, $|G(\xi)| \le \rho$. $\qquad\square$

Proof of Theorem 5.1.1 completed. Thanks to the previous two lemmata, we deduce that for t sufficiently small, there exists a point $\xi = \xi(t)$, a small parameter $\mu = \mu(t)$, and a small bounded function $\phi = \phi(x,t)$ solving (5.3). Change variables and let $\tilde{\phi}(y) = \phi(x)$, where $x = y + t\tilde{\psi}(y)$ for $y \in \Omega_t$. Then,

$$-\Delta_y \tilde{\phi} = \frac{\lambda(t)}{|x-\xi|^2} e^{\tilde{\phi}} + \Delta_y \ln \frac{1}{|x-\xi|^2} \qquad \text{in } \Omega_t.$$

Letting $\tilde{\xi} = \xi + t\psi(\xi)$ and $\tilde{\psi}(y) = \tilde{\phi}(y) + \ln \frac{|y-\tilde{\xi}|^2}{|x-\xi|^2}$, the above equation can be rewritten as

$$-\Delta_y \tilde{\psi} = \frac{\lambda(t)}{|y-\tilde{\xi}|^2} e^{\tilde{\psi}} - \frac{\lambda(0)}{|y-\tilde{\xi}|^2} \qquad \text{in } \Omega_t,$$

where we used the fact that $\Delta_y \ln \frac{1}{|y-\tilde{\xi}|^2} = -\frac{\lambda(0)}{|y-\tilde{\xi}|^2}$. Then, $u = u(t) = \tilde{\psi} + \ln \frac{1}{|y-\tilde{\xi}|^2}$ is a solution to (5.1) associated to $\lambda = \lambda(t)$, and

$$\left\| u(t) - \ln \frac{1}{|y-\tilde{\xi}(t)|^2} \right\|_{L^\infty(\Omega_t)} + |\lambda(t) - 2(N-2)| \to 0, \qquad \text{as } t \to 0. \quad (5.23)$$

It remains to prove that $u(t)$ is the extremal solution to (5.1). By Proposition 3.3.1, it suffices to show that $u(t)$ is stable. Since $N \ge 11$, $2(N-2) < \frac{(N-2)^2}{4}$ and it follows from (5.23) that if t is chosen small enough,

$$\lambda(t) e^{\left\| u - \ln \frac{1}{|y-\tilde{\xi}(t)|^2} \right\|_{L^\infty(\Omega_t)}} < \frac{(N-2)^2}{4}.$$

So, for $\varphi \in C_c^1(\Omega_t)$,

$$\lambda(t) \int_{\Omega_t} e^u \varphi^2 \, dy \leq \frac{(N-2)^2}{4} \int_{\mathbb{R}^N} \frac{\varphi^2}{|y - \xi(t)|^2} \, dy \leq \int_{\mathbb{R}^N} |\nabla \varphi|^2 \, dy,$$

in virtue of Hardy's inequality (Proposition C.1.1). Hence, $u(t)$ is stable, and so $u(t)$ is the extremal solution to (5.1). $\square$

Proof of Corollary 5.1.1. Using the above notation, since $\tilde{\psi}$ is bounded, it follows by Corollary B.5.1 and the fixed point characterization of $\tilde{\psi}$ that $\tilde{\psi}$ is continuous at $y = \tilde{\xi}$. Define the sequence (ψ_n) by

$$\psi_n(y) = \tilde{\psi}\left(\frac{1}{n}(y - \tilde{\xi}) + \tilde{\xi}\right), \qquad \text{for } y \in \Omega^n := n(\Omega_t - \tilde{\xi}) + \tilde{\xi}.$$

Clearly, (ψ_n) converges pointwise to the constant $\tilde{\psi}(\tilde{\xi})$. Also, ψ_n solves

$$-\Delta_y \psi_n = \frac{\lambda(t)}{|y - \tilde{\xi}|^2} e^{\psi_n} - \frac{\lambda(0)}{|y - \tilde{\xi}|^2} \qquad \text{in } \Omega^n. \tag{5.24}$$

Away from $y = \tilde{\xi}$, the right-hand side in the above equality remains bounded. It follows by elliptic regularity that up to a subsequence, (ψ_n) converges to $\tilde{\psi}(\tilde{\xi})$ in the topology of $C^\infty(\mathbb{R}^N \setminus \{\tilde{\xi}\})$. In particular, passing to the limit for $y \neq \tilde{\xi}$ in (5.24), we obtain

$$0 = \frac{\lambda(t)}{|y - \tilde{\xi}|^2} e^{\tilde{\psi}(\tilde{\xi})} - \frac{\lambda(0)}{|y - \tilde{\xi}|^2},$$

hence $\tilde{\psi}(\tilde{\xi}) = \ln \frac{\lambda(0)}{\lambda(t)}$. Since the solution $u(t)$ of (5.1) we constructed is given by $u(t) = \ln \frac{1}{|y-\tilde{\xi}|^2} + \tilde{\psi}$, we have just proved Corollary 5.1.1. $\square$

Exercise 5.1.1 Given $p > 1$, consider the problem

$$\begin{cases} -\Delta u = \lambda(1+u)^p & \text{in } \Omega_t, \\ \quad\;\; u = 0 & \text{on } \partial\Omega_t. \end{cases} \tag{5.25}$$

- When $t = 0$, that is, when the domain is the unit ball, prove that the extremal solution is unbounded and given by $u^* = |x|^{-2/(p-1)} - 1$, if and only if $N \geq 11$ and $p \geq p_c(N)$, where

$$p_c(N) = \frac{(N-2)^2 - 4N + 8\sqrt{N-1}}{(N-2)(N-10)}.$$

Compare to Exercise 4.2.1.

109

- Let $N \geq 11$ and $p > p_c(N)$. Given $t > 0$ small, let $u^*(t)$ denote the extremal solution to (5.25). Show that there exists $t_0 = t_0(N, \psi, p) > 0$ such that if $t < t_0$, $u^*(t)$ is singular. In addition, prove that there exists $\xi(t) \in B_1$ such that $\|u(x, t) - (|x - \xi(t)|^{-2/(p-1)} - 1)\|_{L^\infty(\Omega_t)} \to 0$, as $t \to 0$.

5.2   Flat domains

In the previous section, we proved that the extremal solution to the Gelfand problem (4.14) is singular in any dimension $N \geq 11$ and for any domain close to the ball. In this section, we prove that the situation is completely different when the domain is chosen close to an infinite cylinder with cross-section in $\mathbb{R}^{N_2}$, $N_2 \leq 9$.

Theorem 5.2.1 ([69]) *Let $N = N_1 + N_2 \geq 10$. Let $\Omega \subset \mathbb{R}^N$ denote a smoothly bounded domain. We assume furthermore that Ω is convex and $\partial\Omega$ is uniformly convex, that is, its principal curvatures are bounded away from zero. Given $\varepsilon > 0$, let u_ε^* be the extremal solution to*

$$\begin{cases} -\Delta u = \lambda e^u & \text{in } \Omega_\varepsilon, \\ \qquad u = 0 & \text{on } \partial\Omega_\varepsilon, \end{cases} \tag{5.26}$$

where, writing $\mathbb{R}^N = \mathbb{R}^{N_1} \times \mathbb{R}^{N_2}$ and $x = (x_1, x_2) \in \mathbb{R}^N$ with $x_1 \in \mathbb{R}^{N_1}$, $x_2 \in \mathbb{R}^{N_2}$, we set

$$\Omega_\varepsilon = \{x = (y_1, \varepsilon y_2) : (y_1, y_2) \in \Omega\}. \tag{5.27}$$

Then, if $N_2 \leq 9$, there exists $\varepsilon_0 = \varepsilon_0(N, \Omega) > 0$ such that if $\varepsilon < \varepsilon_0$, u_ε^* is smooth.

Remark 5.2.1 *Let $\Omega = B$ in dimension $N \geq 11$ and let Ω_ε be the ellipsoid defined by (5.27) with $N_2 = 1$. Combining Theorems 5.1.1 and 5.2.1 we can say that for ε close to 1 (round ellipsoid), u^* is singular, while for ε close to 0 (flat ellipsoid), u^* is regular.*

Proof of Theorem 5.2.1. We follow [73]. We assume by contradiction that for a sequence $\varepsilon_j \searrow 0$, we have $u_{\varepsilon_j}^* \notin L^\infty(\Omega_{\varepsilon_j})$. Let $M > 0$ be a constant to be fixed later. By continuity, we can select a number λ_j with $0 < \lambda_j < \lambda_{\varepsilon_j}^*$ such that the minimal solution u_j of (5.26) with parameter λ_j satisfies

$$\max_{\overline{\Omega}_{\varepsilon_j}} u_j = M. \tag{5.28}$$

110

Define

$$v_j(y_1, y_2) = u_j(y_1, \varepsilon_j y_2).$$

Then v_j is defined in $\overline{\Omega}$ and satisfies

$$\begin{cases} -(\varepsilon_j^2 \Delta_{y_1} + \Delta_{y_2})v_j = \varepsilon_j^2 \lambda_j e^{v_j} & \text{in } \Omega, \\ \quad\quad\quad\quad\quad v_j = 0 & \text{on } \partial\Omega, \end{cases} \tag{5.29}$$

where Δ_{y_i} denotes the Laplacian with respect to the variables y_i, $i = 1, 2$.

For some constant C_0 we have

$$\lambda_\varepsilon^* \leq \frac{C_0}{\varepsilon^2}. \tag{5.30}$$

Indeed, let $\mu_\varepsilon = \lambda_1(-\Delta; \Omega_\varepsilon)$ denote the principal eigenvalue for $-\Delta$ in Ω_ε with Dirichlet boundary conditions and $\varphi_\varepsilon > 0$ an associated eigenfunction, that is,

$$\begin{cases} -\Delta\varphi_\varepsilon = \mu_\varepsilon \varphi_\varepsilon & \text{in } \Omega_\varepsilon, \\ \quad\quad \varphi_\varepsilon = 0 & \text{on } \partial\Omega_\varepsilon. \end{cases}$$

We normalize φ_ε so that $\|\varphi_\varepsilon\|_{L^2(\Omega_\varepsilon)} = 1$. Multiplying (5.26) by φ_ε and integrating by parts we find

$$\mu_\varepsilon \int_{\Omega_\varepsilon} u_\varepsilon^* \varphi_\varepsilon \, dx = \lambda_\varepsilon^* \int_{\Omega_\varepsilon} e^{u_\varepsilon^*} \varphi_\varepsilon \, dx.$$

Since $e^u \geq u$ for all $u \in \mathbb{R}$, it follows that $\lambda_\varepsilon^* \leq \mu_\varepsilon$. But by changing variables $(x_1, x_2) = (y_1, \varepsilon y_2)$ we find

$$\mu_\varepsilon = \inf_{\substack{\varphi \in C_c^1(\Omega_\varepsilon) \\ \varphi \neq 0}} \frac{\int_{\Omega_\varepsilon} |\nabla\varphi|^2 \, dx}{\int_{\Omega_\varepsilon} \varphi^2 \, dx} = \inf_{\substack{\psi \in C_c^1(\Omega) \\ \psi \neq 0}} \frac{\int_\Omega |\nabla_{y_1}\psi|^2 + \frac{1}{\varepsilon^2}|\nabla_{y_2}\psi|^2 \, dx}{\int_\Omega \psi^2 \, dx}.$$

Fixing $\psi \in C_c^1(\Omega)$, $\psi \neq 0$ we deduce $\mu_\varepsilon \leq \frac{C_0}{\varepsilon^2}$. Note that $C_0 = C_0(\Omega, N)$ does not depend on M.

Next we show that for some constant C independent of j

$$\|\nabla v_j\|_{L^\infty(\Omega)} \leq C. \tag{5.31}$$

For this, using the uniform convexity of Ω, find $R > 0$ large enough so that for any $y_0 \in \partial\Omega$ there exists $z_0 \in \mathbb{R}^N$ such that the ball $B_R(z_0)$ satisfies $\Omega \subset B_R(z_0)$ and $y_0 \in \partial B_R(z_0)$. For convenience write for $\varepsilon > 0$

$$L_\varepsilon = \varepsilon^2 \Delta_{y_1} + \Delta_{y_2}.$$

Define $\zeta(y) = R^2 - |y - z_0|^2$ so that $\zeta \geq 0$ in Ω and $-L_\varepsilon \zeta = 2\varepsilon N_1 + 2N_2$ (this can be computed easily by shifting so that z_0 is at the origin and writing $|(y_1, y_2)|^2 = |y_1|^2 + |y_2|^2$). From (5.30) we have the uniform bound $\varepsilon_j^2 \lambda_j \leq C$. It follows from (5.29) and the maximum principle that $v_j \leq C\zeta$ with C independent of j and y_0. Since $v_j(y_0) = \zeta(y_0) = 0$, this in turn implies that

$$|\nabla v_j(y_0)| \leq C \quad \forall j, \, y_0 \in \partial\Omega. \tag{5.32}$$

Recall that the minimal solution u_j is strictly stable, that is, $\lambda_1(-\Delta - \lambda_j e^{u_j}; \Omega_{\varepsilon_j}) > 0$. By changing variables, the same holds true for the linearization of (5.29) at v_j, that is, the operator $w \mapsto -L_{\varepsilon_j} w - \varepsilon_j^2 \lambda_j e^{v_j} w$ has a positive principal eigenvalue. This implies that we have the maximum principle in the form: if $w \in C^2(\overline{\Omega})$ satisfies $-L_{\varepsilon_j} w - \varepsilon_j^2 \lambda_j e^{v_j} w = 0$ in Ω then

$$\max_{\overline{\Omega}} |w| \leq \max_{\partial\Omega} |w|.$$

Applying this to the partial derivatives of v_j and using (5.32), we deduce (5.31). By (5.28), (5.31), and (5.30) we can find subsequences, denoted for simplicity (v_j), (ε_j), and (λ_j), such that $v_j \to v$ uniformly in $\overline{\Omega}$ and $\varepsilon_j^2 \lambda_j \to \lambda_0 \geq 0$. Multiplying (5.29) by $\varphi \in C_c^1(\Omega)$ and integrating by parts twice we find

$$-\int_\Omega v_j(\varepsilon_j^2 \Delta_{y_1} \varphi + \Delta_{y_2} \varphi) \, dx = \varepsilon_j^2 \lambda_j \int_\Omega e^{v_j} \varphi \, dx.$$

Letting $j \to \infty$ we obtain

$$-\int_\Omega v \Delta_{y_2} \varphi \, dx = \lambda_0 \int_\Omega e^v \varphi \, dx \quad \forall \varphi \in C_c^1(\Omega).$$

Writing $v_{y_1}(y_2) := v(y_1, y_2)$ for $(y_1, y_2) \in \mathbb{R}^{N_1} \times \mathbb{R}^{N_2} \cap \Omega$, we see that for each nonempty slice

$$\Omega_{y_1} = \{y_2 \in \mathbb{R}^{N_2} : (y_1, y_2) \in \Omega\},$$

112

we have

$$\begin{cases} -\Delta_{y_2} v_{y_1} = \lambda_0 e^{v_{y_1}} & \text{in } \Omega_{y_1} \\ \qquad\quad v_{y_1} = 0 & \text{on } \partial\Omega_{y_1}. \end{cases} \tag{5.33}$$

Let $y_j \in \Omega$ denote the point of maximum of v_j, that is, $v_j(y_j) = \max_{\bar\Omega} v_j = M$. For a subsequence, $y_j \to y_0 \in \bar\Omega$ as $j \to \infty$ and since v_j converges uniformly to v, we have $M = v_j(y_j) \to v(y_0)$. Since $v|_{\partial\Omega} = 0$, we must have $y_0 \in \Omega$.

Write $y_0 = (a, b)$ and observe that Ω_a is nonempty since $y_0 \in \Omega$. Then $v_a(y_2) = v(a, y_2)$ solves (5.33) in Ω_a. Moreover $\max_{\bar\Omega_a} v_a = M$ and v_a is stable, that is,

$$\lambda_0 \int_{\Omega_a} e^{v_a} \varphi^2 \, dx \le \int_{\Omega_a} |\nabla\varphi|^2 \, dx, \quad \forall \varphi \in C_c^1(\Omega_a). \tag{5.34}$$

To see this, let $\varphi \in C_c^1(\Omega_a)$ and $\chi \in C_c^1(\mathbb{R}^{N_1})$ be such that $\chi \equiv 1$ in a neighborhood of a and $\operatorname{supp}(\chi(y_1)\varphi(y_2)) \subset \Omega$. By stability of u_j and changing variables we have

$$\varepsilon_j^2 \lambda_j \int_\Omega e^{v_j} \chi(y_1)^2 \varphi(y_2)^2 \, dx \le \int_\Omega \left(\varepsilon_j^2 \varphi(y_2)^2 |\nabla\chi(y_1)|^2 + \chi(y_1)^2 |\nabla\varphi(y_2)|^2 \right) \, dx.$$

Letting $j \to \infty$ yields

$$\lambda_0 \int_\Omega e^v \chi(y_1)^2 \varphi(y_2)^2 \, dx \le \int_\Omega \chi(y_1)^2 |\nabla\varphi(y_2)|^2 \, dx.$$

Choosing a sequence $\chi_k \in C_c^1(\mathbb{R}^{N_1})$ such that $\chi_k \equiv 1$ in a neighborhood of a and $\operatorname{supp}(\chi_k) \subset B_{1/k}(a)$ we obtain (5.34).

Let $y_{min}^{(1)} = \min\{y_1 : \Omega_{y_1} \ne \emptyset\}$, $y_{max}^{(1)} = \max\{y_1 : \Omega_{y_1} \ne \emptyset\}$. For any $y_{min}^{(1)} < y_1 < y_{max}^{(1)}$ the slice Ω_{y_1} is a smooth open nonempty set and hence for the problem

$$\begin{cases} -\Delta_{y_2} v = \lambda e^v & \text{in } \Omega_{y_1}, \\ \qquad\quad v = 0 & \text{on } \partial\Omega_{y_1}, \end{cases} \tag{5.35}$$

there exists an extremal parameter $0 < \lambda_{y_1}^* < \infty$. In particular,

- If $0 \le \lambda < \lambda_{y_1}^*$ then (5.35) has a unique minimal solution $v_{y_1,\lambda}$. Moreover $v_{y_1,\lambda}$ is smooth and characterized as the unique stable solution to (5.35), that is, the unique solution satisfying

$$\lambda \int_{\Omega_{y_1}} e^{v_{y_1,\lambda}} \varphi^2 \, dx \le \int_{\Omega_{y_1}} |\nabla\varphi|^2 \, dx, \quad \forall \varphi \in C_c^1(\Omega_{y_1}).$$

- If $\lambda > \lambda^*_{y_1}$ then (5.35) has no weak solution.

- If $\lambda = \lambda^*_{y_1}$ then (5.35) has a unique weak solution $v^*_{y_1}$ and $v^*_{y_1} = \lim_{\lambda \nearrow \lambda^*_{y_1}} v_{y_1,\lambda}$.

- If $N_2 \leq 9$ (recall that $\Omega_{y_1} \subset \mathbb{R}^{N_2}$) then $v^*_{y_1}$ is bounded.

We claim that for any $\lambda > 0$ there exists $M_\lambda > 0$ depending only on Ω and λ such that for any $y^{(1)}_{min} < y_1 < y^{(1)}_{max}$ we have

$$\max_{\overline{\Omega}_{y_1}} v_{y_1,\lambda} \leq M_\lambda. \tag{5.36}$$

That is, we assert that if we have some *a priori* control on λ, the boundedness of $v_{y_1,\lambda}$ is uniform when $y^{(1)}_{min} < y_1 < y^{(1)}_{max}$.

Using (5.30) we have the bound $\lambda_0 \leq C_0$. Hence, choosing $M = M_{\lambda_0} + 1$ at the beginning of the proof, (5.36) contradicts (5.28).

Proof of (5.36). The argument is the same as in the proof of Theorem 4.2.1 but we shall emphasize that the bound does not depend on $y^{(1)}_{min} < y_1 < y^{(1)}_{max}$.

For simplicity we write $v = v_{y_1,\lambda}$. Let $0 < \alpha < 2$ and multiply Equation (5.35) by $e^{2\alpha v} - 1$. Integrating in Ω_{y_1} we find

$$2\alpha \int_{\Omega_{y_1}} e^{2\alpha v}|\nabla v|^2\, dx = \lambda \int_{\Omega_{y_1}} \left(e^{(2\alpha+1)v} - e^v\right) dx. \tag{5.37}$$

Using (5.34) with $e^{\alpha v} - 1$ yields

$$\lambda \int_{\Omega_{y_1}} e^v(e^{\alpha v} - 1)^2\, dx \leq \alpha^2 \int_{\Omega_{y_1}} e^{2\alpha v}|\nabla v|^2\, dx. \tag{5.38}$$

Combining (5.37) and (5.38) gives

$$(1 - \frac{\alpha}{2}) \int_{\Omega_{y_1}} e^{(2\alpha+1)v}\, dx \leq 2 \int_{\Omega_{y_1}} e^{(\alpha+1)v}\, dx \leq$$

$$\leq 2 \left[\int_{\Omega_{y_1}} e^{(2\alpha+1)v}\, dx\right]^{\frac{\alpha+1}{2\alpha+1}} |\Omega_{y_1}|^{1 - \frac{\alpha+1}{2\alpha+1}}.$$

For $0 < p < 5$ we deduce the bound

$$\|e^v\|_{L^p(\Omega_{y_1})} \leq C$$

with C independent of $y_{min}^{(1)} < y_1 < y_{max}^{(1)}$.

In dimension $N_2 \leq 9$, we thus have $\|e^v\|_{L^p(\Omega_{y_1})} \leq C$ for some $p > N_2/2$. Recalling (5.35), this shows that $\|v\|_{L^\infty(\Omega_{y_1})} \leq C$ and the constant is independent of $y_{min}^{(1)} < y_1 < y_{max}^{(1)}$, as can be seen using Moser's iteration technique (as in Section B.4) and working on a large ball U such that $\Omega_{y_1} \subset U$ for all $y_{min}^{(1)} < y_1 < y_{max}^{(1)}$, considering all functions on Ω_{y_1} to be extended by zero in $U \setminus \Omega_{y_1}$. $\qquad\square$

5.3 Partial regularity of stable solutions in higher dimensions

In dimension $N \geq 10$, the extremal solution to the Gelfand problem (4.14) can be singular or regular, depending on the geometry of Ω. This is what we have learned in the two previous sections. In case the stable solution is singular, one may wonder how large its singular set can be: a point, a curve, a higher dimensional manifold? This question has recently been addressed in [221]. We present here a partial regularity theorem for the Lane-Emden nonlinearity (see also [220] for related results).

Theorem 5.3.1 ([76]) *Let $N \geq 3$ and $\Omega \subset \mathbb{R}^N$ an open set. Suppose $p > \frac{N+2}{N-2}$ and $u \in H^1(\Omega) \cap L^{p+1}(\Omega)$, $u \geq 0$, is a stable weak solution to*

$$-\Delta u = u^p \qquad \text{in } \Omega. \tag{5.39}$$

Then, $u \in C^\infty(\Omega \setminus \Sigma)$, where Σ is a closed set which the Hausdorff dimension is bounded above by

$$\mathcal{H}_{dim}(\Sigma) \leq N - 2\frac{p+\gamma}{p-1},$$

with $\gamma = 2p + 2\sqrt{p(p-1)} - 1$.

Remark 5.3.1 *In Exercise 4.2.1, we proved that the extremal solution to the problem*

$$\begin{cases} -\Delta u = \lambda(1+u)^p & \text{in } \Omega, \\ u = 0 & \text{on } \partial\Omega, \end{cases}$$

is regular if $1 < p < p_c(N)$, where $p_c(N)$ is given by (4.16). By Theorem 5.3.1, whenever $N \geq 11$ and $p \geq p_c(N)$, the singular set Σ of the extremal solution has *its Hausdorff dimension*

$$\mathcal{H}_{dim}(\Sigma) \leq N - 2\frac{p+\gamma}{p-1},$$

where $\gamma = 2p + 2\sqrt{p(p-1)} - 1$. Note that for $p = p_c(N)$, this implies $\mathcal{H}_{dim}(\Sigma) = 0$. This is precisely the case when Ω is the unit ball (see Exercise 5.1.1): the extremal solution is singular at one point.

5.3.1 Approximation of singular stable solutions

In this section, we extend the approximation procedure discussed in Section 3.2.2, in the context of (5.39).

Lemma 5.3.1 *Suppose $u \in H^1(\Omega) \cap L^{p+1}(\Omega)$, $u \geq 0$, is a stable weak solution to (5.39), that is,*

$$\int_\Omega \nabla u \nabla \varphi \, dx = \int_\Omega u^p \varphi \, dx, \quad \text{for all } \varphi \in C_c^\infty(\Omega),$$

and (1.5) holds. Then, there exists a sequence of stable solutions $u_n \in C^\infty(\Omega)$ to (5.39), such that $u_n \nearrow u$ a.e. and in $H^1(\Omega)$.

Proof. Given $c > 0$, consider the function

$$\phi_c(t) = \left(c + t^{-(p-1)}\right)^{-\frac{1}{p-1}}, \quad \text{defined for } t > 0.$$

We also set $\phi_c(0) = 0$. Then, ϕ_c is increasing, concave, and smooth for $t > 0$. In addition, $\phi_c(t) \nearrow t$ as $c \searrow 0^+$, and $\phi_c(t) \leq t$, for all $t \geq 0$. Also, if $c > 0$, then ϕ_c, ϕ_c' are uniformly bounded. We have

$$\phi_c'(t) = \frac{\phi_c(t)^p}{t^p} \quad \forall t > 0.$$

Let w_c denote the unique solution to

$$\begin{cases} -\Delta w_c = 0 & \text{in } \Omega, \\ w_c = \phi_c(u) & \text{on } \partial\Omega. \end{cases}$$

Then, $w_c \geq 0$, $w_c \in L^\infty(\Omega) \cap H^1(\Omega)$. Moreover, w_c is nonincreasing with respect to c. We claim that $w_c \to w$ in $H^1(\Omega)$ as $c \to 0$, where w is the solution to

$$\begin{cases} -\Delta w = 0 & \text{in } \Omega, \\ w = u & \text{on } \partial\Omega. \end{cases}$$

116

To see this, consider the problem

$$\begin{cases} -\Delta v = (v + w_c)^p & \text{in } \Omega, \\ \quad\; v = 0 & \text{on } \partial\Omega. \end{cases} \tag{5.40}$$

Since $w_c \in L^\infty(\Omega)$, (5.40) has a minimal solution v_c, which can be constructed by the method of sub- and supersolutions, as follows. Note that $\underline{v} = 0$ is a subsolution, since $w_c \geq 0$. Moreover, by Kato's inequality (Lemma 3.2.1), $\overline{v} = \phi_c(u) - w_c$ is a bounded supersolution:

$$-\Delta(\phi_c(u) - w_c) = -\Delta\phi_c(u) \geq -\phi_c'(u)\Delta u = \phi_c(u)^p = (\phi_c(u) - w_c + w_c)^p.$$

In particular, (5.40) has a minimal solution v_c. This minimal solution is bounded and by elliptic regularity, v_c belongs to $C^{1,\alpha}(\overline{\Omega})$. Moreover, v_c is stable in the sense that

$$p \int_\Omega (v_c + w_c)^{p-1} \varphi^2 \, dx \leq \int_\Omega |\nabla\varphi|^2 \, dx, \quad \text{for all } \varphi \in C_c^1(\Omega).$$

Since v_c is minimal and w_c is nonincreasing with respect to c, we deduce that v_c is also nonincreasing with respect to c. It follows that $v(x) = \lim_{c \to 0} v_c(x)$ is well defined for all $x \in \Omega$. Since $v_c \in C^1(\overline{\Omega})$, we have

$$\int_\Omega |\nabla v_c|^2 \, dx = \int_\Omega (v_c + w_c)^p v_c \, dx \leq \int_\Omega u^{p+1} \, dx.$$

In particular, v_c is bounded in $H_0^1(\Omega)$. It follows that $v_c \rightharpoonup v$ weakly in $H_0^1(\Omega)$. Multiplying (5.40) by $\varphi \in C_c^\infty(\Omega)$, integrating, and passing to the limit as $c \to 0$, we see that v is a weak solution to

$$\begin{cases} -\Delta v = (v + w)^p & \text{in } \Omega, \\ \quad\; v = 0 & \text{on } \partial\Omega. \end{cases} \tag{5.41}$$

Let $\varphi_k \in C_c^{0,1}(\Omega)$ be a sequence such that $\varphi_k \to v$ in $H_0^1(\Omega)$. Since $v \geq 0$ we can assume $\varphi_k \geq 0$. We can also assume that $\varphi_k \to v$ a.e. in Ω. Multiplying (5.41) by φ_k and integrating, we obtain

$$\int_\Omega \nabla v \nabla \varphi_k \, dx = \int_\Omega (v + w)^p \varphi_k \, dx.$$

By Fatou's lemma,

$$\int_\Omega (v + w)^p v \, dx \leq \liminf_{k \to \infty} \int_\Omega \nabla v \nabla \varphi_k \, dx = \int_\Omega |\nabla v|^2 \, dx.$$

By monotone convergence,

$$\lim_{c \to 0} \int_\Omega |\nabla v_c|^2 \, dx = \lim_{c \to 0} \int_\Omega (v_c + w_c)^p v_c \, dx = \int_\Omega (v + w)^p v \, dx.$$

Hence,

$$\lim_{c \to 0} \int_\Omega |\nabla v_c|^2 \, dx = \int_\Omega (v + w)^p v \, dx \leq \int_\Omega |\nabla v|^2 \, dx.$$

Since $v_c \rightharpoonup v$ weakly in $H_0^1(\Omega)$, the reverse inequality

$$\int_\Omega |\nabla v|^2 \, dx \leq \liminf_{c \to 0} \int_\Omega |\nabla v_c|^2 \, dx$$

also holds, which proves that $v_c \to v$ in $H_0^1(\Omega)$.

We claim that $u = v + w$, from which Lemma 5.3.1 follows. By construction, $v = \lim v_c \leq \lim(\phi_c(u) - w_c) = u - w$. We need thus only prove that $u \leq v + w$. Note that $\tilde{v} = u - w$ solves (5.41). Let $z = \tilde{v} - v \geq 0$. Then, $z \in H_0^1(\Omega)$, and since u is stable,

$$p \int_\Omega (\tilde{v} + w)^{p-1} (\tilde{v} - v)^2 \, dx \leq \int_\Omega |\nabla(\tilde{v} - v)|^2 \, dx. \tag{5.42}$$

Now, $\tilde{v} - v$ satisfies

$$\int_\Omega \nabla(\tilde{v} - v) \nabla \varphi \, dx = \int_\Omega ((\tilde{v} + w)^p - (v + w)^p) \varphi \, dx, \qquad \forall \varphi \in C_c^\infty(\Omega).$$

We would like to take $\varphi = \tilde{v} - v$. First, we claim that we can take $\varphi \in H_0^1(\Omega) \cap L^\infty(\Omega)$. These functions can be approximated in $H_0^1(\Omega)$ by functions in $C_c^\infty(\Omega)$ with a uniform bound. Then, take $\varphi = \min(\tilde{v} - v, t)$, $t > 0$, which belongs to $H_0^1(\Omega) \cap L^\infty(\Omega)$. We get

$$\int_{[\tilde{v} - v \leq t]} |\nabla(\tilde{v} - v)|^2 \, dx = \int_\Omega ((\tilde{v} + w)^p - (v + w)^p) \min(\tilde{v} - v, t) \, dx.$$

Now let $t \to \infty$. Then,

$$\int_\Omega |\nabla(\tilde{v} - v)|^2 \, dx = \int_\Omega ((\tilde{v} + w)^p - (v + w)^p)(\tilde{v} - v) \, dx.$$

Combined with (5.42) we find

$$\int_\Omega (\tilde{v} - v) \left[p(\tilde{v} + w)^{p-1}(\tilde{v} - v) - (\tilde{v} + w)^p + (v + w)^p \right] \, dx \leq 0.$$

But $p(\tilde{v} + w)^{p-1}(\tilde{v} - v) - (\tilde{v} + w)^p + (v + w)^p \geq 0$ with strict inequality, unless $\tilde{v} \equiv v$. $\qquad \square$

5.3.2 Elliptic regularity in Morrey spaces

The next ingredient in the proof of Theorem 5.3.1 is a so-called ε-regularity result for weak solutions of (5.39) in Morrey spaces.

Definition 5.3.1 *Let Ω be a bounded open set of $\mathbb{R}^N$, $N \geq 1$. Given $p > 1$ and $\lambda \in [0, N]$, the Morrey space $L^{p,\lambda}(\Omega)$ is the set of functions u in $L^p(\Omega)$ such that the following norm is finite:*

$$\|u\|^p_{L^{p,\lambda}(\Omega)} = \sup_{x_0 \in \Omega, r > 0} r^{-\lambda} \int_{B(x_0,r)\cap\Omega} |u|^p \, dx < \infty.$$

Remark 5.3.2 *Observe that $L^{p,0}(\Omega) = L^p(\Omega)$, while $L^{p,N}(\Omega) = L^\infty(\Omega)$.*

We begin by proving the following regularity theorem.

Theorem 5.3.2 ([172]) *Let $N \geq 3$, $p > N/(N-2)$, and let $u \geq 0$ be a weak solution to (5.39). If $u \in L^{p,\lambda}(\Omega)$ for some $\lambda > N - 2 - \frac{2}{p-1}$, then u is regular in Ω.*

The proof relies on two useful technical lemmata.

Lemma 5.3.2 ([172]) *Let $N \geq 3$, $B = B(x,r) \subset \mathbb{R}^N$ and $p > 1$. Assume that $f \in L^{1,\lambda}(B)$ for some $\lambda \in (N - 2 - \frac{2}{p-1}, N - 2)$. For $x \in B$, let*

$$v(x) = \int_B |x-y|^{2-N} f(y) \, dy.$$

Then, there exists a constant $C = C(p,N,\lambda) > 0$, such that

$$\|v\|_{L^p(B)} \leq C r^{\frac{N}{p} - (N-2-\lambda)} \|f\|_{L^{1,\lambda}(B)}. \tag{5.43}$$

Proof. Fix $\mu \in (N - 2 - \frac{2}{p-1}, \lambda)$. Then, for $x \in B$,

$$\left| \int_B |x-y|^{-\mu} f(y) \, dy \right| \leq \sum_{k=0}^{+\infty} \left| \int_{B\cap[2^{-k}r < |x-y| < 2^{-k+1}r]} |x-y|^{-\mu} f(y) \, dy \right|$$

$$\leq \sum_{k=0}^{+\infty} 2^{k\mu} r^{-\mu} \left| \int_{B\cap[|x-y|<2^{-k+1}r]} f(y) \, dy \right|$$

$$\leq 2 \sum_{k=0}^{+\infty} 2^{-k(\lambda-\mu)} r^{\lambda-\mu} \|f\|_{L^{1,\lambda}(B)}$$

$$\leq C r^{\lambda-\mu} \|f\|_{L^{1,\lambda}(B)}.$$

We deduce that

$$|v(x)|^p = \left| \int_B |x-y|^{2-N+\mu} |x-y|^{-\mu} f(y) \, dy \right|^p$$

$$\leq \left(\int_B |x-y|^{(2-N+\mu)p-\mu} |f(y)| \, dy \right) \left(\int_B |x-y|^{-\mu} |f(y)| \, dy \right)^{p-1}$$

$$\leq C \left(\int_B |x-y|^{(2-N+\mu)p-\mu} |f(y)| \, dy \right) r^{(\lambda-\mu)(p-1)} \|f\|_{L^{1,\lambda}(B)}^{p-1}.$$

Integrating over B and recalling that $\mu > N - 2 - 2/(p-1)$, we obtain

$$\|v\|_{L^p(B)}^p \leq C r^{(2-N+\mu)p-\mu+N} \left(\int_B |f(y)| \, dy \right) r^{(\lambda-\mu)(p-1)} \|f\|_{L^{1,\lambda}(B)}^{p-1}$$

$$\leq C r^{(2-N+\lambda)p-\lambda+N} \left(\int_B |f(y)| \, dy \right) \|f\|_{L^{1,\lambda}(B)}^{p-1}$$

$$\leq C r^{(2-N+\lambda)p+N} \|f\|_{L^{1,\lambda}(B)}^p.$$

$\square$

Lemma 5.3.3 ([49]) *Let $0 < \lambda < N$ and $c > 0$. Assume that $\phi : \mathbb{R}^+ \to \mathbb{R}^+$ is a nondecreasing function such that for all $0 < \rho < R$, we have*

$$\phi(\rho) \leq c \left(\frac{\rho^N}{R^N} \phi(R) + R^\lambda \right). \tag{5.44}$$

Then, there exists a constant $C = C(\lambda, N, R, \phi(R))$, such that

$$\phi(\rho) \leq C \rho^\lambda, \qquad \text{for all } \rho \in (0, R). \tag{5.45}$$

Proof. Set $\rho = \tau R$, with $\tau \in (0,1)$. By (5.44),

$$\phi(\tau R) \leq c \left(\tau^N \phi(R) + R^\lambda \right).$$

Now fix $\gamma \in (\lambda, N)$ and $\tau \in (0,1)$ such that $c\tau^N = \tau^\gamma$. Then,

$$\phi(\tau R) \leq \tau^\gamma \phi(R) + cR^\lambda.$$

Iterating the above formula, we deduce that for $k \in \mathbb{N}$,

$$\phi(\tau^k R) \leq \tau^{k\gamma} \phi(R) + c\tau^{\lambda(k-1)} R^\lambda \sum_{j=0}^{k-1} \tau^{j(\gamma-\lambda)} \leq C\tau^{k\lambda}.$$

Now take any $\rho \in (0,R)$. There exists an integer k such that $\tau^{k+1}R < \rho \le \tau^k R$. Since ϕ is nondecreasing, (5.45) follows. $\qquad\square$

Proof of Theorem 5.3.2. Let u be a nonnegative weak solution to (5.39). Fix a ball $B = B(x,R) \subset \Omega$, and let $f(y) = u^p(y)$, for $y \in B$. Also let

$$v(x) = \int_B \Gamma(x-y)f(y)\,dy,$$

where Γ is the fundamental solution to the Laplace equation, given by (A.6). Then,

$$-\Delta(u-v) = 0, \qquad \text{in } B,$$

and so $w = u - v$ is smooth in B. By the mean-value formula (A.10), for all $y \in B/4$, we can write

$$w(y) = \fint_{B(y,R/4)} w(z)\,dz.$$

Integrating over $B(x,\rho)$, $0 < \rho < R/4$, we deduce that

$$\int_{B(x,\rho)} |w|^p\,dy \le C\frac{\rho^N}{R^N} \int_{B(x,R)} |w|^p\,dz.$$

Since $u = v + w$ in B, we deduce that

$$\int_{B(x,\rho)} u^p\,dy \le C \left(\int_{B(x,\rho)} |w|^p\,dy + \int_{B(x,\rho)} v^p\,dy \right)$$

$$\le C \left(\frac{\rho^N}{R^N} \int_{B(x,R)} |w|^p\,dz + \int_{B(x,\rho)} v^p\,dz \right)$$

$$\le C \left(\frac{\rho^N}{R^N} \int_{B(x,R)} u^p\,dz + \int_{B(x,R)} v^p\,dz \right).$$

We claim that $u \in L_{loc}^{p,N-2}(\Omega)$. If $\lambda \ge N - 2$, this is obvious. Otherwise, we may apply (5.43) and obtain for $0 < \rho < R/4$,

$$\int_{B(x,\rho)} u^p\,dy \le C \left(\frac{\rho^N}{R^N} \int_{B(x,R)} u^p\,dz + R^{N-p(N-2-\lambda)} \right).$$

Choosing the constant C larger if necessary, the above inequality remains true for all $\rho \in (0,R)$. Applying Lemma 5.3.3, we deduce that

$$\int_{B(x,\rho)} u^p \, dy \leq C\rho^{N-p(N-2-\lambda)}$$

and so $u \in L^{p,\lambda_1}_{loc}(\Omega)$, with $\lambda_1 = N - p(N-2-\lambda)$. If $\lambda_1 \geq N-2$, the claim follows. Otherwise, we iterate the above estimate and conclude that $u \in L^{p,\lambda_{k+1}}_{loc}(\Omega)$, whenever $\lambda_k < N - 2$ and where, for $k \geq 1$,

$$\lambda_{k+1} = N - p(N - 2 - \lambda_k),$$

that is,

$$\lambda_k = p^k \left(\lambda - \frac{(N-2)p - N}{p-1} \right) + \frac{(N-2)p - N}{p-1}.$$

Since $\lambda > N - 2 - 2/(p-1) = ((N-2)p - N)/(p-1)$, we see that there exists k such that $\lambda_k \geq N - 2$ and the claim follows. Applying once more (5.43) with $\lambda \to (N-2)^-$, we deduce that $v \in L^q(B)$ for all $1 < q < +\infty$, hence $u \in L^q_{loc}(\Omega)$ for all $1 < q < +\infty$. By standard elliptic regularity, we conclude that u is regular. □

Theorem 5.3.2 can be further refined as follows.

Theorem 5.3.3 ([174]) *Let $N \geq 3$, $p > 1$, and $\lambda = N - 2\frac{p+1}{p-1}$. Also let $B(x_0, r_0)$ be a ball. There exists $\varepsilon = \varepsilon(N,p) > 0$ such that for any weak solution $u \in H^1(B(x_0, r_0)) \cap C(\overline{B(x_0, r_0)})$, $u \geq 0$, to (5.39) satisfying*

$$\|u\|_{L^{p+1,\lambda}(B(x_0,r_0))} \leq \varepsilon, \tag{5.46}$$

there holds

$$\|u\|_{L^\infty(B(x_0,r_0/2))} \leq \left(\frac{4}{r_0} \right)^{\frac{2}{p-1}}.$$

Proof. We follow [128]. Without loss of generality, we may assume that $x_0 = 0$. For $y \in B_{r_0}$, let

$$\Phi(y) = (r_0 - |y|)^{\frac{2}{p-1}} u(y).$$

Let $y_0 \in B_{r_0}$ denote a point of maximum of Φ in $\overline{B_{r_0}}$, $\rho_0 = |y_0|$, and y_1 a point of maximum of u in B_{ρ_0}. Then, for $y \in B_{r_0}$, $\Phi(y) \leq \Phi(y_0)$ and so

$$u(y) \leq \left(\frac{r_0 - \rho_0}{r_0 - |y|} \right)^{\frac{2}{p-1}} u(y_0) \leq \left(\frac{r_0 - \rho_0}{r_0 - |y|} \right)^{\frac{2}{p-1}} u(y_1). \tag{5.47}$$

We claim that

$$u(y_1) \le \left(\frac{2}{r_0 - \rho_0}\right)^{\frac{2}{p-1}}, \tag{5.48}$$

from which the theorem follows. Assume by contradiction that (5.48) fails. Then,

$$\rho_1 := u(y_1)^{-\frac{p-1}{2}} \le \frac{r_0 - \rho_0}{2},$$

so that $B(y_1, \rho_1) \subset B(0, (r_0 + \rho_0)/2) \subset B(0, r_0)$. Let

$$v(x) = \rho_1^{\frac{2}{p-1}} u(\rho_1 x + y_1), \qquad \text{for } x \in B_1.$$

Then, v solves (5.39) in B_1, and $v(0) = 1$. In addition, it follows from (5.47) that for all $y \in B(0, (r_0 + \rho_0)/2)$,

$$u(y) \le \left(\frac{r_0 - \rho_0}{r_0 - |y|}\right)^{\frac{2}{p-1}} u(y_1) \le \left(\frac{r_0 - \rho_0}{(r_0 - \rho_0)/2}\right)^{\frac{2}{p-1}} u(y_1),$$

hence

$$\|v\|_{L^\infty(B_1)} \le 2^{\frac{2}{p-1}}. \tag{5.49}$$

Now, Kato's inequality (Lemma 3.2.1) implies that

$$-\Delta v^{p+1} \le (p+1)v^{p-1}v^{p+1} \le 4(p+1)v^{p+1} \quad \text{in } B_1,$$

where we used (5.49) in the last inequality. By the mean-value inequality (Exercise A.1.5), we deduce that

$$1 = v^{p+1}(0) \le C \fint_{B_1} v^{p+1} \, dx = C\rho_1^{-\lambda} \int_{B(y_1, \rho_1)} u^{p+1} \, dx.$$

This contradicts (5.46), provided $\varepsilon^{p+1} < 1/C$. $\qquad\qquad\square$

5.3.3 Measuring singular sets

Theorem 5.3.2 shows that a weak solution $u \ge 0$ of (5.39) is smooth in a neighborhood of a point $x_0 \in \Omega$, provided it belongs to a suitable Morrey space in a neighborhood of x_0. In this section, we estimate the Hausdorff dimension of the set of points where a function $u \in L^1_{loc}(\Omega)$ is locally large, provided it fails to belong to some Morrey space.

Theorem 5.3.4 *Let Ω denote an open set of $\mathbb{R}^N$, $N \geq 1$, u a function in $L^1_{loc}(\Omega)$, and $0 \leq s < N$. Set*

$$E_s = \left\{ x \in \Omega \; : \; \limsup_{r \to 0^+} r^{-s} \int_{B_r(x)} |u(y)|\, dy > 0 \right\}.$$

Then,

$$\mathcal{H}^s(E_s) = 0,$$

where $\mathcal{H}^s$ denotes the Hausdorff measure of dimension s.

The proof of Theorem 5.3.4 relies on the following covering lemma.

Lemma 5.3.4 *Let Σ denote a bounded set of $\mathbb{R}^N$, $N \geq 1$ and let $r = r(x)$ be a function defined on Σ with values in $(0,1)$. Then, there exists a sequence of points $x_i \in \Sigma$ such that*

$$\begin{aligned} B(x_i, r(x_i)) \cap B(x_j, r(x_j)) &= \emptyset \qquad \text{if } i \neq j, \\ \cup_i B(x_i, 3r(x_i)) &\supset \Sigma. \end{aligned} \tag{5.50}$$

Proof. We follow [121]. Consider the family

$$B_{1,1/2} = \{B(x, r(x)) : x \in \Sigma, \; \frac{1}{2} \leq r(x) < 1\}.$$

Since Σ is bounded, there exists a finite subfamily of disjoint balls

$$\overline{B_{1,1/2}} = \{B(x_i, r(x_i)) : \frac{1}{2} \leq r(x_i) < 1 \, 1 \leq i \leq n_1\},$$

which is maximal in the sense that each ball in $B_{1,1/2}$ intersects at least one element in $\overline{B_{1,1/2}}$. Once we have constructed $x_1, \ldots, x_{n_j}$, among the balls $B(x, r(x))$ with $2^{-j-1} \leq r(x) < 2^{-j}$, which do not intersect any of the balls $B(x_i, r(x_i))$, $i = 1, \ldots, n_j$, we can find a finite family of balls, say, $n_{j+1} - n_j$ (possibly void) such that each $B(x, r(x))$ with $2^{-j-2} \leq r(x) < 2^{-j-1}$ intersects at least one of the balls in $\{B(x_i, r(x_i)) : i = 1, \ldots, n_{j+1}\}$. The sequence satisfies (5.50). In fact, the balls $B(x_i, r(x_i))$ are disjoint by construction; for $x \in \Sigma$, there exists x_i such that

$$B(x, r(x)) \cap B(x_i, r(x_i)) \neq \emptyset$$

and $2r(x_i) \geq r(x)$. Hence,

$$|x - x_i| \leq r(x) + r(x_i) \leq 3r(x_i),$$

so that $x \in B(x_i, 3r(x_i))$. □

Proof of Theorem 5.3.4. We follow [121]. It suffices to show that for each compact subset $K \subset\subset \Omega$,

$$\mathcal{H}^s(F) = 0, \qquad \text{where } F = E_s \cap K.$$

For $n \geq 1$, set

$$F^{(n)} = \left\{ x \in F \ : \ \limsup_{r \to 0^+} r^{-s} \int_{B(x,r)} |u(y)| > 1/n \right\}.$$

Obviously, $F = \cup_{n=1}^{+\infty} F^{(n)}$ and it suffices to show that $\mathcal{H}^s(F^{(n)}) = 0$ for all $n \geq 1$. Take a bounded open set Q such that $K \subset Q \subset \overline{Q} \subset \Omega$ and $d = \min(1, d(x, \partial Q))$, where $x \in F^{(n)}$. For all $\varepsilon > 0$, $0 < \varepsilon < d$, and for all $x \in F^{(n)}$, there exists $r(x)$, $0 < r(x) < \varepsilon$, such that

$$r(x)^{-s} \int_{B(x,r(x))} |u(y)| \, dy \geq \frac{1}{2n}.$$

Let x_i be the sequence in Lemma 5.3.4 corresponding to $\Sigma = F^{(n)}$, and let $r_i = r(x_i)$. Then,

$$\sum_i r_i^s \leq 2n \sum_i \int_{B_{r_i}(x_i)} |u(y)| \, dy = 2n \int_{\cup B_{r_i}(x_i)} |u(y)| \, dy. \qquad (5.51)$$

Since $s < N$, this inequality implies

$$\left| \cup B_{r_i}(x_i) \right| \leq |B_1| \sum_i r_i^N \leq C \varepsilon^{N-s} \sum_i r_i^s \leq Cn\varepsilon^{N-s} \int_Q |u(y)| \, dy. \qquad (5.52)$$

From (5.52), it follows that the right-hand side of (5.51) converges to 0 as $\varepsilon \to 0$. Taking into account (5.50) and the definition of the Hausdorff measure $\mathcal{H}^s(F^{(n)})$, we deduce that $\mathcal{H}^s(F^{(n)}) = 0$. □

5.3.4 A monotonicity formula

Consider a weak solution u to (5.3.1), a ball $B(x_0, r_0) \subset \Omega$, and let $\mu = N - 2\frac{p+1}{p-1}$. By the approximation lemma (Lemma 5.3.1) and the ε-regularity theorem (Theorem 5.3.3), u is smooth in a neighborhood of x_0 provided the quantity

$$r^{-\mu} \int_{B(x,r)} u^{p+1} \, dy \qquad (5.53)$$

is uniformly small at every scale $r \in (0, r_0)$ and for all x in a neighborhood of x_0. For fixed x, by Theorem 5.3.4, the above quantity is indeed small for r smaller than some $r(x) > 0$, unless x belongs to a set of zero $\mathscr{H}^\mu$ Hausdorff measure. In this section, we prove a monotonicity formula, which will serve as a bridge between these two results: If (5.53) is small at some point x_0 and some scale r_0, it remains small at all scales $r < r_0$ and for all x near x_0.

Theorem 5.3.5 ([173]) *Let $u \in C^2(\Omega)$, $u \geq 0$, denote a solution to the Lane-Emden equation, Equation (5.39). Given $x \in \Omega$ and $r > 0$ such that $B(x, r) \subset \Omega$, consider the energy*

$$\mathscr{E}_u(x, r) = r^{-\mu} \int_{B(x,r)} \left(\frac{1}{2}|\nabla u|^2 - \frac{1}{p+1} u^{p+1} \right) dx + \frac{r^{-\mu-1}}{p-1} \int_{\partial B(x,r)} u^2 \, d\sigma,$$
$$(5.54)$$

where

$$\mu = N - 2\frac{p+1}{p-1}.$$

Then, $\mathscr{E}_u(x, r)$ is a nondecreasing function of r.

The aforementioned monotonicity formula remains valid for weak solutions.

Corollary 5.3.1 ([173]) *Let $u \in H^1(\Omega) \cap L^{p+1}(\Omega)$, $u \geq 0$, denote a stable weak solution to the Lane-Emden equation (5.39). For $x \in \Omega$, $r > 0$, such that $B(x, r) \subset \Omega$, consider the energy $\mathscr{E}_u(x, r)$ given by (5.54). Then,*

- *$\mathscr{E}_u(x, r)$ is nondecreasing in r.*

- *$\mathscr{E}_u(x, r)$ is continuous in $x \in \Omega$ and $r > 0$.*

Proof. Using Theorem 5.3.5 and the approximation lemma (Lemma 5.3.1), we easily see that $\mathscr{E}_u(x, r)$ is nondecreasing in r. Also, the continuity of $\mathscr{E}_u(x, r)$ reduces to that of $\int_{\partial B(x,r)} u^2 \, d\sigma$. Let $u_n \in C^\infty(\Omega)$ denote the approximating sequence given by Lemma 5.3.1. Multiply (5.3.1) by u_n and integrate on $B(x, r) \subset \Omega$:

$$\int_{\partial B(x,r)} u_n \frac{\partial u_n}{\partial n} \, d\sigma = \int_{B(x,r)} (|\nabla u_n|^2 - u_n^{p+1}) \, dy.$$

In other words,

$$\frac{\partial}{\partial r} \int_{\partial B(x,r)} u_n^2 \, d\sigma = 2 \int_{B(x,r)} (|\nabla u_n|^2 - u_n^{p+1}) \, dy. \tag{5.55}$$

126

Fix $x_0 \in \Omega$ and $r_0 > 0$ such that $B(x_0, 2r_0) \subset \Omega$. Integrate (5.55) between r and r_0. Then, for $x \in B(x_0, r_0)$,

$$\fint_{\partial B(x,r)} u_n^2 \, d\sigma = \fint_{\partial B(x,r_0)} u_n^2 \, d\sigma + 2 \int_{r_0}^r \fint_{B(x,t)} (|\nabla u_n|^2 - u_n^{p+1}) \, dy \, dt.$$

Passing to the limit as $n \to +\infty$, we deduce that

$$\fint_{\partial B(x,r)} u^2 \, d\sigma = \fint_{\partial B(x,r_0)} u^2 \, d\sigma + 2 \int_{r_0}^r \fint_{B(x,t)} (|\nabla u|^2 - u^{p+1}) \, dy \, dt,$$

and so we just need to prove that $x \to \fint_{\partial B(x,r_0)} u^2 \, d\sigma$ is continuous at x_0. We may always assume that $x_0 = 0$. Then,

$$
\begin{aligned}
\int_{\partial B(x,r_0)} u_n^2 \, d\sigma &= \int_{\partial B(0,r_0)} u_n^2 \, d\sigma + \int_0^1 \frac{d}{dt} \left(\int_{\partial B(tx,r_0)} u_n^2 \, d\sigma \right) dt \\
&= \int_{\partial B(0,r_0)} u_n^2 \, d\sigma + 2 \int_0^1 \left(\int_{\partial B(tx,r_0)} u_n \nabla u_n \cdot x \, d\sigma \right) dt \\
&= \int_{\partial B(0,r_0)} u_n^2 \, d\sigma + 2|x| \int_0^1 \left(\int_{\partial B(tx,r_0)} u_n \frac{\partial u_n}{\partial n} \, d\sigma \right) dt \\
&= \int_{\partial B(0,r_0)} u_n^2 \, d\sigma + 2|x| \int_0^1 \left(\int_{B(tx,r_0)} (|\nabla u_n|^2 - u_n^{p+1}) \, dy \right) dt.
\end{aligned}
$$

Passing to the limit as $n \to +\infty$, we deduce that

$$\int_{\partial B(x,r_0)} u^2 \, d\sigma = \int_{\partial B(0,r_0)} u^2 \, d\sigma + 2|x| \int_0^1 \left(\int_{B(tx,r_0)} (|\nabla u|^2 - u^{p+1}) \, dy \right) dt,$$

which is a continuous quantity of x, as desired. $\qquad\square$

Proof of Theorem 5.3.5. To simplify notation, we write $B(x,r) = B_r$. We begin by applying Pohozaev's identity. By (8.8):

$$\int_{B_r} \Delta u (x \cdot \nabla u) \, dx = \frac{N-2}{2} \int_{B_r} |\nabla u|^2 \, dx - \frac{r}{2} \int_{\partial B_r} |\nabla u|^2 \, d\sigma + r \int_{\partial B_r} \left(\frac{\partial u}{\partial r} \right)^2 d\sigma.$$

Since u solves (5.39), we also have

$$\int_{B_r} \Delta u (x \cdot \nabla u) \, dx = -\int_{B_r} u^p (x \cdot \nabla u) \, dx =$$

$$N \int_{B_r} \frac{u^{p+1}}{p+1} \, dx - r \int_{\partial B_r} \frac{u^{p+1}}{p+1} \, d\sigma.$$

Combining the above two equalities, we obtain the Pohozaev identity:

$$\frac{N-2}{2}\int_{B_r}|\nabla u|^2\,dx-\frac{r}{2}\int_{\partial B_r}|\nabla u|^2\,d\sigma+r\int_{\partial B_r}\left(\frac{\partial u}{\partial r}\right)^2\,d\sigma=$$

$$\frac{N}{p+1}\int_{B_r}u^{p+1}\,dx-\frac{r}{p+1}\int_{\partial B_r}u^{p+1}\,d\sigma. \quad (5.56)$$

Now multiply the Lane-Emden equation (5.39) by u and integrate over B_r. Then,

$$\int_{B_r}|\nabla u|^2\,dx=\int_{B_r}u^{p+1}\,dx-\int_{\partial B_r}u\frac{\partial u}{\partial r}\,d\sigma. \quad (5.57)$$

Differentiating the above equation, we obtain

$$\int_{\partial B_r}|\nabla u|^2\,dx=\int_{\partial B_r}u^{p+1}\,dx-\frac{d}{dr}\int_{\partial B_r}u\frac{\partial u}{\partial r}\,d\sigma. \quad (5.58)$$

Plugging (5.57) and (5.58) in (5.56), we deduce that

$$\frac{1}{r}\left(\frac{N-2}{2}-\frac{N}{p+1}\right)\int_{B_r}u^{p+1}\,dx-\left(\frac{1}{2}-\frac{1}{p+1}\right)\int_{\partial B_r}u^{p+1}\,d\sigma$$

$$=\frac{1}{2}\frac{d}{dr}\left\{\int_{\partial B_r}u\frac{\partial u}{\partial r}\,d\sigma\right\}-\int_{\partial B_r}\left(\left(\frac{\partial u}{\partial r}\right)^2+\frac{N-2}{2}\frac{u}{r}\frac{\partial u}{\partial r}\right)\,d\sigma.$$

Letting $\mu=N-2\frac{p+1}{p-1}$, this identity can be rewritten as

$$\frac{p-1}{p+1}\frac{d}{dr}\left\{r^{-\mu}\int_{B_r}u^{p+1}\,dx\right\}+\frac{1}{2}r^{-\mu}\frac{d}{dr}\left\{\int_{\partial B_r}u\frac{\partial u}{\partial r}\,d\sigma\right\}$$

$$=r^{-\mu}\int_{\partial B_r}\left(\left(\frac{\partial u}{\partial r}\right)^2+\frac{N-2}{2}\frac{u}{r}\frac{\partial u}{\partial r}\right)\,d\sigma. \quad (5.59)$$

The second term on the left-hand side of this identity can be rewritten as

$$r^{-\mu}\frac{d}{dr}\left\{\int_{\partial B_r}u\frac{\partial u}{\partial r}\,d\sigma\right\}=\frac{d}{dr}\left\{r^{-\mu}\int_{\partial B_r}u\frac{\partial u}{\partial r}\,d\sigma\right\}+\mu r^{-\mu-1}\int_{\partial B_r}u\frac{\partial u}{\partial r}\,d\sigma.$$

$$(5.60)$$

In addition,

$$\frac{d}{dr}\left\{r^{-\mu}\int_{\partial B_r} u\frac{\partial u}{\partial r}\,d\sigma\right\} = \frac{1}{2}\frac{d}{dr}\left\{r^{-\mu}|\partial B_r|\frac{d}{dr}\fint_{\partial B_r} u^2\,d\sigma.\right\}$$

$$= \frac{1}{2}\frac{d^2}{dr^2}\left\{r^{-\mu}\int_{\partial B_r} u^2\,d\sigma\right\} - \frac{p+3}{p-1}r^{-\mu}\int_{\partial B_r}\left(\frac{2}{p-1}\frac{u^2}{r^2} + \frac{u}{r}\frac{\partial u}{\partial r}\right)d\sigma$$

Collecting this identity and (5.60) and plugging them in (5.59), we find

$$\frac{1}{2}\frac{p-1}{p+1}\frac{d}{dr}\left\{r^{-\mu}\int_{B_r} u^{p+1}\,dx\right\} + \frac{1}{4}\frac{d^2}{dr^2}\left\{r^{-\mu}\int_{\partial B_r} u^2\,d\sigma\right\}$$

$$= r^{-\mu}\int_{\partial B_r}\left(\left(\frac{\partial u}{\partial r}\right)^2 + \frac{p+7}{2(p-1)}\frac{u}{r}\frac{\partial u}{\partial r} + \frac{p+3}{(p-1)^2}\frac{u^2}{r^2}\right)d\sigma.$$

This can be rewritten as

$$\frac{d}{dr}\left\{\frac{p-1}{2(p+1)}r^{-\mu}\int_{B_r} u^{p+1}\,dx+\right.$$

$$\left. + \frac{1}{4}\frac{d}{dr}\left(r^{-\mu}\int_{\partial B_r} u^2\,d\sigma\right) - \frac{1}{4}r^{-\mu-1}\int_{\partial B_r} u^2\,d\sigma\right\}$$

$$= r^{-\mu}\int_{\partial B_r}\left(\frac{\partial u}{\partial r} + \frac{2}{p-1}\frac{u}{r}\right)^2 d\sigma \geq 0.$$

So, the quantity

$$\mathscr{F}_u(x,r) =$$

$$\frac{p-1}{2(p+1)}r^{-\mu}\int_{B_r} u^{p+1}\,dx + \frac{1}{4}\frac{d}{dr}\left\{r^{-\mu}\int_{\partial B_r} u^2\,d\sigma\right\} - \frac{1}{4}r^{-\mu-1}\int_{\partial B_r} u^2\,d\sigma$$

$$\tag{5.61}$$

is nondecreasing. Using (5.57) and (5.58), we see that $\mathscr{E}_u(x,r) = \mathscr{F}_u(x,r)$. $\square$

Remark 5.3.3 *Eliminating the terms $r^{-\mu-1}\int_{\partial B_r} u^2\,d\sigma$ between (5.54) and*

(5.61), *we obtain the equivalent formulation for the energy* $\mathcal{E}_u(x,r)$:

$$\mathcal{E}_u(x,r) =$$
$$\frac{p-1}{p+3} r^{-\mu} \int_{B(x,r)} \left(\frac{1}{2}|\nabla u|^2 + \frac{1}{p+1} u^{p+1} \right) dy + \frac{1}{p+3} \frac{d}{dr} \left(r^{-\mu} \int_{\partial B(x,r)} u^2 \, d\sigma \right).$$
$$(5.62)$$

5.3.5   Proof of Theorem 5.3.1

Our last ingredient is the following crucial capacitary estimate.

Proposition 5.3.1 ([97]) *Let* Ω *be an open set of* $\mathbb{R}^N$, $p > 1$. *Let* $u \in C^2(\Omega)$ *denote a stable solution to*

$$-\Delta u = |u|^{p-1} u \qquad \text{in } \Omega.$$

Then, for any $\gamma \in [1, 2p + 2\sqrt{p(p-1)} - 1)$, *any* $\psi \in C_c^1(\Omega)$, $0 \le \psi \le 1$, *and any integer* $m \ge \max \left\{ \frac{p+\gamma}{p-1}, 2 \right\}$, *there exists a constant* $C_{p,m,\gamma} > 0$ *such that*

$$\int_\Omega \left(\left| \nabla \left(|u|^{\frac{\gamma-1}{2}} u \right) \right|^2 + |u|^{p+\gamma} \right) \psi^{2m} \, dx \le C_{p,m,\gamma} \int_\Omega |\nabla \psi|^{2\left(\frac{p+\gamma}{p-1} \right)} \, dx.$$

Proof. Following [97], we split the proof into steps.
Step 1. For any $\varphi \in C_c^1(\Omega)$, we have

$$\int_\Omega \left| \nabla \left(|u|^{\frac{\gamma-1}{2}} u \right) \right|^2 \varphi^2 \, dx = \frac{(\gamma+1)^2}{4\gamma} \int_\Omega \left(|u|^{p+\gamma} \varphi^2 - \nabla \frac{|u|^{\gamma+1}}{\gamma+1} \cdot \nabla \varphi^2 \right) dx.$$
$$(5.63)$$

Multiply the equation by $|u|^{\gamma-1} u \varphi^2$ and integrate by parts. Then,

$$\int_\Omega \gamma |\nabla u|^2 |u|^{\gamma-1} \varphi^2 \, dx + \int_\Omega |u|^{\gamma-1} u \nabla u \cdot \nabla \varphi^2 \, dx = \int_\Omega |u|^{p+\gamma} \varphi^2 \, dx,$$

and therefore

$$\frac{\gamma}{\left(\frac{\gamma+1}{2} \right)^2} \int_\Omega \left| \nabla \left(|u|^{\frac{\gamma-1}{2}} u \right) \right|^2 \varphi^2 \, dx + \int_\Omega \nabla \frac{|u|^{\gamma+1}}{\gamma+1} \cdot \nabla \varphi^2 \, dx = \int_\Omega |u|^{p+\gamma} \varphi^2 \, dx.$$

Identity (5.63) then follows by multiplying the above equation by the factor $\frac{\left(\frac{\gamma+1}{2}\right)^2}{\gamma}$.

Step 2. For any $\varphi \in C_c^1(\Omega)$ and any $\varepsilon > 0$, there exists a constant $C = C(\varepsilon, \gamma)$ such that

$$\left(p - \frac{(\gamma+1)^2}{4\gamma} - \varepsilon\right) \int_\Omega |u|^{p+\gamma} \varphi^2 \, dx \leq C \int_\Omega |u|^{\gamma+1} |\nabla \varphi|^2 \, dx. \qquad (5.64)$$

The function $\psi = |u|^{\frac{\gamma-1}{2}} u \varphi$ belongs to $C_c^1(\Omega)$, and thus it can be used as a test function in the stability inequality (1.5).

$$p \int_\Omega |u|^{p+\gamma} \varphi^2 \, dx \leq \int_\Omega \left|\nabla(|u|^{\frac{\gamma-1}{2}} u)\right|^2 \varphi^2 \, dx + \int_\Omega \left(|u|^{\frac{\gamma-1}{2}} u\right)^2 |\nabla \varphi|^2 \, dx +$$
$$\int_\Omega 2\nabla(|u|^{\frac{\gamma-1}{2}} u) \cdot \nabla \varphi |u|^{\frac{\gamma-1}{2}} u \varphi \, dx. \qquad (5.65)$$

By Young's inequality, given $\eta > 0$, there exists a constant $C = C(\eta)$ such that

$$\int_\Omega 2\nabla(|u|^{\frac{\gamma-1}{2}} u) \cdot \nabla \varphi |u|^{\frac{\gamma-1}{2}} u \varphi \, dx$$
$$\leq \eta \int_\Omega \left|\nabla(|u|^{\frac{\gamma-1}{2}} u)\right|^2 \varphi^2 \, dx + C \int_\Omega |u|^{\gamma+1} |\nabla \varphi|^2 \, dx. \qquad (5.66)$$

Using this in (5.65), we obtain

$$p \int_\Omega |u|^{p+\gamma} \varphi^2 \, dx \leq (1+\eta) \int_\Omega \left|\nabla(|u|^{\frac{\gamma-1}{2}} u)\right|^2 \varphi^2 \, dx + C \int_\Omega |u|^{\gamma+1} |\nabla \varphi|^2 \, dx. \qquad (5.67)$$

Using (5.63) and (5.66), we also have

$$\int_\Omega \left|\nabla\left(|u|^{\frac{\gamma-1}{2}} u\right)\right|^2 \varphi^2 \, dx \leq$$
$$(1+\eta)\frac{(\gamma+1)^2}{4\gamma} \int_\Omega |u|^{p+\gamma} \varphi^2 \, dx + C \int_\Omega |u|^{\gamma+1} |\nabla \varphi|^2 \, dx. \qquad (5.68)$$

From the above inequality and (5.67), we obtain

$$p \int_\Omega |u|^{p+\gamma} \varphi^2 \, dx \leq (1+\eta)^2 \frac{(\gamma+1)^2}{4\gamma} \int_\Omega |u|^{p+\gamma} \varphi^2 \, dx + C \int_\Omega |u|^{\gamma+1} |\nabla \varphi|^2 \, dx,$$

which, letting $\eta = c\varepsilon$, with $c > 0$ small, gives the identity (5.64).

Step 3. For any $\gamma \in [1, 2p + 2\sqrt{p(p-1)} - 1)$ and any integer $m \geq$ $\max\left(\frac{p+\gamma}{p-1}, 2\right)$, there exists a constant $C = C(p, m, \gamma)$, such that

$$\int_\Omega \left(|u|^{p+\gamma} + |\nabla(|u|^{\frac{\gamma-1}{2}}u)|^2 \right) \psi^{2m} \, dx \leq C \int_\Omega |\nabla\psi|^{2\frac{p+\gamma}{p-1}} \, dx, \qquad (5.69)$$

for all test functions $\psi \in C_c^1(\Omega)$ satisfying $|\psi| \leq 1$ in Ω. Note that for $\gamma \in [1, 2p + 2\sqrt{p(p-1)} - 1)$, there exists $\varepsilon > 0$ sufficiently small so that $\left(p - \frac{(\gamma+1)^2}{4\gamma} - \varepsilon \right) > 0$. Apply (5.64) with $\varphi = \psi^m$, $m \geq 1$. Then,

$$\int_\Omega |u|^{p+\gamma}\psi^{2m} \, dx \leq C \int_\Omega |u|^{\gamma+1}\psi^{2m-2}|\nabla\psi|^2 \, dx$$

$$\leq C \left(\int_\Omega |u|^{p+\gamma}\psi^{(2m-2)\frac{p+\gamma}{\gamma+1}} \, dx \right)^{\frac{\gamma+1}{p+\gamma}} \left(\int_\Omega |\nabla\psi|^{2\frac{p+\gamma}{p-1}} \, dx \right)^{\frac{p-1}{p+\gamma}}.$$

At this point, we notice that $m \geq \max\left(\frac{p+\gamma}{p-1}, 2\right)$ implies $(2m - 2)\frac{p+\gamma}{\gamma+1} \geq 2m$. Since $|\psi| \leq 1$ in Ω, we deduce that

$$\int_\Omega |u|^{p+\gamma}\psi^{2m} \, dx \leq C \int_\Omega |\nabla\psi|^{2\frac{p+\gamma}{p-1}} \, dx.$$

Going back to (5.68), (5.69) follows. $\qquad\qquad\qquad\qquad\qquad\qquad \square$

Proof of Theorem 5.3.1. Suppose $u \in H^1(\Omega) \cap L^{p+1}(\Omega)$, $u \geq 0$ is a stable weak solution to (5.39). Given $\varepsilon > 0$, define

$$\Sigma_\varepsilon = \left\{ x \in \Omega : \forall r > 0 \quad \int_{B(x,r)} (u^{p+1} + |\nabla u|^2) \, dx \geq \varepsilon r^{N - 2\frac{p+1}{p-1}} \right\}.$$

Step 1. There exists a fixed value of $\varepsilon > 0$ such that for every $x \notin \Sigma_\varepsilon$, u is bounded (hence smooth) in a neighborhood of x.

To see this, let $x_0 \notin \Sigma_\varepsilon$: there exists $r_0 > 0$ such that

$$r_0^{-\mu} \int_{B(x_0,r_0)} (u^{p+1} + |\nabla u|^2) \, dx < \varepsilon,$$

where $\mu = N - 2\frac{p+1}{p-1}$. By (5.54), for $r < r_0$,

$$\mathcal{E}_u(x_0, r) \leq r^{-\mu} \int_{B(x_0,r)} \frac{1}{2} |\nabla u|^2 \, dy + \frac{r^{-\mu-1}}{p-1} \int_{\partial B(x_0,r)} u^2 \, d\sigma$$

$$\leq r^{-\mu} \int_{B(x_0,r_0)} \frac{1}{2} |\nabla u|^2 \, dy + \frac{r^{-\mu-1}}{p-1} \int_{\partial B(x_0,r)} u^2 \, d\sigma$$

$$\leq \frac{\varepsilon}{2} \left(\frac{r}{r_0}\right)^{-\mu} + \frac{r^{-\mu-1}}{p-1} \int_{\partial B(x_0,r)} u^2 \, d\sigma.$$

Integrating between $r = r_0/2$ and r_0, and recalling that $\mathcal{E}_u(x, r)$ is nondecreasing in r, we deduce that

$$\frac{r_0}{2} \mathcal{E}_u(x_0, r_0/2) \leq 2^{\mu-2} \varepsilon r_0 + \frac{1}{p-1} \int_{r_0/2}^{r_0} r^{-\mu-1} \left(\int_{\partial B(x_0,r)} u^2 \, d\sigma\right) dr$$

$$\leq C\varepsilon r_0 + C r_0^{-\mu-1} \int_{B(x_0,r_0)} u^2 \, dy$$

$$\leq C\varepsilon r_0 + C r_0^{-\mu-1} \left(\int_{B(x_0,r_0)} u^{p+1} \, dy\right)^{\frac{2}{p+1}} r_0^{N(1-\frac{2}{p+1})}$$

$$< C\varepsilon r_0.$$

Hence,

$$\mathcal{E}_u(x_0, r_0/2) < C\varepsilon.$$

Since $\mathcal{E}_u$ is continuous in x, there exists $r_1 < r_0/2$ such that $\mathcal{E}_u(x, r_0/2) < 2C\varepsilon$, for $x \in B(x_0, r_1)$. Since $\mathcal{E}_u$ is nonincreasing in r, we deduce that for all $x \in B(x_0, r_1)$ and all $r < r_1$,

$$\mathcal{E}_u(x, r) < 2C\varepsilon. \tag{5.70}$$

Now take an approximating sequence u_n given by Lemma 5.3.1. Integrating (5.62) between 0 and $r_2 < r_1$, we find

$$\frac{p-1}{p+3} \int_0^{r_2} r^{-\mu} \left(\int_{B(x,r)} \left(\frac{1}{2} |\nabla u_n|^2 + \frac{1}{p+1} u_n^{p+1}\right) dy\right) dr + \frac{r_2^{-\mu}}{p+3} \int_{\partial B(x,r_2)} u_n^2 \, d\sigma$$
$$\leq r_2 \mathcal{E}_{u_n}(x, r_2).$$

It follows that

$$Cr_2 \mathscr{E}_u(x, r_2) \geq \int_0^{r_2} \left(r^{-\mu} \int_{B(x,r)} u^{p+1} \, dy \right) dr$$

$$\geq \int_{r_2/2}^{r_2} \left(r^{-\mu} \int_{B(x,r)} u^{p+1} \, dy \right) dr.$$

By the fundamental theorem of calculus, we deduce that there exists $r_3 \in (r_2/2, r_2)$ such that

$$C \mathscr{E}_u(x, r_2) \geq r_3^{-\mu} \int_{B(x,r_3)} u^{p+1} \, dy \geq r_2^{-\mu} \int_{B(x,r_2/2)} u^{p+1} \, dy.$$

Now apply (5.70). Then,

$$r^{-\mu} \int_{B(x,r)} u^{p+1} \, dy \leq C\varepsilon,$$

for all $x \in B(x_0, r_1)$ and all $r < r_1/2$. Taking ε sufficiently small, it follows from Theorem 5.3.3 that (u_n) is uniformly bounded near x_0, and so u is smooth in a neighborhood of x_0.

Step 2. For all $\gamma \geq 1$, there exists $\varepsilon' > 0$ such that

$$\Sigma_\varepsilon \subseteq \tilde{\Sigma}_{\varepsilon'} := \left\{ x \in \Omega : \forall r > 0 \quad \int_{B(x,r)} u^{p+\gamma} \, dx \geq \varepsilon' r^{N-2\frac{p+\gamma}{p-1}} \right\}.$$

Indeed, suppose $x \notin \tilde{\Sigma}_{\varepsilon'}$. Then,

$$\int_{B(x,r)} u^{p+\gamma} \, dx < \varepsilon' r^{N-2\frac{p+\gamma}{p-1}}$$

for some $r > 0$. By Hölder's inequality,

$$\int_{B(x,r)} u^{p+1} \, dx \leq C \left(\int_{B(x,r)} u^{p+\gamma} \, dx \right)^{\frac{p+1}{p+\gamma}} r^{N(1-\frac{p+1}{p+\gamma})}$$

$$< C \left(\varepsilon' r^{N-2\frac{p+\gamma}{p-1}} \right)^{\frac{p+1}{p+\gamma}} r^{N(1-\frac{p+1}{p+\gamma})} = C(\varepsilon')^{\frac{p+1}{p+\gamma}} r^{N-2\frac{p+1}{p-1}}. \quad (5.71)$$

Take a function $\varphi \in C_c^2(\Omega)$ and multiply the Lane-Emden equation (5.39) by $u\varphi^2$. Then,

$$\int_\Omega |\nabla u|^2 \varphi^2 \, dx + \int_\Omega u \nabla u \cdot \nabla \varphi^2 \, dx = \int_\Omega u^{p+1} \varphi^2 \, dx,$$

that is,

$$\int_\Omega |\nabla u|^2 \varphi^2 \, dx = \int_\Omega u^{p+1} \varphi^2 \, dx + \frac{1}{2} \int_\Omega u^2 \Delta \varphi^2 \, dx.$$

Now choose φ such that $\varphi = 1$ in $B(x, r/2)$, $\varphi = 0$ outside $B(x,r)$, and $|\Delta \varphi^2| \leq C/r^2$. Then,

$$\int_{B(x,r/2)} |\nabla u|^2 \, dx \leq C \int_{B(x,r)} u^{p+1} \, dx + \frac{C}{r^2} \int_{B(x,r)} u^2 \, dx.$$

We estimate

$$\frac{1}{r^2} \int_{B(x,r)} u^2 \, dx \leq \frac{C}{r^2} \left(\int_{B(x,r)} u^{p+\gamma} \, dx \right)^{\frac{2}{p+\gamma}} r^{1 - \frac{2}{p+\gamma}}$$

$$< \frac{C}{r^2} \left(\varepsilon' r^{n - 2\frac{p+\gamma}{p-1}} \right)^{\frac{2}{p+\gamma}} r^{1 - \frac{2}{p+\gamma}} = C(\varepsilon')^{\frac{2}{p+\gamma}} r^{N - 2\frac{p+1}{p-1}}.$$

Using (5.71), we deduce that

$$\int_{B(x,r/2)} (u^{p+1} + |\nabla u|^2) \, dx < C(\varepsilon')^{\frac{2}{p+\gamma}} r^{N - 2\frac{p+1}{p-1}}.$$

Choosing ε' such that $C(\varepsilon')^{\frac{2}{p+\gamma}} \leq \varepsilon$, we deduce that $x \notin \Sigma_\varepsilon$. And so, $\tilde{\Sigma}_{\varepsilon'} \supset \Sigma_\varepsilon$.
Step 3. By the capacitary estimate (Proposition 5.3.1), $u \in L^{p+\gamma}_{loc}(\Omega)$ if $\gamma \in [1, 2p + 2\sqrt{p(p-1)} - 1)$. By Theorem 5.3.4, it follows that for $\varepsilon' > 0$ small,

$$\mathcal{H}^{N - 2\frac{p+\gamma}{p+1}}(\tilde{\Sigma}_{\varepsilon'}) = 0.$$

This being true for all $\gamma \in [1, 2p + 2\sqrt{p(p-1)} - 1)$, Theorem 5.3.1 follows.
$\square$

Chapter 6

Liouville theorems for stable solutions

In this chapter, we explore Equation (1.4) when $\Omega = \mathbb{R}^N$. We begin with the study of radial solutions.

6.1 Classifying radial stable entire solutions

In this section, we focus our attention on bounded radial stable solutions of

$$-\Delta u = f(u) \qquad \text{in } \mathbb{R}^N. \tag{6.1}$$

The following Liouville-type theorem is sharp (see, for example, [97, Theorem 5] for counter-examples).

Theorem 6.1.1 ([38, 216]) *Let* $1 \leq N \leq 10$ *and* $f \in C^1(\mathbb{R})$. *Then, every bounded radial stable solution* $u \in C^2(\mathbb{R}^N)$ *of* (6.1) *is constant.*

The proof of Theorem 6.1.1 uses two technical lemmata, which we present next.

Lemma 6.1.1 ([216]) *Let* $N \geq 1$ *and* u *be a stable nonconstant radial solution to* (6.1). *Then, there exists* $K > 0$ *such that*

$$\int_r^{+\infty} \frac{ds}{s^{N-1} u_r^2(s)} \leq K r^{-2\sqrt{N-1}}, \qquad \text{for all } r \geq 1. \tag{6.2}$$

Proof. We prove the lemma for $N \geq 2$ and refer the reader to [216] for the case $N = 1$.

Step 1 ([38]) $|u_r| > 0$ in $\mathbb{R}^N \setminus \{0\}$.

Assume by contradiction that $u_r(R) = 0$ for some $R > 0$. Differentiate (6.1) with respect to r. Then, u_r solves

$$-\Delta u_r + \frac{N-1}{r^2} u_r = f'(u)u_r, \qquad \text{for } r > 0. \tag{6.3}$$

Take a cutoff function $\zeta_\varepsilon \in H^1(\mathbb{R}^N)$ defined by

$$\zeta_\varepsilon(x) = \begin{cases} 0 & \text{if } |x| < \varepsilon^2, \\ 2 - \dfrac{\ln|x|}{\ln\varepsilon} & \text{if } \varepsilon^2 \leq |x| < \varepsilon, \\ 1 & \text{if } |x| \geq \varepsilon, \end{cases}$$

if $N = 2$, and $\zeta_\varepsilon(x) = \zeta(x/\varepsilon)$, with $\zeta \in C^1(\mathbb{R}^N)$ such that $\zeta \equiv 0$ in B_1 and $\zeta \equiv 1$ in $\mathbb{R}^N \setminus B_2$, if $N \geq 3$. One easily checks that $\zeta_\varepsilon \to 1$ in $H^1_{loc}(\mathbb{R}^N)$. Multiply (6.3) with $\zeta_\varepsilon u_r$, integrate, and let $\varepsilon \to 0$. We deduce

$$Q_u(u_r \chi_{B_R}) = \int_{B_R} \left(|\nabla u_r|^2 - f'(u)u_r^2 \right) dx = -(N-1) \int_{B_R} \frac{u_r^2}{r^2} \, dx < 0,$$

a contradiction with the stability of u, since $u_r \chi_{B_R} \in H^1_0(B_R)$ is a valid test function in (1.5) (see Remark 1.1.1).

Step 2 ([38]) For all $\eta \in H^1(\mathbb{R}^N)$,

$$(N-1) \int_{\mathbb{R}^N} \frac{u_r^2 \eta^2}{|x|^2} \, dx \leq \int_{\mathbb{R}^N} u_r^2 |\nabla \eta|^2 \, dx. \tag{6.4}$$

This is a restatement of the geometric Poincaré formula, Equation (4.25), in the radial setting. By density, it suffices to consider $\eta \in C^1_c(\mathbb{R}^N \setminus \{0\})$. We calculate $Q_u(u_r \eta)$:

$$0 \leq Q_u(u_r \eta) = \int_{\mathbb{R}^N} \left(|\nabla(u_r \eta)|^2 - f'(u)(u_r \eta)^2 \right) dx$$

$$= \int_{\mathbb{R}^N} \left(u_r^2 |\nabla \eta|^2 + \nabla \eta^2 \cdot u_r \nabla u_r + \eta^2 |\nabla u_r|^2 - f'(u)u_r^2 \eta^2 \right) dx$$

$$= \int_{\mathbb{R}^N} \left(u_r^2 |\nabla \eta|^2 - \eta^2 \nabla \cdot (u_r \nabla u_r) + \eta^2 |\nabla u_r|^2 - f'(u)u_r^2 \eta^2 \right) dx$$

$$= \int_{\mathbb{R}^N} \left(u_r^2 |\nabla \eta|^2 - \eta^2 (u_r \Delta u_r + f'(u)u_r^2) \right) dx.$$

Using (6.3), we obtain (6.4).

Step 3. Fix $R > r \geq 1$ and consider the function

$$\eta(t) = \begin{cases} 1 & \text{if } 0 \leq t \leq 1, \\ t^{-\sqrt{N-1}} & \text{if } 1 < t \leq r, \\ \dfrac{r^{-\sqrt{N-1}}}{\int_r^R \frac{ds}{s^{N-1}u_r^2(s)}} \displaystyle\int_t^R \frac{ds}{s^{N-1}u_r^2(s)} & \text{if } r \leq t \leq R, \\ 0 & \text{if } R < t. \end{cases}$$

In particular,

$$(N-1)\int_0^{+\infty} u_r^2 \eta^2\, t^{N-3}\, dt \geq (N-1)\int_0^1 u_r^2 t^{N-3}\, dt +$$
$$+ (N-1)\int_1^r u_r^2 t^{-2\sqrt{N-1}+N-3}\, dt.$$

In addition,

$$\int_0^{+\infty} u_r^2 (\eta')^2 t^{N-1}\, dt = (N-1)\int_1^r u_r^2 t^{-2\sqrt{N-1}+N-3}\, dt + \frac{r^{-2\sqrt{N-1}}}{\int_r^R \frac{ds}{s^{N-1}u_r^2}}.$$

Applying Step 2, we conclude that

$$(N-1)\int_0^1 u_r^2 t^{N-3}\, dt \leq \frac{r^{-2\sqrt{N-1}}}{\int_r^R \frac{ds}{s^{N-1}u_r^2}}.$$

Finally, since $R > r$ is arbitrarily large, we obtain (6.2) with $K = \left((N-1)\int_0^1 u_r^2 t^{N-3}\, dt \right)^{-1}$. $\qquad\square$

Lemma 6.1.2 ([216]) *Let $N \geq 2$ and let u denote a nonconstant stable radial solution to (6.1). Then, there exists a constant $M > 0$ such that*

$$|u(2r) - u(r)| \geq M r^{-N/2 + \sqrt{N-1} + 2}, \qquad \text{for all } r \geq 1. \tag{6.5}$$

Proof. Fix $r \geq 1$ and consider the functions

$$\alpha(s) = s^{\frac{1-N}{3}} \left| u_r(s) \right|^{-\frac{2}{3}}, \qquad\qquad s \in (r, 2r),$$
$$\beta(s) = \left| u_r(s) \right|^{\frac{2}{3}}, \qquad\qquad s \in (r, 2r).$$

By Lemma 6.1.1, we have

$$\|\alpha\|_{L^3(r,2r)} \leq K^{\frac{1}{3}} r^{-\frac{2}{3}\sqrt{N-1}},$$

where $K > 0$ is independent of r. In addition, since $|u_r| > 0$ in $\mathbb{R}^N \setminus \{0\}$, it follows that

$$\|\beta\|_{L^{\frac{3}{2}}(r,2r)} = |u(2r) - u(r)|^{\frac{2}{3}}.$$

Applying Hölder's inequality to α and β, we deduce that

$$r^{\frac{4-N}{3}} \int_1^2 t^{\frac{1-N}{3}} \, dt = \|s^{\frac{1-N}{3}}\|_{L^1(r,2r)} \leq \|\alpha\|_{L^3(r,2r)} \|\beta\|_{L^{\frac{3}{2}}(r,2r)}$$

$$\leq K^{\frac{1}{3}} r^{-\frac{2}{3}\sqrt{N-1}} |u(2r) - u(r)|^{\frac{2}{3}}.$$

$\square$

We may now prove Theorem 6.1.1.

Proof.
Case 1. $2 \leq N \leq 9$. Let $r \geq 1$. There exists $m \in \mathbb{N}$ and $1 \leq r_1 < 2$ such that $r = 2^{m-1} r_1$. From the monotonicity of u and Lemma 6.1.2, we obtain

$$|u(r)| \geq |u(r) - u(r_1)| - |u(r_1)| = \sum_{k=1}^{m-1} |u(2^k r_1) - u(2^{k-1} r_1)| - |u(r_1)| \geq$$

$$\geq M \sum_{k=1}^{m-1} (2^{k-1} r_1)^{-N/2+\sqrt{N-1}+2} - |u(r_1)| =$$

$$= M \left(\frac{r^{-N/2+\sqrt{N-1}+2} - r_1^{-N/2+\sqrt{N-1}+2}}{2^{-N/2+\sqrt{N-1}+2} - 1} - |u(r_1)| \right).$$

Since $r_1 \in [1,2)$, u is continuous and $-N/2 + \sqrt{N-1} + 2 > 0$, we finally have

$$|u(r)| \geq M r^{-N/2+\sqrt{N-1}+2} - M_2,$$

contradicting the assumption that u is bounded.
Case 2. $N = 10$. This time, $-N/2 + \sqrt{N-1} + 2 = 0$ and working as above, we obtain

$$|u(r)| \geq M(m-1) - |u(r_1)| = \frac{M(\ln r - \ln r_1)}{\ln 2} - |u(r_1)|,$$

contradicting again the assumption that u is bounded. $\square$

6.2 Classifying stable entire solutions

6.2.1 The Liouville equation

We now turn to possible nonradial solutions. At the time of writing of this book, the theory is yet incomplete. We begin our presentation with the non-linearity $f(u) = e^u$ (see, for example, [88], [89] for more general results). That is, we study stable solutions of

$$-\Delta u = e^u \qquad \text{in } \mathbb{R}^N, \tag{6.6}$$

which we shall refer to as the Liouville equation. Stable solutions are classified as follows.

Theorem 6.2.1 ([95]) *For $1 \le N \le 9$, there is no stable solution $u \in C^2(\mathbb{R}^N)$ of (6.6).*

Remark 6.2.1 *The theorem is sharp: For every $N \ge 10$, there exists a radial stable solution to (6.6) (see Chapter 2).*

The proof of Theorem 6.2.1 relies on the following capacitary estimate.

Proposition 6.2.1 ([95]) *Assume that $N \ge 1$ and let Ω denote an open set of $\mathbb{R}^N$. Let $u \in C^2(\Omega)$ denote a stable solution to*

$$-\Delta u = e^u \qquad \text{in } \Omega.$$

Then, for any integer $m \ge 5$ and any $\alpha \in (0,2)$, there exists a constant $C = C(m, \alpha) > 0$ such that

$$\int_\Omega e^{(2\alpha+1)u} \psi^{2m} \, dx \le C \int_\Omega |\nabla\psi|^{2(2\alpha+1)} \, dx, \tag{6.7}$$

for all $\psi \in C_c^1(\Omega)$ such that $0 \le \psi \le 1$.

Proof. The first step is very similar to the proof of Theorem 4.2.1: given $\varphi \in C_c^1(\Omega)$, we multiply the equation by $e^{2\alpha u}\varphi^2$ on the one hand, and use the stability inequality (1.5) with test function $e^{\alpha u}\varphi$ on the other hand. The former computation is developed next.

$$\int_\Omega \nabla u \nabla \left(e^{2\alpha u}\right) \varphi^2 \, dx + \int_\Omega e^{2\alpha u} \nabla u \nabla \left(\varphi^2\right) \, dx = \int_\Omega e^{(2\alpha+1)u}\varphi^2 \, dx.$$

Hence,

$$\int_\Omega e^{(2\alpha+1)u}\varphi^2\,dx = \frac{2}{\alpha}\int_\Omega |\nabla(e^{\alpha u})|^2\,\varphi^2\,dx + \frac{2}{\alpha}\int_\Omega \varphi\nabla(e^{\alpha u})e^{\alpha u}\nabla\varphi\,dx$$
$$\geq \left(\frac{2}{\alpha}-\varepsilon\right)\int_\Omega |\nabla(e^{\alpha u})|^2\,\varphi^2\,dx - C_\varepsilon\int_\Omega e^{2\alpha u}|\nabla\varphi|^2\,dx,$$

where we used the Cauchy-Schwarz inequality and where $\varepsilon > 0$ is chosen so small that $2/\alpha - \varepsilon > 1$. So, there exists $\beta = (2/\alpha - \varepsilon)^{-1} < 1$ such that

$$\int_\Omega |\nabla(e^{\alpha u})|^2\,\varphi^2\,dx \leq \beta\int_\Omega e^{(2\alpha+1)u}\varphi^2\,dx + C\int_\Omega e^{2\alpha u}|\nabla\varphi|^2\,dx. \tag{6.8}$$

Next, we use stability (1.5) with test function $e^{\alpha u}\varphi$ and obtain

$$\int_\Omega e^{(2\alpha+1)u}\varphi^2\,dx \leq \int_\Omega |\nabla(e^{\alpha u})|^2\,\varphi^2\,dx + \int_\Omega |\nabla\varphi|^2 e^{2\alpha u}\,dx +$$
$$+ 2\int_\Omega e^{\alpha u}\nabla(e^{\alpha u})\varphi\nabla\varphi\,dx \leq$$
$$\leq (1+\varepsilon)\int_\Omega |\nabla(e^{\alpha u})|^2\,\varphi^2\,dx + C_\varepsilon\int_\Omega |\nabla\varphi|^2 e^{2\alpha u}\,dx,$$

where $\varepsilon > 0$ is chosen so small that $\gamma = \beta(1+\varepsilon) < 1$. Plugging (6.8) in the latter inequality yields

$$\int_\Omega e^{(2\alpha+1)u}\varphi^2\,dx \leq \gamma\int_\Omega e^{(2\alpha+1)u}\varphi^2\,dx + C\int_\Omega |\nabla\varphi|^2 e^{2\alpha u}\,dx.$$

That is,

$$\int_\Omega e^{(2\alpha+1)u}\varphi^2\,dx \leq C\int_\Omega |\nabla\varphi|^2 e^{2\alpha u}\,dx. \tag{6.9}$$

The next step consists in iterating the above formula: set $\varphi = \psi^m$, where $\psi \in C_c^1(\Omega)$, $0 \leq \psi \leq 1$ and $m \geq 5$. We obtain

$$\int_\Omega e^{(2\alpha+1)u}\psi^{2m}\,dx \leq C\int_\Omega |\nabla\psi|^2\psi^{2(m-1)}e^{2\alpha u}\,dx$$
$$\leq C\left(\int_\Omega |\nabla\psi|^{2(2\alpha+1)}\,dx\right)^{\frac{1}{2\alpha+1}}\left(\int_\Omega \psi^{2(m-1)\frac{2\alpha+1}{2\alpha}}e^{(2\alpha+1)u}\,dx\right)^{\frac{2\alpha}{2\alpha+1}}.$$

Notice that for $m \geq 5$ and $\alpha \in (0,2)$, there holds $(m-1)\frac{2\alpha+1}{\alpha} \geq 2m$, hence $\psi^{(m-1)\frac{2\alpha+1}{\alpha}} \leq \psi^{2m}$ and (6.7) follows. $\qquad\square$

We are now ready for the proof of the theorem.

Proof of Theorem 6.2.1. Assume by contradiction that equation (6.6) admits a stable solution for some $N \leq 9$. Fix an integer $m \geq 5$ and choose $\alpha \in (0,2)$ such that $N - 2(2\alpha + 1) < 0$. This is always possible since $N \leq 9$. Consider a standard cutoff ϕ_R, defined for $R > 0$ and $x \in \mathbb{R}^N$ by $\phi_R(x) = \phi(|x|/R)$, where $\phi \in C_c^1(\mathbb{R})$ is such that $0 \leq \phi \leq 1$ and

$$\phi(t) = \begin{cases} 1 & \text{if } |t| \leq 1, \\ 0 & \text{if } |t| \geq 2. \end{cases} \qquad (6.10)$$

Apply (6.7) with $\psi = \phi_R$ and $\Omega = \mathbb{R}^N$ to get

$$\int_{B_R} e^{(2\alpha+1)u}\,dx \leq CR^{N-2(2\alpha+1)},$$

where $C > 0$ is independent of R. Letting $R \to +\infty$, we obtain $\int_{\mathbb{R}^N} e^{(2\alpha+1)u}\,dx = 0$, a contradiction. $\qquad\square$

Exercise 6.2.1 ([97]) In this exercise, we classify stable solutions to the Lane-Emden equation

$$-\Delta u = |u|^{p-1}u \qquad \text{in } \mathbb{R}^N. \qquad (6.11)$$

Using the capacitary estimate in Proposition 5.3.1, prove that if $u \in C^2(\mathbb{R}^N)$ is a stable solution to (6.11), with $1 < p < p_c(N)$, where $p_c(N)$ is given by (4.16), then $u \equiv 0$.

6.2.2 Dimension $N = 2$

Is Theorem 6.2.1 true for general nonlinearities? In addition to stability, we must impose further restrictions on u in order to obtain a Liouville theorem for

$$-\Delta u = f(u) \qquad \text{in } \mathbb{R}^N. \qquad (6.12)$$

Indeed, in the simple case $f \equiv 0$, all solutions are stable, but the Liouville theorem fails without assuming, say, that u is bounded. This being said, we have the following statement.

Definition 6.2.1 *A function $u : \mathbb{R}^N \to \mathbb{R}$ is said to be one-dimensional if up to a rotation of space, u is a function of x_N only, that is, if there exists $\tau \in \mathbb{S}^{N-1}$ and $g : \mathbb{R} \to \mathbb{R}$ such that*

$$u(x) = g(\tau \cdot x) \qquad \text{for all } x \in \mathbb{R}^N.$$

Theorem 6.2.2 ([99]) *Let $f \in C^1(\mathbb{R})$ and let $u \in C^2(\mathbb{R}^2)$ be a stable solution to (6.12) such that $|\nabla u| \in L^\infty(\mathbb{R}^2)$. Then, u is one-dimensional.*

Remark 6.2.2 *Theorem 6.2.2 remains valid if f is only assumed to be locally Lipschitz continuous (but the proof is quite different). See [99]. A difficult open problem is to decide whether Theorem 6.2.2 still holds in dimension $N = 3$.*

Proof. Since $|\nabla u|$ is bounded and $N = 2$, there exists a constant $C > 0$ independent of $R > 0$ such that

$$\int_{B_R} |\nabla u|^2 \, dx \leq CR^2. \tag{6.13}$$

Since u is stable, there exists a solution $v > 0$ to the linearized equation

$$-\Delta v = f'(u)v \qquad \text{in } \mathbb{R}^2. \tag{6.14}$$

Note that $\partial u / \partial x_j$ also solves (6.14). Now recall that the principal eigenvalue of an elliptic operator on a bounded domain is simple. By analogy, we can hope that $\partial u / \partial x_j$ is a constant multiple of v. Next, we prove that this is indeed the case. Let

$$\sigma_j = \frac{1}{v} \frac{\partial u}{\partial x_j} \tag{6.15}$$

for $j = 1, 2$. Then, since v and $\partial u / \partial x_j$ both solve the linearized equation (6.14), it follows that

$$-\nabla \cdot \left(v^2 \nabla \sigma_j \right) = 0 \qquad \text{in } \mathbb{R}^2. \tag{6.16}$$

Proposition 6.2.2 ([17], [7]) *Let $N \geq 1$ and let $v \in L^\infty_{loc}(\mathbb{R}^N)$ denote a positive function. Suppose that $\sigma \in H^1_{loc}(\mathbb{R}^N)$ satisfies*

$$-\sigma \nabla \cdot \left(v^2 \nabla \sigma \right) \leq 0 \qquad \text{in } \mathscr{D}'(\mathbb{R}^N). \tag{6.17}$$

Assume that there exists a constant $C > 0$ such that for every $R > 1$,

$$\int_{B_R} v^2 \sigma^2 \, dx \leq CR^2. \tag{6.18}$$

Then, σ is constant.

Proof. Take a standard cutoff function φ_0 such that $\varphi_0 = 1$ on B_1 and $\varphi_0 = 0$ on $\mathbb{R}^N \setminus B_2$. Test (6.17) with φ^2, where $\varphi(x) = \varphi_0(x/R)$. We obtain

$$\int_{\mathbb{R}^N} \varphi^2 v^2 |\nabla \sigma|^2 \, dx \leq -2 \int_{\mathbb{R}^N} \varphi v^2 \sigma \nabla \varphi \nabla \sigma \, dx$$

$$\leq 2 \left(\int_{[R<|x|<2R]} \varphi^2 v^2 |\nabla \sigma|^2 \, dx \right)^{1/2} \left(\int_{\mathbb{R}^N} v^2 \sigma^2 |\nabla \varphi|^2 \, dx \right)^{1/2}$$

$$\leq C \left(\int_{[R<|x|<2R]} \varphi^2 v^2 |\nabla \sigma|^2 \, dx \right)^{1/2} \left(\frac{1}{R^2} \int_{B_{2R}} v^2 \sigma^2 \, dx \right)^{1/2}.$$

By (6.18), it follows that

$$\int_{\mathbb{R}^N} \varphi^2 v^2 |\nabla \sigma|^2 \, dx \leq C \left(\int_{[R<|x|<2R]} \varphi^2 v^2 |\nabla \sigma|^2 \, dx \right)^{1/2}. \tag{6.19}$$

In particular, $\int_{\mathbb{R}^N} \varphi^2 v^2 |\nabla \sigma|^2 \, dx \leq C^2$ and letting $R \to +\infty$, we obtain that

$$\int_{\mathbb{R}^N} v^2 |\nabla \sigma|^2 \, dx < +\infty.$$

Hence, the right-hand side of (6.19) converges to zero as $R \to +\infty$ and so,

$$\int_{\mathbb{R}^N} v^2 |\nabla \sigma|^2 \, dx = 0.$$

Since v is positive, we conclude that σ is constant. $\qquad\square$

Thanks to (6.13), we may apply Proposition 6.2.2 to σ_j given by (6.15), and deduce that σ_j is constant, that is, there exists a constant C_j such that

$$\frac{\partial u}{\partial x_j} = C_j v.$$

In particular, the gradient of u points in a fixed direction, that is, u is one-dimensional. $\qquad\square$

6.2.3 Dimensions $N = 3, 4$

In higher dimensions, only partial information is currently available. We present here a Liouville theorem that applies to any positive nonlinearity.

145

Theorem 6.2.3 ([88]) *Assume that $f \in C^1(\mathbb{R})$, $f \geq 0$ and $1 \leq N \leq 4$. Assume that $u \in C^2(\mathbb{R}^N)$ is a bounded, stable solution to (6.12). Then, u is constant.*

Proof. For $R > 0$, let B_R denote the ball of radius R centered at the origin. We begin by proving that there exists a constant $C > 0$ independent of $R > 0$ such that

$$\int_{B_R} |\nabla u|^2 \, dx \leq CR^{N-2}. \tag{6.20}$$

Let $M \geq \|u\|_\infty$, $\varphi \in C_c^2(\mathbb{R}^N)$, $\varphi \geq 0$, and multiply (6.12) by $(u - M)\varphi$:

$$\int_{\mathbb{R}^N} -\Delta u(u - M)\varphi \, dx = \int_{\mathbb{R}^N} f(u)(u - M)\varphi \, dx.$$

Integrating by parts and recalling that $f \geq 0$, it follows that

$$\int_{\mathbb{R}^N} |\nabla u|^2 \varphi \, dx + \int_{\mathbb{R}^N} (u - M)\nabla u \nabla \varphi \, dx = \int_{\mathbb{R}^N} f(u)(u - M)\varphi \, dx$$
$$\leq 0,$$

hence,

$$\int_{\mathbb{R}^N} |\nabla u|^2 \varphi \, dx \leq - \int_{\mathbb{R}^N} \frac{1}{2} \nabla (u - M)^2 \nabla \varphi \, dx = \int_{\mathbb{R}^N} \frac{(u - M)^2}{2} \Delta \varphi \, dx$$
$$\leq 2M^2 \int_{\mathbb{R}^N} |\Delta \varphi| \, dx.$$

Let φ_0 denote any nonnegative test function such that $\varphi_0 = 1$ on B_1 and apply the above inequality with $\varphi(x) = \varphi_0(x/R)$. We obtain (6.20).

Since u is stable, there exists a solution $v > 0$ of the linearized equation

$$-\Delta v = f'(u)v \qquad \text{in } \mathbb{R}^N. \tag{6.21}$$

Let $\sigma_j = \frac{1}{v} \frac{\partial u}{\partial x_j}$ for $j = 1, \dots, N$. Then, since v and $\partial u/\partial x_j$ both solve the linearized equation (6.21), it follows that

$$-\nabla \cdot \left(v^2 \nabla \sigma_j\right) = 0 \qquad \text{in } \mathbb{R}^N. \tag{6.22}$$

Apply Proposition 6.2.2 to (6.22). By (6.20), we deduce that if $N \leq 4$, then σ_j is constant, that is, there exists a constant C_j such that

$$\frac{\partial u}{\partial x_j} = C_j v.$$

In particular, the gradient of u points in a fixed direction, that is, u is one-dimensional and solves

$$-u'' = f(u) \qquad \text{in } \mathbb{R}.$$

Since $f \geq 0$ and u is bounded, this is possible only if u is constant and $f(u) = 0$. $\qquad\qquad\square$

6.3 Classifying solutions that are stable outside a compact set

Much less is known about the classification of solutions that are stable outside a compact set. Power nonlinearities were first treated by [97], followed by the exponential nonlinearity by [71] and more general convex nonlinearities by [88]. In this section, we return to the Liouville equation, Equation (6.6). The results are substantially different in the cases $N = 2$ (critical), $3 \leq N \leq 9$ (supercritical range) and $N \geq 10$ (twice supercritical).

6.3.1 The critical case

Theorem 6.3.1 ([95]) *Let $N = 2$. Let $u \in C^2(\mathbb{R}^2)$ denote a solution to (6.6). Then, u is stable outside a compact set if and only if u is of the form*

$$u(x) = \ln\left(\frac{32\lambda^2}{\left(4 + \lambda^2 |x - x_0|^2\right)^2}\right), \qquad \lambda > 0, \quad x_0 \in \mathbb{R}^2. \qquad (6.23)$$

Proof.
Step 1. Any function given by (6.23) is stable outside a large ball of $\mathbb{R}^2$. Clearly, it is not enough to prove this claim for $x_0 = 0$ and $\lambda > 0$. To this end, we observe that there exists $R = R(\lambda) > 1$ such that

$$e^{u(x)} \leq \frac{1}{4|x|^2 \ln^2(|x|)}$$

for $|x| > R$, and that for all $\psi \in C_c^1(\mathbb{R}^2 \setminus \overline{B(0,R)})$ we have

$$\int_{[|x|>R]} \left(|\nabla \psi|^2 - \frac{\psi^2}{4|x|^2 \ln^2(|x|)} \right) dx \geq 0.$$

The latter follows immediately from the fact that the function $\varphi(x) = \ln^{1/2}(|x|) \in C^2(\mathbb{R}^2 \setminus \overline{B(0,1)})$ is a positive solution to $-\Delta \varphi - \frac{1}{4|x|^2 \ln^2(|x|)} \varphi = 0$ in $\mathbb{R}^2 \setminus \overline{B(0,1)}$. Combining these two properties, we obtain the desired conclusion.

Step 2. Conversely, if u is stable outside a compact set, there exists $R_0 > 0$ such that u is stable in $\Omega = \mathbb{R}^N \setminus \overline{B(0, R_0)}$. For every $R > R_0 + 3$ and every $x \in \mathbb{R}^2$, consider a function $\psi_R \in C_c^1(\mathbb{R}^2 \setminus \overline{B(0, R_0)})$ satisfying

$$\psi_R(x) = \begin{cases} \xi(x) & \text{if } |x| \leq R_0 + 3, \\ \phi_R(x) & \text{if } |x| \geq R_0 + 3, \end{cases} \qquad (6.24)$$

where $\phi_R(x) = \phi(|x|/R)$, with $\phi \in C_c^1(\mathbb{R})$, $0 \leq \phi \leq 1$, satisfying (6.10), and where ξ is any function belonging to $C^1(\mathbb{R}^2)$ such that $0 \leq \xi \leq 1$ in $\mathbb{R}^2$, $\xi \equiv 0$ inside the ball $B(0, R_0+1)$ and $\xi \equiv 1$ outside the ball $B(0, R_0+2)$. u is stable on the support of ψ_R, so $Q_u(\psi_R) \geq 0$. In particular, letting $R \to +\infty$, we deduce that

$$\int_{\mathbb{R}^2} e^u \, dx < +\infty.$$

We then conclude that u is of the form (6.23), using Theorem 6.3.2 below.

$\square$

Theorem 6.3.2 ([58]) *Let $N = 2$. Let $u \in C^2(\mathbb{R}^2)$ denote a solution to (6.6). If*

$$\int_{\mathbb{R}^2} e^u \, dx < +\infty, \qquad (6.25)$$

then u is of the form (6.23).

The proof of Theorem 6.3.2 uses the moving-planes device and the following information on the asymptotics of solutions.

Lemma 6.3.1 ([58]) *Assume that $u \in C^2(\mathbb{R}^2)$ is a solution to (6.6) satisfying the integrability condition (6.25). Then,*

$$\lim_{|x| \to +\infty} \frac{u(x)}{\ln |x|} = -\frac{1}{2\pi} \int_{\mathbb{R}^2} e^u \, dx \leq -4.$$

Proof.
Step 1 ([30]). u is bounded above.

148

Since u satisfies the integral condition (6.25), there exists $R > 0$ such that for every $|x_0| \geq R$,

$$\int_{B(x_0,1)} e^u \, dx \leq \pi/2. \tag{6.26}$$

In the ball $B(x_0, 1)$, $u = v + w$, where v, w, respectively solve

$$\begin{cases} -\Delta v = e^u & \text{in } B(x_0, 1), \\ v = 0 & \text{on } \partial B(x_0, 1), \end{cases} \tag{6.27}$$

and

$$\begin{cases} -\Delta w = 0 & \text{in } B(x_0, 1), \\ w = u & \text{on } \partial B(x_0, 1). \end{cases} \tag{6.28}$$

We begin by estimating w. Since u is superharmonic, $w \leq u$ in $B(x_0, 1)$. By the mean-value formula (Proposition A.1.2), for any $x \in B(x_0, 1/2)$,

$$w(x) = \fint_{B(x,1/2)} w \, dy \leq \fint_{B(x,1/2)} u \, dy \leq \fint_{B(x,1/2)} e^u \, dy \leq 2, \tag{6.29}$$

where we used (6.26). Next, we estimate v. Using the maximum principle (Proposition A.2.2) and the expression of the fundamental solution of the Laplace operator (A.7), we first note that $v \leq \tilde{v}$ in $B(x_0, 1)$, where

$$\tilde{v}(x) = -\frac{1}{2\pi} \int_{\mathbb{R}^2} \ln|x - y| f(y) \, dy,$$

and $f(y) = e^{u(y)}$ for $y \in B(x_0, 1)$, $f(y) = 0$ otherwise. Now,

$$\int_{B(x_0,1)} e^{2v} \, dx \leq \int_{B(x_0,1)} e^{2\tilde{v}} \, dx =$$

$$\int_{B(x_0,1)} \exp\left(-\frac{1}{\pi} \int_{\mathbb{R}^2} \ln|x - y| f(y) \, dy\right) dx =$$

$$\int_{B(x_0,1)} \exp\left(\frac{\|f\|_{L^1(\mathbb{R}^2)}}{\pi} \int_{B(x_0,1)} -\ln|x - y| \frac{f(y)}{\|f\|_{L^1(\mathbb{R}^2)}} \, dy\right) dx.$$

Set $\delta = \|f\|_{L^1(\mathbb{R}^2)}/\pi$ and note that $\delta \leq 1/2$, by (6.26). Using Jensen's inequality with probability measure $d\mu = f/\|f\|_{L^1(\mathbb{R}^2)} dy$, it follows that

$$\int_{B(x_0,1)} e^{2v} \, dx \leq \int_{B(x_0,1)\times B(x_0,1)} |x - y|^{-\delta} \frac{f(y)}{\|f\|_{L^1(\mathbb{R}^2)}} \, dy \, dx$$

$$\leq \int_{B(0,2)} |x|^{-\delta} \, dx < +\infty. \tag{6.30}$$

Combining (6.29) and (6.30), we deduce that there exists a constant $C > 0$ independent of x_0 such that $\int_{B(x_0,1/2)} e^{2u} \, dx \leq C$ and $\int_{B(x_0,1)} v \, dx \leq C$. By elliptic regularity applied to (6.27), we deduce that v is bounded in $B(x_0, 1/4)$ by a constant C independent of x_0. By (6.29), we deduce that u is bounded above in $B(x_0, 1/4)$ by a constant C independent of x_0. Step 1 follows.
Step 2 ([86]).

$$\int_{\mathbb{R}^2} e^u \, dx \geq 8\pi. \tag{6.31}$$

For $t \in \mathbb{R}$, let $\Omega_t = \{x \in \mathbb{R}^N : u(x) > t\}$. Then,

$$\int_{\Omega_t} e^u \, dx = -\int_{\Omega_t} \Delta u = \int_{\partial \Omega_t} |\nabla u| \, d\sigma.$$

Using the coarea formula, we have for almost every t

$$-\frac{d}{dt} |\Omega_t| = \int_{\partial \Omega_t} \frac{d\sigma}{|\nabla u|}.$$

Applying the Cauchy-Schwarz inequality and the isoperimetric inequality, we also have

$$\int_{\partial \Omega_t} \frac{d\sigma}{|\nabla u|} \int_{\partial \Omega_t} |\nabla u| \, d\sigma \geq |\partial \Omega_t|^2 \geq 4\pi |\Omega_t|.$$

Hence,

$$-\frac{d}{dt} |\Omega_t| \left(\int_{\Omega_t} e^u \, dx \right) \geq 4\pi |\Omega_t|.$$

And so, using the coarea formula once more,

$$\frac{d}{dt} \left(\int_{\Omega_t} e^u \, dx \right)^2 = 2e^t \left(\frac{d}{dt} |\Omega_t| \right) \left(\int_{\Omega_t} e^u \, dx \right) \leq -8\pi e^t |\Omega_t|.$$

Integrating from $-\infty$ to $+\infty$ gives

$$-\left(\int_{\mathbb{R}^2} e^u \, dx \right)^2 \leq -8\pi \int_{\mathbb{R}^2} e^u \, dx,$$

which implies (6.31).
Step 3.

$$\lim_{|x| \to +\infty} \frac{w(x)}{\ln |x|} = \frac{1}{2\pi} \int_{\mathbb{R}^2} e^u \, dx,$$

150

where

$$w(x) = \frac{1}{2\pi} \int_{\mathbb{R}^2} \left(\ln|x - y| - \ln|y| \right) e^{u(y)} \, dy, \qquad \text{for } x \in \mathbb{R}^2.$$

Since e^u is bounded and $\int_{\mathbb{R}^2} e^u \, dx < +\infty$, w is well defined for all $x \in \mathbb{R}^2$. We want to show that

$$I := \int_{\mathbb{R}^2} \frac{\ln|x - y| - \ln|y| - \ln|x|}{\ln|x|} e^u \, dy \to 0, \qquad \text{as } |x| \to +\infty.$$

Given $\varepsilon > 0$, choose $R > 0$ so large that

$$\int_{|y| > R} e^u \, dy < \varepsilon.$$

Write $I = I_1 + I_2 + I_3$, where I_1, I_2, and I_3 are the integrals on the respective regions

$$D_1 = \{ y \in \mathbb{R}^2 : |y - x| \le 1 \},$$
$$D_2 = \{ y \in \mathbb{R}^2 : |y - x| > 1 \text{ and } |y| \le R \},$$
$$D_3 = \{ y \in \mathbb{R}^2 : |y - x| > 1 \text{ and } |y| > R \}.$$

I_1 is estimated as follows

$$|I_1| \le \frac{1}{\ln|x|} \int_{|x-y| \le 1} |\ln|x - y|| e^u \, dy + C \int_{|x-y| \le 1} e^u \, dy$$

$$\le \frac{\|e^u\|_{L^\infty(\mathbb{R}^2)}}{\ln|x|} \int_{B(0,1)} |\ln|x|| \, dx + C \int_{|y| \ge |x| - 1} e^u \, dy \to 0, \quad \text{as } |x| \to +\infty.$$

To estimate I_2, we note that $|\ln|x - y| - \ln|x|| \le C$ for $|y| \le R$ and $\int_{|y| \le R} |\ln|y|| \, dy < +\infty$. We deduce that

$$|I_2| \le \frac{\|e^u\|_{L^\infty(\mathbb{R}^2)}}{\ln|x|} \int_{|y| \le R} \left(|\ln|x - y| - \ln|x|| + |\ln|y|| \right) dy \to 0, \text{ as } |x| \to +\infty.$$

To estimate I_3, we note that for $|x - y| > 1$ and $|y| > R$,

$$\left| \frac{\ln|x - y| - \ln|y| - \ln|x|}{\ln|x|} \right| \le C.$$

Hence,
$$|I_3| \leq C \int_{|y|>R} e^u \, dy \leq C\varepsilon.$$

We deduce that $\limsup_{|x|\to+\infty} |I| \leq C\varepsilon$. Since $\varepsilon > 0$ is arbitrary, Step 3 follows.
Step 4. End of proof.
Clearly, $\Delta w = e^u$ in $\mathbb{R}^2$. So, the function $v = u + w$ is harmonic and by Steps 1 and 3,
$$v(x) \leq C + C_1 \ln(|x| + 1),$$
for some constants $C, C_1 > 0$. This implies
$$|v(x)| \leq |v(x) - C + C_1 \ln(|x| + 1)| + C + C_1 \ln(|x| + 1)$$
$$\leq 2(C + C_1 \ln(|x| + 1)) - v(x).$$

Since v is harmonic, so is $\partial_i v$. Using the mean-value formula (Proposition A.2.2), we deduce that for any $R > 0$,
$$|\partial_i v(x)| = \left| \fint_{B(x,R)} \partial_i v \, dy \right| = \frac{1}{|B_R|} \left| \int_{\partial B(x,R)} v n_i \, d\sigma \right| \leq \frac{C}{R} \fint_{\partial B(x,R)} |v| \, d\sigma$$
$$\leq \frac{C}{R} \fint_{\partial B(x,R)} \left[2(C + C_1 \ln(|y| + 1)) - v \right] d\sigma(y)$$
$$= \frac{C}{R} \left(\fint_{\partial B(x,R)} \left[2(C + C_1 \ln(|y| + 1)) \right] d\sigma(y) - v(x) \right)$$
$$\leq C \frac{(1 + \ln(|x| + R + 1)) - v(x)}{R},$$

where we used the mean-value formula again in the penultimate line. Letting $R \to +\infty$, we deduce that v is constant, so that
$$\lim_{|x|\to+\infty} \frac{u(x)}{\ln|x|} = \frac{1}{2\pi} \int_{\mathbb{R}^2} e^u \, dx.$$

Applying (6.31), the lemma follows. $\square$

Proof of Theorem 6.3.2. By Lemma 6.3.1, u achieves its maximum at some point $x_0 \in \mathbb{R}^2$. Replacing u by $u(x + x_0)$ if needed, we may always assume that $x_0 = 0$. Our goal is to apply the moving-planes device to prove that u is radially symmetric. Since for given $\alpha \in \mathbb{R}$, the solution to the ODE
$$\begin{cases} -u'' - \dfrac{1}{r}u' = e^u, \\ u'(0) = 0, \\ u(0) = \alpha, \end{cases}$$

is unique, (6.23) will follow. So, we are left with proving that u is radially symmetric. It suffices to prove that u is symmetric in any given direction. Choose one. Rotating space if necessary, we may work with the x_1 direction and show only that

$$u(x_1, x_2) \geq u(-x_1, x_2), \qquad \text{whenever } x_1 \leq 0. \qquad (6.32)$$

Given $\lambda > 0$, set

$$T_\lambda = \{x = (x_1, x_2) \in \mathbb{R}^2 : x_1 = \lambda\} \qquad \text{and} \qquad \Sigma_\lambda = \{x \in \mathbb{R}^2 : x_1 < \lambda\}.$$

For $x = (x_1, x_2) \in \mathbb{R}^2$, let x^λ denote the reflection of x with respect to the plane T_λ, that is, $x^\lambda = (2\lambda - x_1, x_2)$.
Step 1. There exists $R_1 > 0$ such that $u(x) > u(x^\lambda)$ in Σ_λ for $\lambda \geq R_1$.
  To see this, given $\lambda > 0$, we note that $v_\lambda(x) = u(x_\lambda) - u(x)$ satisfies

$$-\Delta v_\lambda - V_\lambda(x) v_\lambda = 0 \qquad \text{in } \Sigma_\lambda, \qquad (6.33)$$

where $V_\lambda(x) = \int_0^1 e^{tu(x_\lambda) + (1-t)u(x)}\, dt$. Since $|x_\lambda| > |x|$ for all $x \in \Sigma_\lambda$, Lemma 6.3.1 implies that for some $R_0 > 1$ independent of λ, there holds

$$V_\lambda(x) \leq \frac{1}{|x|(1 + |x|^2)\ln(|x| - 1)}, \qquad \text{in } \Sigma_\lambda \setminus \overline{B_{R_0}}. \qquad (6.34)$$

In particular, $w(x) := \ln(|x| - 1)$ satisfies

$$\begin{cases} -\Delta w - V_\lambda(x) w \geq 0 & \text{in } \Sigma_\lambda \setminus \overline{B_{R_0}}, \\ \quad\quad\quad\quad\quad\;\; w > 0 & \text{in } \Sigma_\lambda \setminus \overline{B_{R_0}}. \end{cases} \qquad (6.35)$$

By Lemma 6.3.1, $\lim_{|x| \to +\infty} u(x) = -\infty$. Then there exists $R_1 > R_0$ such that $\max_{|x| \geq R_1} u < \min_{|x| \leq R_0} u$. And so, for $\lambda \geq R_1$,

$$v_\lambda > 0 \qquad \text{in } \overline{B_{R_0}}. \qquad (6.36)$$

It remains to prove that $v_\lambda \geq 0$ in $\Sigma_\lambda \setminus \overline{B_{R_0}}$. To this end, we observe that

$$\begin{cases} -\Delta v_\lambda - V_\lambda(x) v_\lambda = 0 & \text{in } \Sigma_\lambda \setminus \overline{B_{R_0}}, \\ \quad\quad\quad\quad\quad\quad v_\lambda > 0 & \text{on } \partial B_{R_0}. \\ \displaystyle \lim_{|x| \to +\infty} \frac{v_\lambda(x)}{w(x)} = 0. \end{cases} \qquad (6.37)$$

The first equation follows from (6.33), the second from (6.36), and the third from the expression of w and Lemma 6.3.1. Using the maximum principle (Proposition A.7.2) and the strong maximum principle (Corollary A.5.1), Step 1 follows.

Step 2. For $\lambda \in \Lambda = \{\lambda > 0 \; : \; \forall \mu \geq \lambda, \; u(x) > u(x^{\mu}) \text{ in } \Sigma_{\mu}\}$, there holds

$$\frac{\partial u}{\partial x_1} < 0 \qquad \text{in } \mathbb{R}^2 \setminus \Sigma_{\lambda}. \tag{6.38}$$

Indeed, for $\mu \geq \lambda$, v_{μ} solves

$$\begin{cases} -\Delta v_{\mu} = V_{\mu}(x)v_{\mu} \geq 0 & \text{in } \Sigma_{\mu}, \\ v_{\mu} > 0 & \text{in } \Sigma_{\mu}, \\ v_{\mu} = 0. & \text{on } T_{\mu} = \partial \Sigma_{\mu}. \end{cases}$$

By the boundary point lemma (Proposition A.4.1), we deduce that

$$2\frac{\partial u}{\partial x_1}(\mu, x_2) = \frac{\partial v_{\mu}}{\partial x_1}(\mu, x_2) < 0,$$

as desired.

Step 3. $\inf \Lambda = 0$.

Assume by contradiction that $\lambda_0 = \inf \Lambda > 0$. By continuity of u, $v_{\lambda_0} \geq 0$ in Σ_{λ_0} and by the strong maximum principle, either $v_{\lambda_0} \equiv 0$, or $v_{\lambda_0} > 0$ in Σ_{λ_0}. In the former case, the function $v(x) = u(\lambda_0 + x)$ is even. Using Step 2, we deduce that $\partial u/\partial x_1(0) = -\partial u/\partial x_1(2\lambda_0, 0) > 0$. This contradicts our initial assumption that 0 is a point of maximum of u.

So, $v_{\lambda_0} > 0$ in Σ_{λ_0}. By definition of λ_0, there exists a sequence $\lambda_n \nearrow \lambda_0$ and a sequence $x_n \in \Sigma_{\lambda_n}$ such that $v_{\lambda_n}(x_n) \leq 0$. It follows that $v_{\lambda_n}(y_n) \leq 0$ for some sequence $y_n \in \Sigma_{\lambda_n} \cap \overline{B_{R_0}}$. Otherwise, we would have $v_{\lambda_n} > 0$ in $\Sigma_{\lambda_n} \cap \overline{B_{R_0}}$ and by the strong maximum principle applied to (6.37) with $\lambda = \lambda_n$, we would deduce that $v_{\lambda_n} > 0$ in Σ_{λ_n}, contradicting $v_{\lambda_n}(x_n) \leq 0$. Since $v_{\lambda_n}(y_n) \leq 0$, there exists a point z_n in the segment joining y_n to $y_n^{\lambda_n}$ such that $\partial u/\partial x_1(z_n) \geq 0$. Finally, since (y_n) is bounded, so is (z_n) and a subsequence $z_{n_k} \to z_0$, with $z_0 \in T_{\lambda_0}$, $v_{\lambda_0}(z_0) = 0$, and $\partial u/\partial x_1(z_0) \geq 0$. This contradicts the boundary point lemma. Equation (6.32) follows. $\qquad \square$

6.3.2 The supercritical range

Theorem 6.3.3 ([71]) *Let* $3 \leq N \leq 9$. *There is no solution* $u \in C^2(\mathbb{R}^N)$ *of (6.6), which is stable outside a compact set.*

Remark 6.3.1 *As observed in Remark 6.2.1, the above theorem ceases to hold for $N \geq 10$.*

Proof. Take a standard cutoff function $\phi \in C_c^1(\mathbb{R})$, that is, $0 \leq \phi \leq 1$ in $\mathbb{R}$ and (6.10) holds. For $s > 0$, also let $\theta_s \in C_c^1(\mathbb{R})$, such that $0 \leq \theta_s \leq 1$ in $\mathbb{R}$ and

$$\theta_s(t) = \begin{cases} 0 & \text{if } |t| \leq s+1, \\ 1 & \text{if } |t| \geq s+2. \end{cases} \tag{6.39}$$

The proof of the theorem is by contradiction and proceeds in four steps. Let us suppose that the equation admits a C^2 solution, which is stable outside a compact set. Then,

Step 1. There exists $R_0 > 0$ such that

- For every $\alpha \in (0,2)$ and every $r > R_0 + 3$ we have

$$\int_{[R_0+2<|x|<r]} e^{(2\alpha+1)u} \, dx \leq A + Br^{N-2(2\alpha+1)}, \tag{6.40}$$

  where A, B are positive constants depending only on α, N, R_0, but not on r.

- For every $\alpha \in (0,2)$ and every open ball $B = B(y, 2R)$ such that $B \subset \{x \in \mathbb{R}^N : |x| > R_0\}$, we have

$$\int_B e^{(2\alpha+1)u} \, dx \leq CR^{N-2(2\alpha+1)}, \tag{6.41}$$

  where C is a positive constant depending only on α, N, R_0, but neither on R nor y.

To prove this, since u is stable outside a compact set, there exists $R_0 > 0$ such that Proposition 6.2.1 holds in $\Omega = \mathbb{R}^N \setminus \overline{B(0, R_0)}$. We fix $m = 5$ and, for every $r > R_0 + 3$, consider the test function $\xi_r \in C_c^1(\mathbb{R}^N)$ defined by

$$\xi_r(x) = \begin{cases} \theta_{R_0}(|x|) & \text{if } |x| \leq R_0 + 3, \\ \phi\left(\dfrac{|x|}{r}\right) & \text{if } |x| \geq R_0 + 3. \end{cases} \tag{6.42}$$

Applying Proposition 6.2.1 with $\psi = \xi_r$ yields

$$\int_{[R_0+2<|x|<r]} e^{(2\alpha+1)u}\,dx \leq \int_{\mathbb{R}^N \backslash \overline{B(0,R_0)}} e^{(2\alpha+1)u}\,dx \leq$$

$$\leq C \int_{\mathbb{R}^N} \left|\nabla \xi_r\right|^{2(2\alpha+1)}\,dx \leq$$

$$\leq C \left[\int_{[|x|\leq R_0+2]} \left|\nabla \theta_{R_0}\right|^{2(2\alpha+1)}\,dx + \int_{[r\leq|x|\leq 2r]} \left|\nabla \xi_r\right|^{2(2\alpha+1)}\,dx \right]$$

$$\leq C_1 + C_2 r^{N-2(2\alpha+1)},$$

for all $r > R_0 + 3$. Equation (6.40) follows. The estimate (6.41) is obtained similarly by using the test function $\psi_{R,y}(x) = \phi\left(\frac{|x-y|}{R}\right)$ in Proposition 6.2.1.

Step 2. Let $\eta > 0$. Then, there exists $R_1 = R_1(N, \eta, u) > 0$ such that

$$\int_{[|x|>R_1]} e^{\frac{N}{2}u}\,dx \leq \eta^{\frac{N}{2}}. \tag{6.43}$$

Let $\alpha_1 = \frac{N-2}{4} \in (0,2)$. By (6.40) we infer that, for all $r > R_0 + 3$,

$$\int_{[R_0+2<|x|<r]} e^{\frac{N}{2}u}\,dx = \int_{[R_0+2<|x|<r]} e^{(2\alpha_1+1)u}\,dx \leq A + B r^{N-2(2\alpha_1+1)}.$$

So, $\int_{[|x|>R_0+2]} e^{\frac{N}{2}u}\,dx < +\infty$, which immediately yields (6.43).

Step 3.

$$\lim_{|x|\to+\infty} |x|^2 e^{u(x)} = 0.$$

Set $\epsilon = 1/10$ and observe that $\frac{N}{2-\epsilon} \in (1,5)$. Thus, there exists $\alpha_2 = \alpha_2(N) \in (0,2)$ such that $2\alpha_2 + 1 = \frac{N}{2-\epsilon}$. Here we have used the assumption $3 \leq N \leq 9$. Next, we fix $\eta > 0$ and observe that $w = e^u$ satisfies

$$-\Delta w - e^u w \leq 0 \qquad \text{in } B(y, 2R).$$

The Serrin-Trudinger inequality (Theorem B.4.2) for positive subsolutions to the equation $-\Delta w - e^u w = 0$ asserts that, for any $t > 1$,

$$\|w\|_{L^\infty(B(y,R))} \leq C_{ST} R^{-\frac{N}{t}} \|w\|_{L^t(B(y,2R))}, \tag{6.44}$$

where C_{ST} is a positive constant depending on N, t and

$$R^\epsilon \|e^u\|_{L^{\frac{N}{2-\epsilon}}(B(y,2R))}.$$

In order to apply the above result, we consider points $y \in \mathbb{R}^N$ such that $|y| > 10R_1$ and set $R = |y|/4$, $t = N/2 > 1$. Here, $R_1 > R_0$ is defined by (6.43) of Step 2. This choice yields

$$B(y, 2R) \subset \{x \in \mathbb{R}^N \, : \, |x| > R_1\} \subset \{x \in \mathbb{R}^N \, : \, |x| > R_0\}.$$

In addition, (6.43) holds. So,

$$R^\epsilon \|e^u\|_{L^{\frac{N}{2-\epsilon}}(B(y,2R))} = R^\epsilon \left(\int_{B(y,2R)} e^{\frac{N}{2-\epsilon}u} \right)^{\frac{2-\epsilon}{N}} =$$

$$= R^\epsilon \left(\int_{B(y,2R)} e^{(2\alpha_2+1)u} \right)^{\frac{2-\epsilon}{N}} \le R^\epsilon \left[CR^{N-2(2\alpha_2+1)} \right]^{\frac{2-\epsilon}{N}} \le C'R^\epsilon R^{2-\epsilon}R^{-2} = C',$$

where in the latter we have used (6.41). This proves that the constant C_{ST} is independent of both y and R. Actually, it depends only on N and R_0. Now, using $t = N/2$ in (6.44) and Step 2, we are led to

$$\left| e^{u(y)} \right| \le C_{ST} R^{-2} \|e^u\|_{L^{N/2}(B(y,2R))} \le 16 C_{ST} \left| y \right|^{-2} \|e^u\|_{L^{N/2}(B(y,2R))} \le C_2 \left| y \right|^{-2} \eta,$$

which proves Step 3.

Step 4. End of the proof. By Step 3, there exists $R_2 > 0$ such that the function $v = v(|x|)$, defined as the mean-value of the solution u over spheres of radii $|x| > 0$, satisfies

$$-\Delta v \le \frac{1}{2r^2} \qquad \text{for all } r > R_2.$$

Hence, the radial function v verifies

$$v'(r) \ge \frac{C(N)}{r^{N-1}} - \frac{1}{2(N-2)r} \qquad \text{for all } r > R_2$$

and thus

$$v'(r) \ge -\frac{1}{r} \qquad \text{for all } r > R_3,$$

for some $R_3 > R_2$. Integrating the latter and taking the exponential, we get

$$r^2 e^{v(r)} \ge Cr \qquad \text{for all } r > R_3, \tag{6.45}$$

where C is a positive constant independent of r. Finally, we observe that (6.45) contradicts Step 3. Indeed, by Jensen's inequality, we have for all $r > R_3$,

$$\max_{|x|=r} |x|^2 e^{u(x)} = r^2 \max_{|x|=r} e^{u(x)} \geq \frac{r^2}{|[|x|=r]|} \int_{[|x|=r]} e^u \, d\sigma \geq r^2 e^{v(r)} \geq cr,$$

which clearly contradicts the conclusion of Step 3. $\qquad\square$

6.3.3 Flat nonlinearities

The previous classification results remain valid for a greater class of nonlinearities (see [97] for the Lane-Emden nonlinearity, as well as [88] for more general results). However, the following example teaches us that no Liouville theorem can hold for general f (even in the restricted class of convex functions), if the solution u is only assumed to be stable outside a compact set.

Example 6.3.1 ([89]) Let $N \geq 3$ and $1 < p < \frac{N+2}{N-2}$. For every $R > 0$ there exists $\beta = \beta(R, p, N) > 0$ such that the equation

$$-\Delta u = [(u - \beta)^+]^p \qquad in \quad \mathbb{R}^N \qquad (6.46)$$

admits a solution $u = u_R \in C^2(\mathbb{R}^N)$ satisfying:

(i) u is positive, radially symmetric, and strictly radially decreasing.

(ii) $u(R) = \beta$, $\lim_{r \to +\infty} u(r) = 0$, where $r = |x|$.

(iii) The Morse index of u is finite and positive.

Furthermore one has that $\beta(R, p, N) = \beta(1, p, N) R^{-\frac{2}{p-1}}$ where $\beta(1, p, N) = \frac{1}{N-2} \int_0^1 \phi_1^p(s) s^{N-1} ds > 0$ and ϕ_1 is the unique positive radial solution to

$$\begin{cases} -\Delta \phi_1 = \phi_1^p & in \quad B_1, \\ \phi_1(1) = 0. \end{cases}$$

In particular, for every $\beta > 0$, Equation (6.46) admits a solution u satisfying the above properties.

Proof. Fix $R > 0$. Since $N \geq 3$ and $1 < p < \frac{N+2}{N-2}$, there exists a unique radial function $\phi_R = \phi_R(r) \in C^2(\overline{B_R})$ solution to

$$\begin{cases} -\Delta\phi_R = \phi_R^p & \text{in } B_R, \\ \phi_R(R) = 0, \\ \phi_R > 0 & \text{in } B_R, \\ \phi_R' < 0 & \text{in } \overline{B_R} \setminus \{0\}. \end{cases} \tag{6.47}$$

Now we consider the function

$$v_R = v_R(r) = \begin{cases} \phi_R(r) & \text{for } 0 \leq r \leq R, \\ \alpha r^{2-N} - \beta & \text{for } r \geq R, \end{cases}$$

with $\alpha = \frac{1}{N-2} \int_0^R \phi_R^p(s) s^{N-1} ds > 0$ and $\beta = \alpha R^{2-N}$. Set $h_R(r) = \alpha r^{2-N} - \beta$. The above function v_R is clearly continuous, nonnegative, radially symmetric, strictly radially decreasing, and satisfies $\lim_{r \to +\infty} v_R(r) = -\beta < 0$. Furthermore v_R belongs to $C^2(\mathbb{R}^N)$ and solves the equation

$$-\Delta v = (v^+)^p \qquad \text{in } \mathbb{R}^N. \tag{6.48}$$

To check that v_R is of class C^2 we observe that, by integrating the ODE satisfied by ϕ_R, we get

$$\phi_R'(R) = -R^{1-N} \int_0^R \phi_R^p(s) s^{N-1} ds = (2 - N)\alpha R^{1-N} = h_R'(R).$$

Moreover,

$$\phi_R''(R) = -\phi_R^p(R) - \frac{(N-1)}{R}\phi_R'(R) = 0 + (1-N)(2-N)\alpha R^{-N} = h''(R).$$

Therefore v_R is in $C^2(\mathbb{R}^N)$ and solves Equation (6.48).

The desired function u is then obtained by setting $u := u_R = v_R + \beta$. Then, u belongs to $C^2(\mathbb{R}^N)$, solves Equation (6.46) and, by making use of the properties of v_R, it satisfies (i) and (ii). To prove (iii) we first show that the Morse index of u_R is at least 1. To this end we multiply Equation (6.46) by the function $v_R^+ \in C_c^{0,1}(\mathbb{R}^N)$ (note that $v_R^+ = (u - \beta)^+ > 0$) and integrate by parts to obtain:

$$\int_{\mathbb{R}^N} |\nabla v_R^+|^2 \, dx = \int_{\mathbb{R}^N} \nabla u \nabla v_R^+ \, dx = \int_{\mathbb{R}^N} [(u-\beta)^+]^{p+1} \, dx =$$

$$= \int_{\mathbb{R}^N} [(u-\beta)^+]^{p-1}(v_R^+)^2 \, dx < \int_{\mathbb{R}^N} p[(u-\beta)^+]^{p-1}(v_R^+)^2 \, dx. \tag{6.49}$$

Hence,

$$Q_u(v_R^+) = \int_{\mathbb{R}^N} |\nabla v_R^+|^2 \, dx - \int_{\mathbb{R}^N} p[(u-\beta)^+]^{p-1}(v_R^+)^2 \, dx < 0.$$

The latter proves that the Morse index of u is at least 1. To prove that u has finite Morse index, we work as follows. Since

$$\text{ind}_{\mathbb{R}^N}(u) = \sup_{n \in \mathbb{N}^\star} \text{ind}_{B_n}(u), \tag{6.50}$$

it is enough to bound, independently of n, each one of the quantities $\text{ind}_{B_n}(u)$. To do so, we first observe that $\text{ind}_{B_n}(u)$ is the number of negative Dirichlet eigenvalues of the operator $-\Delta - p[(u-\beta)^+]^{p-1}$ in B_n and next we invoke the Cwikel-Lieb-Rozenbljum formula (see [64], [142], [187], as well as [141] for the version of the formula used here). This formula says that the number of negative Dirichlet eigenvalues of the operator $-\Delta - V$ in B_n is bounded by

$$\left(\frac{4e}{N(N-2)}\right)^{\frac{N}{2}} (\omega_{N-1})^{-1} \int_{B_n} V^{\frac{N}{2}} \, dx, \tag{6.51}$$

provided $N \geq 3$ and V is a nonnegative potential belonging to $L^{\frac{N}{2}}(B_n)$. The desired conclusion then follows by observing that

$$\forall \, n > R \quad \int_{B_n} p^{\frac{N}{2}}[(u-\beta)^+]^{\frac{N(p-1)}{2}} = \int_{B_R} p^{\frac{N}{2}} \phi_R^{\frac{N(p-1)}{2}}$$

$$= C = C(p, N, \phi_R) < +\infty, \tag{6.52}$$

where C is independent of n since $(u-\beta)^+$ is supported in $\overline{B_R}$.

To conclude the proof we observe that $\phi_R(r) = R^{-\frac{2}{p-1}}\phi_1(R^{-1}r)$ and that $\beta = \beta(R,p,N) = R^{2-N}\frac{1}{N-2}\int_0^R \phi_R^p(r)r^{N-1}dr = \beta(1,p,N)R^{-\frac{2}{p-1}}$ by making the change of variable $s = Rr$.

$\square$

Remark 6.3.2 *(i) The function $f(t) = [(t-\beta)^+]^p$, $\beta > 0$ is nonnegative, nondecreasing, convex, and of class C^1. It is of class C^2 for $p > 2$. The latter is always possible in dimension $N = 3, 4, 5$ (recall that p is subcritical).*
(ii) Let $\beta > 0$. Any solution to

$$\begin{cases} -\Delta w = [(w-\beta)^+]^p & \text{in } \mathbb{R}^N, \\ w > 0 & \text{in } \mathbb{R}^N, \\ \lim_{|x|\to+\infty} w(x) = 0, \end{cases} \tag{6.53}$$

must be one of the functions u_R built in Example 6.3.1 (up to translation).

Proof. Indeed, (up to translation) by a theorem of Gidas, Ni, and Nirenberg, w must be radially symmetric and strictly radially decreasing. Let $R = R(\beta) > 0$, the only value for which $w(R) = \beta$ (such a value always exists since a bounded nonconstant solution to (6.53) must satisfy: $\sup w > \beta$, otherwise it would be harmonic and hence constant). Now, it is clear that $w = u_{R(\beta)}$. Indeed, $w - \beta$ is radial and harmonic for $r \geq R(\beta)$ and hence $w - \beta = Ar^{2-N} + B$, where A and B are constants. The assumption $\lim_{|x| \to +\infty} w(x) = 0$ implies $B = \beta$ and $w = Ar^{2-N}$ for $r \geq R(\beta)$ and $A > 0$. In addition, $w - \beta$ is a solution to (6.47) and thus it must be equal to $\phi_{R(\beta)}$ on $B_{R(\beta)}$. Combining this information and using the continuity of w and w', we have $w = u_{R(\beta)}$.   □

(iii) In particular, for every $\beta > 0$ problem (6.53) admits a unique solution (up to translation). This solution necessarily coincides with $u_{R(\beta)}$. Clearly the value $R(\beta)$ can be explicitly calculated by using the formula for β given in Example 6.3.1.

(iv) Note that any solution to (6.46) (not necessarily positive) converging uniformly to some constant $\gamma < \beta$ is automatically stable outside a compact set of $\mathbb{R}^N$. This follows by observing that f (and thus f') is zero on $(-\infty, \beta)$. Arguing as in the above proof one can prove that such u must have finite Morse index.

Chapter 7

A conjecture of De Giorgi

7.1 Statement of the conjecture

In this chapter, we discuss a celebrated conjecture, due to De Giorgi ([81]):

Conjecture 7.1.1 ([81]) *Let $N \geq 1$. Let $u \in C^2(\mathbb{R}^N, [-1, 1])$ satisfy*

$$-\Delta u = u - u^3 \qquad and \qquad \frac{\partial u}{\partial x_N} > 0 \qquad in \ all \ of \ \mathbb{R}^N. \qquad (7.1)$$

Then, the level sets of u are hyperplanes, at least if $N \leq 8$.

Note that if the level sets of u are hyperplanes, then they must be parallel (no two level sets can cross). So the conclusion of the conjecture is equivalent to requesting that u is one-dimensional (that is, u is a function of one variable only; see Definition 6.2.1). One-dimensional solutions of (7.1) are easily classified, as the following proposition demonstrates.

Proposition 7.1.1 *Let $N \geq 1$ and $u \in C^2(\mathbb{R}^N, [-1, 1])$. Assume that u is a one-dimensional solution to (7.1). Then, there exists a unit vector $\tau \in \mathscr{S}^{N-1}$ and a constant $\alpha \in \mathbb{R}$ such that*

$$u(x) = g_0(\tau \cdot x + \alpha), \qquad for \ all \ x \in \mathbb{R}^N,$$

where $g_0(s) = \tanh(s/\sqrt{2})$.

Proof. Since u is one-dimensional, there exists a vector $\tau \in \mathscr{S}^{N-1}$ and a function $h \in C^2(\mathbb{R})$ such that $u(x) = h(\tau \cdot x)$. In addition,

$$-h'' = h - h^3 \quad and \quad h' > 0 \quad in \ \mathbb{R}. \qquad (7.2)$$

Note that since h is increasing, there exists constants $m_-, m_+ \in [-1, 1]$ such that $\lim_{\pm\infty} h = m_\pm$. In particular, $\liminf_{\pm\infty} h' = 0$. By the mean-value theorem, we deduce that $\liminf_{\pm\infty} h'' = 0$. Using (7.2), we deduce that

$$m_\pm - m_\pm^3 = 0. \tag{7.3}$$

Multiply (7.2) by $2h'$ and integrate: there exists a constant $c \in \mathbb{R}$ such that

$$(h')^2 = \frac{1}{2}\left(1 - h^2\right)^2 - c \tag{7.4}$$

and so $c = \frac{1}{2}\left(1 - m_\pm^2\right)^2$. In particular, $m_+^2 = m_-^2$ and since h is increasing, it follows, using (7.3), that $m_- = -1$, $m_+ = 1$ and so $c = 0$. Let $m = h(0)$. By direct integration, we see that the unique increasing solution to (7.4) such that $h(0) = m$ is given by

$$\sqrt{2} \int_m^{h(t)} \frac{ds}{1 - s^2} = t.$$

One can easily check that $h(t) = g_0(t + \alpha)$, where $\alpha \in \mathbb{R}$ is such that $g_0(\alpha) = m$ is the desired solution. $\qquad \square$

7.2 Motivation for the conjecture

In this section, we give a possible motivation for the conjecture of De Giorgi 7.1.1. We warn the reader that full proofs will not be given; a thorough investigation would be beyond the scope of this book. This should however provide us enough insight on the conjecture to move on to its proof in dimensions $N = 2$ and $N = 3$, which, as we are about to discover, uses stability as a central tool.

7.2.1 Phase transition phenomena

The Allen-Cahn equation appearing in (7.1) arises as a crude model in the theory of phase transitions. Consider a pure body, contained in a bounded region Ω of space, which state may change from one (say, thermodynamical) phase to another. To each of these phases, we assign a given value, say $u = -1$ and $u = +1$, while the transient state of the body will be assigned a value $u \in (-1, +1)$. In the informal discussion that follows, we are interested in the description of the interface between these two states. We argue that such an interface should be "close" to a surface of minimal area.

The system tends to minimize an energy functional, which should favor the states ± 1. Take W, a double-well potential having minimal value at $u = \pm 1$, for example,

$$W(u) = \frac{1}{4}(1 - u^2)^2. \tag{7.5}$$

We might be tempted at first to model our problem by minimizing an energy of the form $E(u) = \int_{\Omega} W(u)\, dx$ among all density distributions u of prescribed total mass $\int_{\Omega} u\, dx = m$. One quickly realizes that any solution to such a minimization problem must be piecewise constant. In addition, there are infinitely many solutions, with no restriction on the interface between the sets $[u = -1]$ and $[u = +1]$. In particular, there is no way to recover the physically reasonable criterion that the interface has minimal area.

Accordingly, an interfacial energy must be added. This can be done by considering the Ginzburg-Landau energy

$$\mathscr{E}_{\Omega}(u) = \frac{1}{2} \int_{\Omega} |\nabla u|^2 \, dx + \int_{\Omega} W(u)\, dx, \tag{7.6}$$

which the Euler-Lagrange equation is given by

$$-\Delta u = u - u^3 \qquad \text{in } \Omega. \tag{7.7}$$

It can be argued that the interfacial energy should be relatively small, compared to the potential term, that is, we should rather consider the rescaled energy

$$\mathscr{E}_{\varepsilon,\Omega}(u) = \frac{\varepsilon}{2} \int_{\Omega} |\nabla u|^2 \, dx + \frac{1}{\varepsilon} \int_{\Omega} W(u)\, dx. \tag{7.8}$$

Note that u minimizes $\mathscr{E}_{\Omega/\varepsilon}$ if and only if $u_\varepsilon(x) := u(x/\varepsilon)$ minimizes $\mathscr{E}_{\varepsilon,\Omega}$. Using Young's inequality and the coarea formula, there holds

$$\mathscr{E}_{\varepsilon,\Omega}(u_\varepsilon) \geq \int_{\Omega} \sqrt{2W(u_\varepsilon)}\,|\nabla u_\varepsilon|\, dx = \int_{-1}^{1} \sqrt{2W(s)}\mathscr{H}^{N-1}([u_\varepsilon = s])\, ds,$$

for any $u_\varepsilon \in C^2(\overline{\Omega}, [-1, 1])$. In addition, the inequality is an equality, provided

$$|\nabla u_\varepsilon| = \frac{1}{\varepsilon}\sqrt{2W(u_\varepsilon)}. \tag{7.9}$$

So, heuristically, if (7.9) holds and if the level sets $[u_\varepsilon = s]$ are surfaces of minimal area, then u_ε should be a local minimizer of (7.8). Now, (7.9) implies that ∇u_ε has constant length along any given level set and so the level sets of u_ε must be parallel. Assume that the level set $\Gamma = [u_\varepsilon = 0]$ is a smooth hypersurface and let d_Γ denote the distance to Γ. Writing $u_\varepsilon(x) = g(d_\Gamma(x))$, (7.9) reduces to

$$g' = \frac{1}{\varepsilon}\sqrt{2W(g)}.$$

Working as in Proposition 7.1.1, we deduce that $g(s) = g_0\left(\frac{s}{\varepsilon} + \alpha\right)$ for some $\alpha \in \mathbb{R}$, where $g_0(s) = \tanh(s/\sqrt{2})$.

Unfortunately, the above discussion is not completely correct: the level sets of u_ε need not be surfaces of minimal area. It can be shown however that these level sets are close to a surface of minimal area: for a sequence (u_{ε_k}), there exists a set E of minimal perimeter such that $u_{\varepsilon_k} \to \chi_E - \chi_{E^c}$ in L^1_{loc}, as $\varepsilon_k \to 0$ (see [156]).

7.2.2 Monotone solutions and global minimizers

In the previous section, we saw how the Allen-Cahn equation (7.7) appears naturally in the description of phase transition phenomena.

But why did De Giorgi state his conjecture for monotone solutions? To gain further insight, let us return to the study of minimizers of $\mathscr{E}_{\varepsilon,\Omega}$. Recall that u minimizes $\mathscr{E}_{\Omega/\varepsilon}$, given by Equation (7.6), if and only if $u_\varepsilon(x) := u(x/\varepsilon)$ minimizes $\mathscr{E}_{\varepsilon,\Omega}$, given by Equation (7.8). So, to find a minimizer of $\mathscr{E}_{\varepsilon,\Omega}$ for arbitrary $\varepsilon > 0$, it suffices to exhibit a function $u \in C^2(\mathbb{R}^N)$ that minimizes $\mathscr{E}_{\Omega/\varepsilon}$ for all $\varepsilon > 0$.

Definition 7.2.1 *A function $u \in C^2(\mathbb{R}^N)$ is said to be a global minimizer of the Ginzburg-Landau energy functional if*

$$\mathscr{E}_\Omega(u) \leq \mathscr{E}_\Omega(u + \varphi)$$

for every $\varphi \in C^1_c(\Omega)$ and for every bounded domain $\Omega \subset \mathbb{R}^N$, where $\mathscr{E}_\Omega$ is given by (7.6).

Global minimizers and monotone solutions are related through the following theorem.

Theorem 7.2.1 ([3]) *Let $N \geq 1$ and let $u \in C^2(\mathbb{R}^N; [-1,1])$ denote a monotone solution to (7.1). In addition, assume that*

$$\lim_{x_N \to \pm\infty} u(x', x_N) = \pm 1, \qquad \text{for every } x' \in \mathbb{R}^{N-1}. \tag{7.10}$$

Then, u is a global minimizer of $\mathscr{E}_\Omega$, defined by (7.6).

Remark 7.2.1 *Note that the additional requirement (7.10) is compatible with the De Giorgi conjecture (Conjecture 7.1.1). Note also that in (7.10), we do not require the limits to be uniform in the x' variable.*

Proof. We use a foliation argument of Weierstraß (see [122]). Consider the family of functions u_τ defined for $\tau \in \mathbb{R}$ by

$$u_\tau(x) = u(x', x_N + \tau), \qquad \text{for all } x = (x', x_N) \in \mathbb{R}^N.$$

Since u is monotone, the graphs of the family $(u_\tau)_\tau$ form a foliation of $\mathbb{R}^N \times [-1, 1]$, that is, the graphs are (strictly) ordered and they fill $\mathbb{R}^N \times [-1, 1]$. Also observe that by standard elliptic regularity, (u_τ) converges locally uniformly to 1, as $\tau \to +\infty$.

Now fix a bounded domain $\Omega \subset \mathbb{R}^N$ and let us prove that u is the unique solution to

$$\begin{cases} -\Delta v = v - v^3 & \text{in } \Omega, \\ v = u & \text{on } \partial\Omega. \end{cases} \qquad (7.11)$$

The following exercise will then imply that u is a global minimizer.
Exercise 7.2.1

- Prove that every solution to (7.11) satisfies $-1 < v < 1$ in $\overline{\Omega}$.

- Check that there exists an absolute minimizer of the energy (7.6) subject to the boundary condition $v = u$ on $\partial\Omega$.

It remains to be proven that u is the unique solution to (7.11). Assume by contradiction that there exists a solution $v \neq u$ of (7.11). Then, $-1 < v < 1$ in $\overline{\Omega}$ and so the set

$$\Lambda = \{\tau > 0 : v \leq u_\tau \quad \text{in } \overline{\Omega}\}$$

is nonempty. Let

$$\tau_0 = \inf \Lambda.$$

Since $v \neq u$ but $v = u$ on $\partial\Omega$, there exists a point $x_0 \in \Omega$ such that $v(x_0) \neq u(x_0)$. Say $v(x_0) > u(x_0)$ (the reverse inequality can be treated similarly). In particular, $\tau_0 > 0$. In addition, by definition of τ_0, $v \leq u_{\tau_0}$ in $\overline{\Omega}$, and there exists a point $x_1 \in \overline{\Omega}$ such that $v(x_1) = u_{\tau_0}(x_1)$. Since $u_{\tau_0}|_{\partial\Omega} > u|_{\partial\Omega} = v|_{\partial\Omega}$, we deduce that $x_1 \in \Omega$. By the strong maximum principle, we must have $v \equiv u_{\tau_0}$, which contradicts $u_{\tau_0}|_{\partial\Omega} > v|_{\partial\Omega}$. $\qquad \square$

Theorem 7.2.1 can be combined to useful energy estimates that hold for global minimizers.

Corollary 7.2.1 *Let $N \geq 1$. Assume that $u \in C^2(\mathbb{R}^N; [-1,1])$ is a global minimizer of the Ginzburg-Landau energy (7.6). Then, there exists a constant $C > 0$, depending on N only, such that given any $R > 1$,*

$$\mathscr{E}_{B_R}(u) = \int_{B_R} \left(\frac{1}{2} |\nabla u|^2 + \frac{1}{4}(1-u^2)^2 \right) dx \leq CR^{N-1}. \tag{7.12}$$

Proof. Given $R > 1$, take a cutoff function $\varphi_R \in C_c^1(B_{R+1})$ such that $\varphi_R \equiv 1$ in B_R, $0 \leq \varphi_R \leq 1$ and $|\nabla \varphi_R| \leq C$. Let $v = \varphi_R + u(1 - \varphi_R)$. Since u is a global minimizer,

$$\mathscr{E}_{B_R}(u) \leq \mathscr{E}_{B_R}(v) = \int_{B_R \backslash B_{R-1}} \left(\frac{1}{2} |\nabla v|^2 + \frac{1}{4}(1-v^2)^2 \right) dx \leq CR^{N-1}.$$

$\square$

Note that Corollary 7.2.1 is optimal: inequality (7.12) is an equality if u is a one-dimensional solution. The following monotonicity formula completes the description of the energy $\mathscr{E}_{B_R}(u)$.

Theorem 7.2.2 ([157]) *Let $N \geq 1$. Let $u \in C^3(\mathbb{R}^N)$ denote a bounded solution to the Allen-Cahn equation*

$$-\Delta u = u - u^3 \qquad in \ \mathbb{R}^N. \tag{7.13}$$

Then,

$$\Phi(R) = R^{1-N} \mathscr{E}_{B_R}(u) \tag{7.14}$$

is a nondecreasing function of R.

Proof. We exploit Pohozaev's identity. By (8.8),

$$\int_{B_R} \Delta u(x \cdot \nabla u) \, dx = \frac{N-2}{2} \int_{B_R} |\nabla u|^2 \, dx - \frac{R}{2} \int_{\partial B_R} |\nabla u|^2 \, d\sigma + R \int_{\partial B_R} \left(\frac{\partial u}{\partial r} \right)^2 d\sigma.$$

Since u solves (7.13), we also have

$$\int_{B_R} \Delta u(x \cdot \nabla u) \, dx = -\int_{B_R} (u - u^3)(x \cdot \nabla u) \, dx =$$

$$-N \int_{B_R} W(u) \, dx + R \int_{\partial B_R} W(u) \, d\sigma,$$

where $W(u)$ is given by (7.5). Combining the above two equalities, we obtain the Pohozaev identity:

$$\int_{B_R} \left((N-2)|\nabla u|^2 + 2NW(u) \right) dx =$$
$$R\int_{\partial B_R} \left(|\nabla u|^2 + 2W(u) \right) d\sigma - 2R\int_{\partial B_R} \left(\frac{\partial u}{\partial r} \right)^2 d\sigma. \quad (7.15)$$

Now differentiate Φ given by (7.14):

$$2\Phi'(R) =$$
$$-(N-1)R^{-N} \int_{B_R} \left(|\nabla u|^2 + 2W(u) \right) dx + R^{1-N} \int_{\partial B_R} \left(|\nabla u|^2 + 2W(u) \right) d\sigma.$$

Using (7.15), we deduce that

$$2R^N \Phi'(R) = \int_{B_R} \left(2W(u) - |\nabla u|^2 \right) dx + 2R\int_{\partial B_R} \left(\frac{\partial u}{\partial r} \right)^2 d\sigma.$$

To complete the proof, we apply Theorem 7.2.3 below. $\qquad\qquad\square$

Theorem 7.2.3 ([158]) *Let $N \geq 1$. Let $u \in C^3(\mathbb{R}^N)$ denote a bounded solution to the Allen-Cahn equation (7.13). Then,*

$$|\nabla u|^2 \leq 2W(u) \qquad in \ \mathbb{R}^N, \quad (7.16)$$

where $W(u)$ is given by (7.5).

Proof. We want to show that the function $P(x) := |\nabla u|^2 - 2W(u)$ is nonpositive. Since u is bounded, it follows that P is bounded (by standard elliptic regularity) and that $\inf_{\mathbb{R}^N} |\nabla u| = 0$. In particular, given $\delta > 0$, we may assume (up to a translation of space) that

$$|\nabla u|^2(0) < \delta. \quad (7.17)$$

Step 1. Set $f(u) = -W'(u) = u - u^3$. Then, P satisfies the inequality

$$|\nabla u|^2 \Delta P \geq \frac{1}{2}|\nabla P|^2 - 2f(u)\nabla u \cdot \nabla P \qquad in \ \mathbb{R}^N. \quad (7.18)$$

169

Indeed, by definition of P, we have for $i = 1, \ldots, N$,

$$\partial_i P = 2 \sum_j \partial_j u \partial_{ij} u + 2 f(u) \partial_i u. \tag{7.19}$$

In particular,

$$\sum_i (\partial_i P - 2f(u)\partial_i u)^2 = 4 \sum_i \left(\sum_j \partial_j u \partial_{ij} u \right)^2$$

$$\leq 4 \sum_j (\partial_j u)^2 \sum_{i,j} (\partial_{ij} u)^2$$

$$= 4 |\nabla u|^2 |Hu|^2, \tag{7.20}$$

where Hu denotes the Hessian matrix of u. Differentiating once more (7.19) and using (7.9), we obtain

$$\Delta P = 2 \sum_{i,j} (\partial_{ij} u)^2 + 2 \sum_j \partial_j u \partial_j \Delta u + 2 f'(u) |\nabla u|^2 + 2 f(u) \Delta u$$

$$= 2 |Hu|^2 - 2 f(u)^2.$$

Using (7.20), it follows that

$$|\nabla u|^2 \Delta P \geq \frac{1}{2} \sum_i (\partial_i P - 2f(u)\partial_i u)^2 - 2f(u)^2 |\nabla u|^2$$

$$= \frac{1}{2} |\nabla P|^2 - 2f(u)\nabla u \cdot \nabla P,$$

as claimed.

Step 2. We assume temporarily that given $\varepsilon > 0, R > 0$, there exists a radial cutoff function $\eta(x) = \eta_{\varepsilon,R}(|x|) \in C^2(\mathbb{R}^N)$ having the following properties:

$$\eta(R) = 1, \quad \eta > 0, \quad \eta' < 0 \quad \text{and} \quad \lim_{r \to +\infty} \eta(r) = 0. \tag{7.21}$$

$$\lim_{\varepsilon \to 0^+} \eta(r) = 1 \quad \text{for all } r \geq R \tag{7.22}$$

$$\frac{\eta^2}{(\eta')^2} \left(\frac{2\eta'}{\eta} - \frac{M}{\varepsilon} \eta' - \eta'' - \frac{(N-1)\eta'}{r} \right) \leq \frac{\varepsilon}{L}, \quad \text{for } r \geq R, \tag{7.23}$$

where

$$M = \sup_{\mathbb{R}^N} 2 |f(u)| |\nabla u|, \quad L = \sup_{\mathbb{R}^N} 2 |\nabla u|^2. \tag{7.24}$$

Step 3. Set $v = \eta P$. Then,

$$v(x) \leq \max(\varepsilon, \max_{|x|=R} P), \qquad \text{for all } |x| \geq R. \tag{7.25}$$

Equation (7.25) is obvious if $\sup_{|x| \geq R} v \leq 0$, so we may assume that v is positive somewhere. Since P is bounded and $\lim_{+\infty} \eta = 0$, $\lim_{|x| \to +\infty} v = 0$. So, either v reaches its maximum on $|x| = R$ and then (7.25) follows, or v achieves its maximum at some point x_0 such that $|x_0| > R$. At x_0, $0 = \nabla v = \eta \nabla P + P \nabla \eta$, hence $\nabla P = -P \frac{\nabla \eta}{\eta}$. Using (7.18), we deduce that at x_0,

$$|\nabla u|^2 \Delta v \geq |\nabla u|^2 P \Delta \eta + 2 |\nabla u|^2 \nabla P \cdot \nabla \eta + \frac{1}{2} \eta |\nabla P|^2 - 2 \eta f(u) \nabla u \cdot \nabla P$$

$$= \left[|\nabla u|^2 \Delta \eta - 2 \frac{|\nabla u|^2 |\nabla \eta|^2}{\eta} + 2 f(u) \nabla u \cdot \nabla \eta \right] P + \frac{P^2 |\nabla \eta|^2}{2 \eta}.$$

Now, $\Delta v(x_0) \leq 0$ since x_0 is an interior point of maximum. Furthermore, $P(x_0) > 0$, since $\eta(x_0) > 0$ and $v(x_0) > 0$. So, at x_0,

$$\frac{P |\nabla u|^2}{2 \eta} \leq 2 \frac{|\nabla u|^2 |\nabla \eta|^2}{\eta} - 2 f(u) \nabla u \cdot \nabla \eta - |\nabla u|^2 \Delta \eta. \tag{7.26}$$

If $|\nabla u|^2(x_0) \leq \varepsilon$, then since $\eta \leq 1$ and $W \geq 0$, we have $v(x) \leq v(x_0) \leq P(x_0) \leq |\nabla u|^2(x_0) \leq \varepsilon$ for $|x| \geq R$ and so (7.25) holds.

If $|\nabla u|^2(x_0) > \varepsilon$, since $\eta' < 0$, (7.24) and (7.26) imply that at x_0,

$$v = \eta P \leq \frac{2 \eta^2}{|\nabla \eta|^2} |\nabla u|^2 \left(\frac{2 |\nabla \eta|^2}{\eta} + \frac{M |\nabla \eta|}{|\nabla u|^2} \right) - \Delta \eta$$

$$\leq L \frac{\eta^2}{|\nabla \eta|^2} \left(\frac{2 |\nabla \eta|^2}{\eta} + \frac{M |\nabla \eta|}{\varepsilon} - \Delta \eta \right).$$

Recalling (7.23), we conclude that $v(x_0) \leq \varepsilon$ and (7.25) is proven.
Step 4. Now, we may let $\varepsilon \to 0$ in (7.25). Then,

$$P(x) \leq \max(0, \max_{|x|=R} P), \qquad \text{for } |x| \geq R.$$

Now let $R \to 0^+$. Using (7.17), we obtain

$$P(x) \leq \max(0, P(0)) < \delta \qquad \text{for all } x \in \mathbb{R}^N.$$

Since $\delta > 0$ is arbitrary, $P \leq 0$ as requested. So, we are left with the proof of Step 2. Set

$$g_\varepsilon(t) = \int_t^1 \frac{e^{-\varepsilon/Ls}}{s^2}\, ds, \qquad \text{for } 0 \leq t \leq 1$$

and

$$h_{\varepsilon,R}(t) = \int_R^t \frac{e^{-(M/\varepsilon)s}}{s^{N-1}}\, ds, \qquad \text{for } t \geq R.$$

Take at last

$$\eta(r) = \eta_{\varepsilon,R}(r) = g_\varepsilon^{-1}(c\, h_{\varepsilon,R}(r)), \qquad \text{for } r \geq R,$$

where $c = g_\varepsilon(0)/h_{\varepsilon,R}(+\infty)$. Then, (7.21) is obvious. For (7.22), it suffices to observe that $h_{\varepsilon,R} \to 0$ pointwise as $\varepsilon \to 0$ and $g_\varepsilon^{-1}(0^+) \to 1$.

For (7.23), differentiate, take the log and differentiate again with respect to r the equality

$$\int_\eta^1 \frac{e^{-\varepsilon/Ls}}{s^2}\, ds = \int_R^r \frac{e^{-(M/\varepsilon)s}}{s^{N-1}}\, ds.$$

$\square$

Thanks to Theorem 7.2.1, it is natural to study monotone entire solutions of the Allen-Cahn equation, Equation (7.1). However, it is not clear at this stage that the conclusion of Conjecture 7.1.1 has a chance to hold. Also, why should we restrict to dimension $N \leq 8$? We discuss this in the next section.

7.2.3 From Bernstein to De Giorgi

In this section, we continue our discussion of the De Giorgi conjecture 7.1.1 in its weak form. That is, we make the additional assumption that (7.10) holds. The following result shows that the level sets of u are "flat at infinity."

Theorem 7.2.4 ([3]) *Let* $1 \leq N \leq 8$ *and let* $u \in C^2(\mathbb{R}^N; [-1,1])$ *denote a monotone solution to (7.1) such that (7.10) hold. Then, there exists a sequence* (ε_k) *converging to zero and a unit vector* $\tau \in \mathscr{S}^{N-1}$ *such that*

$$\lim_{k\to+\infty} \varepsilon_k^{N-1} \int_{B_{1/\varepsilon_k}} \left(|\nabla u|^2 - |\partial_\tau u|^2 \right) dx = 0. \qquad (7.27)$$

Moreover, there exists a half-space $E \subset \mathbb{R}^N$ *orthogonal to* τ *such that*

$$u_{\varepsilon_k}(x) = u(x/\varepsilon_k) \to \chi_E - \chi_{E^c} \qquad (7.28)$$

in $L^1_{loc}(\mathbb{R}^N)$, *as* $k \to +\infty$.

To prove this theorem, one is led to establishing first that (7.28) holds for some set E that is a local minimizer of perimeter. This is where Theorem 7.2.1 and the energy estimates Corollary 7.2.1 and Theorem 7.2.2 are used. In addition, since u is monotone, it follows from the implicit function theorem that any level set $E_\lambda := \{x \in \Omega : u(x) > \lambda\}$ lies above the graph of a function $\psi_\lambda : \Omega_\lambda \subset \mathbb{R}^{N-1} \to \mathbb{R}$, that is,

$$E_\lambda = \{(x', x_N) \in \Omega_\lambda \times \mathbb{R} : x_N > \psi_\lambda(x')\}.$$

Hence, E should lie above an entire graph $\psi : \mathbb{R}^{N-1} \to \mathbb{R}$, that is,

$$E = \{(x', x_N) \in \mathbb{R}^N \times \mathbb{R} : x_N > \psi(x')\}.$$

Since E has a locally minimal perimeter, ψ solves (in a weak sense) the minimal surface equation in all of $\mathbb{R}^{N-1}$. Since global minimal graphs are flat in dimension $N - 1 \le 7$ due to Bernstein-type theorems (see [127]), one finally obtains the desired conclusion.

In fact, much more can be said. The following difficult results, which we state without proof, completely settle the weak form of the De Giorgi conjecture.

Theorem 7.2.5 ([194]) *Let $1 \le N \le 8$. Assume that $u \in C^2(\mathbb{R}^N; [-1, 1])$ is a monotone solution to (7.1) such that (7.10) holds. Then, u is one-dimensional.*

Theorem 7.2.6 ([84]) *Let $N \ge 9$. There exists a monotone solution $u \in C^2(\mathbb{R}^N; [-1, 1])$ of (7.1) such that (7.10) holds and u is not one-dimensional.*

Let us turn now to the full De Giorgi conjecture, which we shall prove in dimensions $N = 2$ and $N = 3$. For the state of the art on what is known in dimensions $4 \le N \le 8$, we refer the reader to [104].

7.3   Dimension $N = 2$

Theorem 7.3.1 ([120]) *Conjecture 7.1.1 holds true if $N = 2$.*

Proof. Let u denote a solution to (7.1). Since u is monotone, u is stable. Since u is bounded, so is $|\nabla u|$ (by applying elliptic regularity in any given ball $B(x_0, 1)$, $x_0 \in \mathbb{R}^N$). We then simply apply Theorem 6.2.2. $\qquad\square$

7.4   Dimension $N = 3$

Theorem 7.4.1 ([7])   *Conjecture 7.1.1 holds true if $N = 3$.*

Proof. We follow [96].
Step 1. Assume that $u = u(x_1, x_2, x_3)$ satisfies (7.1). Then, the function

$$\bar{u}(x_1, x_2) = \lim_{x_3 \to +\infty} u(x_1, x_2, x_3) \tag{7.29}$$

is a bounded stable solution to

$$-\Delta u = u - u^3 \quad \text{in } \mathbb{R}^2. \tag{7.30}$$

For every $t \in \mathbb{R}$, consider the function u^t defined by

$$u^t(x_1, x_2, x_3) = u(x_1, x_2, x_3 + t) \quad \text{for all } (x_1, x_2, x_3) \in \mathbb{R}^3. \tag{7.31}$$

Clearly, u^t solves (7.1) and by standard elliptic regularity, $\|u^t\|_{C^2(\mathbb{R}^3)} \leq C$, for some constant C independent of t. Since $\partial u / \partial x_3 > 0$, we deduce that $u^t \to \bar{u}$ in $C^1_{loc}(\mathbb{R}^3)$, as $t \to +\infty$. Therefore, $\bar{u}$ is a bounded solution to (7.30).

To see that $\bar{u}$ is stable, recall that u^t is monotone, hence stable. In particular, taking $\varphi \in C^1_c(\mathbb{R}^3)$ of the form $\varphi(x_1, x_2, x_3) = \phi(x_1, x_2)\psi_R(x_3)$, with $\phi \in C^1_c(\mathbb{R}^2)$ and $\psi_R \in C^1_c(\mathbb{R})$ such that $0 \leq \psi_R \leq 1$, $0 \leq \psi'_R \leq 2$, $\psi_R = 0$ in $\mathbb{R} \setminus [R, 2R+2]$ and $\psi_R = 1$ in $[R+1, 2R+1]$, the stability of u^t implies that

$$\int_{\mathbb{R}^3} \left| \nabla(\phi(x_1, x_2)\psi_R(x_3)) \right|^2 dx \geq \int_{\mathbb{R}^3} (1 - 3(u^t)^2)\phi(x_1, x_2)^2 \psi_R(x_3)^2 \, dx. \tag{7.32}$$

Now,

$$\left| \nabla(\phi(x_1, x_2)\psi_R(x_3)) \right|^2 = \psi_R^2 \left| \nabla \phi \right|^2 + (\psi'_R)^2 \phi^2$$

and so

$$\int_{\mathbb{R}^3} \left| \nabla(\phi(x_1, x_2)\psi_R(x_3)) \right|^2 dx =$$

$$\left(\int_{[R+1,2R+1]} \psi_R^2 \, dx_3 + \int_{[R,R+1] \cup [2R+1,2R+2]} \psi_R^2 \, dx_3 \right) \int_{\mathbb{R}^2} \left| \nabla \phi \right|^2 dx +$$

$$+ \int_{[R,R+1] \cup [2R+1,2R+2]} (\psi'_R)^2 dx_3 \int_{\mathbb{R}^2} \phi^2 \, dx$$

$$\leq (R+2) \int_{\mathbb{R}^2} \left| \nabla \phi \right|^2 dx + 8 \int_{\mathbb{R}^2} \phi^2 \, dx. \tag{7.33}$$

Similarly,

$$\int_{\mathbb{R}^3} (1 - 3(u^t)^2) \phi(x_1, x_2)^2 \psi_R(x_3)^2 \, dx =$$

$$= \int_{[R+1 \leq x_3 \leq 2R+1]} + \int_{[R<x_3<R+1] \cup [2R+1<x_3<2R+2]} \geq$$

$$\int_{[R+1 \leq x_3 \leq 2R+1]} (1 - 3(u^t)^2) \phi(x_1, x_2)^2 \psi_R(x_3)^2 \, dx - 4 \int_{\mathbb{R}^2} \phi^2 \, dx. \quad (7.34)$$

Collecting (7.32), (7.33), and (7.34), we deduce that

$$\frac{R+2}{R} \int_{\mathbb{R}^2} |\nabla \phi|^2 \, dx + \frac{12}{R} \int_{\mathbb{R}^2} \phi^2 \, dx \geq$$

$$\int_{\mathbb{R}^2} \left(\fint_{[R+1 \leq x_3 \leq 2R+1]} (1 - 3(u^t)^2) \, dx_3 \right) \phi^2 \, dx. \quad (7.35)$$

Since for every $(x_1, x_2) \in \mathbb{R}^2$, we have

$$\left| \fint_{[R+1 \leq x_3 \leq 2R+1]} (1 - 3(u^t)^2) \, dx_3 - (1 - 3\bar{u}^2) \right| \leq$$

$$\leq 3 \fint_{[R+1 \leq x_3 \leq 2R+1]} \left| \bar{u}^2 - (u^t)^2 \right| \, dx_3 \leq$$

$$\leq 6[\bar{u}(x_1, x_2) - u^t(x_1, x_2, R)] \to 0 \qquad \text{as } R \to +\infty,$$

we may pass to the limit $R \to +\infty$ in (7.35), by dominated convergence. We deduce that $\bar{u}$ is stable, as claimed.

Step 2. There exists $\tau = (\tau_1, \tau_2) \in \mathscr{S}^1$ and $h \in C^2(\mathbb{R})$ such that

$$\bar{u}(x_1, x_2) = h(\tau_1 x_1 + \tau_2 x_2), \qquad \text{for all } (x_1, x_2) \in \mathbb{R}^2. \quad (7.36)$$

Furthermore, either $h \equiv 1$ or there exists $\alpha \in \mathbb{R}$ such that $h(t) = \tanh(t/\sqrt{2} + \alpha)$ for every $t \in \mathbb{R}$.

To see this, we apply Theorem 7.3.1. Thus, (7.36) holds and h is a monotone solution to

$$-h'' = h - h^3 \qquad \text{in } \mathbb{R}. \quad (7.37)$$

Using the strong maximum principle (Corollary A.5.1), we see that either h is constant or h is a strictly monotone solution to (7.37). If h is constant, then $h \in \{-1, 0, 1\}$. The case $h = 0$ is excluded since $\bar{u}$ is stable. The case $h = -1$ is also ruled out by the monotonicity assumption $\partial u / \partial x_3 > 0$. Hence, $h = 1$.

When h is not constant, it is easily seen that h can always be chosen strictly increasing (if h were strictly decreasing, simply replace τ by $-\tau$ and $h(t)$ by $h(-t)$ in (7.36)). To conclude the proof of Step 2, simply apply Proposition 7.1.1.

Step 3. The Dirichlet energy of u satisfies

$$\int_{B_R} |\nabla u|^2 \, dx \le CR^2, \qquad \text{for all } R > 1, \tag{7.38}$$

for some positive constant C independent of R.

For every $R > 1$, set

$$\mathscr{E}_R(u^t) = \int_{B_R} \left[\frac{1}{2} |\nabla u^t|^2 + \frac{1}{4}((u^t)^2 - 1)^2 \right] \, dx,$$

where u^t is the function defined in (7.31). Since $u^t \to \bar{u}$ in $C^1_{loc}(\mathbb{R}^3)$ we get

$$\lim_{t \to +\infty} \mathscr{E}_R(u^t) = \mathscr{E}_R(\bar{u}), \qquad \text{for all } R > 1. \tag{7.39}$$

In addition,

$$\partial_t \mathscr{E}_R(u^t) = \int_{B_R} \nabla u^t \cdot \nabla(\partial_t u^t) \, dx + \int_{B_R} ((u^t)^3 - u^t)\partial_t u^t \, dx.$$

Since $\partial u/\partial x_3 > 0$, we have $\partial_t u^t > 0$. Since $\|u^t\|_{C^1 \mathbb{R}^3} \le C$ for some C independent of t, it follows that

$$\partial_t \mathscr{E}_R(u^t) = \int_{\partial B_R} \frac{\partial u^t}{\partial n} \partial_t u^t \, d\sigma \ge -C \int_{\partial B_R} \partial_t u^t \, d\sigma.$$

Hence, for every $T > 0$ and every $R > 1$, we have

$$\mathscr{E}_R(u) = \mathscr{E}_R(u^T) - \int_0^T \partial_t \mathscr{E}_R(u^t) \, dt$$

$$\le \mathscr{E}_R(u^T) + C \int_0^T \left(\int_{\partial B_R} \partial_t u^t \, d\sigma \right) dt$$

$$= \mathscr{E}_R(u^T) + C \int_{\partial B_R} \int_0^T \partial_t u^t \, dt \, d\sigma$$

$$= \mathscr{E}_R(u^T) + C \int_{\partial B_R} (u^T - u) d\sigma \le \mathscr{E}_R(u^T) + C_1 R^2,$$

for some constant C_1 independent of T and R.

Letting $T \to +\infty$ and using (7.39), we obtain

$$\mathcal{E}_R(u) \le \mathcal{E}_R(\bar{u}) + C_1 R^2$$

$$\le C_2 R^2 \int_{-R}^{R} \left[\frac{1}{2}(h')^2 + \frac{1}{4}(h^2 - 1)^2 \right] \, dt + C_1 R^2,$$

where C_2 is independent of R. Due to the expression of h found in Step 2,

$$\int_{\mathbb{R}} \left[\frac{1}{2}(h')^2 + \frac{1}{4}(h^2 - 1)^2 \right] \, dt < +\infty$$

and the desired conclusion follows.

Step 4. Let $\sigma_j = \frac{\partial u}{\partial x_j} / \frac{\partial u}{\partial x_3}$ for $j = 1, 2$. Then, since $\partial u / \partial x_3$ and $\partial u / \partial x_j$ both solve the linearized equation

$$-\Delta v = (1 - 3u^2)v \qquad \text{in } \mathbb{R}^3,$$

it follows that

$$-\nabla \cdot \left(v^2 \nabla \sigma_j \right) = 0 \qquad \text{in } \mathbb{R}^3. \tag{7.40}$$

Apply Proposition 6.2.2 for (7.40). By (7.38), we deduce that σ_j is constant, that is, there exists a constant C_j such that

$$\frac{\partial u}{\partial x_j} = C_j \frac{\partial u}{\partial x_3}.$$

In particular, the gradient of u points in a fixed direction, that is, u is one-dimensional. $\qquad \square$

Chapter 8

Further readings

8.1 Stability versus geometry of the domain

So far, we have mostly dealt with stable solutions of (1.3) for two specific types of domains: Ω is bounded or $\Omega = \mathbb{R}^N$. In this section, we review a number of results applying to other geometries.

8.1.1 The half-space

The Lane-Emden nonlinearity

We begin discussing (1.3) for $\Omega = \mathbb{R}^N_+$ in the case of the Lane-Emden nonlinearity.

$$\begin{cases} -\Delta u = |u|^{p-1}u & \text{in } \mathbb{R}^N_+, \\ u = 0 & \text{on } \partial\mathbb{R}^N_+. \end{cases} \qquad (8.1)$$

When p is subcritical, no positive solution exists.

Theorem 8.1.1 ([124]) *Let $N \geq 3$ and $1 < p \leq p_S(N) = \frac{N+2}{N-2}$ (or $N = 2$ and $p < +\infty$). If $u \in C^2(\overline{\mathbb{R}^N_+})$ is a nonnegative solution to (8.1), then, $u \equiv 0$.*

When restricting to *bounded* nonnegative solutions, the above theorem can be extended to any exponent below the second critical exponent $p_c(N-1)$, defined by Equation (4.16), in dimension $N-1$.

Theorem 8.1.2 ([97]) *Let $N \geq 2$ and $1 < p < p_c(N-1)$ (where $p_c(N)$ is given by (4.16)) . If $u \in C^2(\overline{\mathbb{R}^N_+})$ is a nonnegative bounded solution to (8.1), then, $u \equiv 0$.*

179

Proof. By the strong maximum principle either $u = 0$ or $u > 0$ in $\mathbb{R}^N_+$. Let us prove that the second possibility does not happen. Suppose to the contrary that $u > 0$ in $\mathbb{R}^N_+$. Then, u is monotone (using a similar—but more delicate— strategy as that of Lemma 1.2.1, see [70]), hence stable. The boundedness of u, standard elliptic estimates, and the monotonicity of u with respect to the variable x_N, imply that the function

$$v(x_1, \ldots, x_{N-1}) := \lim_{x_N \to +\infty} u(x)$$

is a *positive* smooth solution to the Lane-Emden equation in $\mathbb{R}^{N-1}$. Furthermore, mimicking Step 1 in the proof of Theorem 7.4.1, we deduce that v is stable. At this point, an application of Exercise 6.2.1 to the solution v in $\mathbb{R}^{N-1}$, gives $v = 0$ in $\mathbb{R}^{N-1}$. This result clearly contradicts $v > 0$ in $\mathbb{R}^{N-1}$. Hence, $u = 0$, which completes the proof. $\qquad\square$

In the proof above, we used the fact that bounded positive solutions of the equation are monotone (hence stable). A similar classification result is available for solutions that are stable outside a compact set (possibly unbounded and/or sign-changing), but for a smaller range of p.

Theorem 8.1.3 ([97]) *Let $N \geq 2$ and $p > 1$. Assume that $u \in C^2(\overline{\mathbb{R}^N_+})$ is a solution to (8.1) that is stable outside a compact set and $1 < p < p_c(N)$, where $p_c(N)$ is given by Equation (4.16). Then, $u \equiv 0$.*

None of the previous three theorems are known to be optimal.

General nonlinearity

In this section, we consider bounded positive solutions to

$$\begin{cases} -\Delta u = f(u) & \text{in } \mathbb{R}^N_+, \\ \quad u > 0 & \text{in } \mathbb{R}^N_+, \\ \quad u = 0 & \text{on } \mathbb{R}^N_+, \end{cases} \qquad (8.2)$$

where $f \in C^1(\mathbb{R})$. A useful criterion for obtaining Liouville-type theorems in this context is the following.

Theorem 8.1.4 ([16]) *Let $N \geq 2$, $f \in C^1(\mathbb{R})$ and let $u \in C^2(\overline{\mathbb{R}^N_+})$ be a solution to (8.2). Assume that*

$$f(M) \leq 0, \qquad \text{where } M = \sup_{\mathbb{R}^N_+} u.$$

Then, u is monotone and one-dimensional.

As a consequence, we obtain the following.

Theorem 8.1.5 ([16, 17, 70, 102]) *Let* $N \geq 2$, $f \in C^1(\mathbb{R})$ *and let* $u \in C^2(\overline{\mathbb{R}^N_+})$ *be a solution to (8.2).*

- *If* $N = 2$, *then* u *is monotone and one-dimensional.*

- *If* $N = 3$ *and* $f(0) \geq 0$, *then* u *is monotone and one-dimensional.*

- *If* $N \leq 5$ *and* $f \geq 0$, *then* u *is monotone and one-dimensional.*

- *If* $N \geq 2$ *and* $f(0) \geq 0$, *then* u *is monotone.*

Remark 8.1.1 *It is not known whether the assumption* $f(0) \geq 0$ *is necessary to conclude that (strictly) positive solutions are monotone. Note, however, that* $u = 1 - \cos(x_N)$ *is a nonnegative solution to (8.2) for* $f(u) = u - 1$, *but* u *is clearly not monotone.*

Proof of Theorem 8.1.5. Using a similar strategy as that of Lemma 1.2.1, see [17], u must be monotone. The boundedness of u, standard elliptic estimates, and the monotonicity of u with respect to the variable x_N, imply that the function

$$v(x_1, \ldots, x_{N-1}) := \lim_{x_N \to +\infty} u(x)$$

solves

$$-\Delta v = f(v) \qquad \text{in } \mathbb{R}^{N-1}.$$

Furthermore, working as in Step 1 in the proof of Theorem 7.4.1, we deduce that v is stable. Using Theorem 6.2.2 if $N - 1 \leq 2$, or Theorem 6.2.3 if $N - 1 \leq 4$, we deduce that $v = 0$ and so

$$0 = \Delta v = f(v) = f\left(\sup_{\mathbb{R}^N_+} u\right).$$

We may then apply Theorem 8.1.4. □

8.1.2 Domains with controlled volume growth

We return to the model Lane-Emden equation, posed this time in an arbitrary proper domain $\Omega \subset \mathbb{R}^N$.

$$\begin{cases} -\Delta u = |u|^{p-1}u & \text{in } \Omega, \\ u = 0 & \text{on } \partial\Omega. \end{cases} \tag{8.3}$$

181

The following theorem shows that Liouville theorems for stable solutions extend to any domain having controlled volume growth.

Theorem 8.1.6 ([97]) *Let $p > 1$ and let Ω denote a proper $C^{2,\alpha}$ domain of $\mathbb{R}^N$. Let $u \in C^2(\overline{\Omega})$ be a stable solution to (8.3). Suppose that there exists a point $x_0 \in \mathbb{R}^N$ and $\gamma \in [1, 2p + 2\sqrt{p(p-1)} - 1)$ such that*

$$\liminf_{R \to +\infty} \frac{\left| \Omega \cap B(x_0, R) \right|}{R^{2\left(\frac{p+\gamma}{p-1}\right)}} = 0. \tag{8.4}$$

Then, $u \equiv 0$.

It is worth observing that the volume growth condition (8.4) is automatically satisfied in many interesting cases.

Proposition 8.1.1 ([97]) *Condition (8.4) is satisfied in any of the following cases*

- $N \geq 2$, $p > 1$ *and Ω has finite volume $|\Omega| < +\infty$.*

- $N \leq 10$, $p > 1$ *and Ω is any domain of $\mathbb{R}^N$.*

- $N \geq 11$, $1 < p < p_c(N)$ *and Ω is any domain of $\mathbb{R}^N$.*

- $N \geq 11$, $p > 1$ *and $\Omega \subset \mathbb{R}^K \times \omega$, where $1 \leq K \leq 10$, $\omega \subset \mathbb{R}^{N-K}$ is any domain with finite $(N - K)$ dimensional Lebesgue measure.*

For the classification of solutions that are stable outside a compact set, the volume growth condition is no longer sufficient: just think of the case of a bounded domain, where all (classical) solutions are stable outside a compact set. However, if Ω is a smooth unbounded proper domain, satisfying

$$\exists X \in \mathbb{R}^N, \quad |X| = 1 \quad : \quad n(x) \cdot X \geq 0, \quad n(x) \cdot X \not\equiv 0 \quad \text{on } \partial\Omega, \tag{8.5}$$

where n is the normal unit vector to $\partial\Omega$ pointing outward, for example, if Ω is a smooth epigraph, then the following result holds.

Theorem 8.1.7 ([97]) *Let Ω be a proper unbounded $C^{2,\alpha}$ domain of $\mathbb{R}^N$, $N \geq 2$ satisfying condition (8.5). Let $u \in C^2(\overline{\Omega})$ be a solution to (8.3) that is stable outside a compact set, with $1 < p \leq p_S(N) = \frac{N+2}{N-2}$ ($p < +\infty$ if $N = 2$). Then, $u \equiv 0$.*

8.1.3 Exterior domains

As the next theorem demonstrates, even when working in unbounded domains, geometric conditions on Ω such as (8.5) cannot be avoided to obtain a Liouville-type result for solutions that are stable outside a compact set.

Theorem 8.1.8 ([72]) *Let $N \geq 3$ and let $\mathscr{D} \subset \mathbb{R}^N$ be a smoothly bounded open set, such that $\Omega = \mathbb{R}^N \setminus \overline{\mathscr{D}}$ is connected. Then, there exists a number $p_0 > \frac{N+2}{N-2}$ such that for any $\frac{N+2}{N-2} < p < p_0$, there exists a solution $u \in C^2(\overline{\Omega})$ to*

$$
\begin{cases}
-\Delta u = u^p & \text{in } \Omega, \\
u > 0 & \text{in } \Omega, \\
u = 0 & \text{on } \partial\Omega,
\end{cases}
$$

having fast decay at infinity, that is, $u(x) = \mathcal{O}(|x|^{2-N})$, as $|x| \to +\infty$. In particular, u is stable outside a compact set.

Remark 8.1.2 *When $\mathscr{D}$ is a ball, the theorem remains valid for any $p > \frac{N+2}{N-2}$. In [72], the authors also construct slow-decay solutions, for any $p > \frac{N+2}{N-2}$. More precisely, such solutions satisfy*

$$
u = c_p |x|^{-\frac{2}{p-1}}(1 + o(1)), \quad \text{as } |x| \to +\infty,
$$

where $c_p = \left(\frac{2}{p-1}\left(N - 2 - \frac{2}{p-1}\right)\right)^{\frac{1}{p-1}}$. By the optimality of Hardy's inequality, every such solution is unstable outside every compact set if $p < p_c(N)$.

One should also note that if a solution decays at least like $|x|^{-2/(p-1)}$, then it must satisfy the following alternative.

Theorem 8.1.9 ([20]) *Let $N \geq 3$, $p \in (1, +\infty) \setminus \{\frac{N+2}{N-2}\}$, and $\Omega = \mathbb{R}^N \setminus B(0,1)$. Let $u \in C^2(\Omega)$ denote a solution to*

$$
\begin{cases}
-\Delta u = u^p & \text{in } \Omega, \\
u > 0 & \text{in } \Omega,
\end{cases}
$$

such that $u(x) = \mathcal{O}(|x|^{-\frac{2}{p-1}})$, as $|x| \to +\infty$. Then, either there exists $\gamma > 0$ such that

$$
\lim_{|x| \to +\infty} |x|^{N-2} u(x) = \gamma,
$$

or there exists a positive solution $w \in C^2(\mathscr{S}^{N-1})$ to

$$-\Delta_{\mathscr{S}^{N-1}} w = w^p - \lambda w \qquad in \ \mathscr{S}^{N-1},$$

with $\lambda = c_p^{p-1} = \frac{2}{p-1}\left(N - \frac{2p}{p-1}\right)$, such that

$$\lim_{r \to +\infty} r^{\frac{2}{p-1}} u(r,\cdot) = w(\cdot),$$

in the $C^k(\mathscr{S}^{N-1})$ topology, for any $k \in \mathbb{N}$.

8.2 Symmetry of stable solutions

We have seen in Proposition 1.3.4 that stable solutions defined on the unit ball are always radial. One may wonder what symmetry properties stable solutions have when working in more general domains. Also, what can be said of solutions with a higher Morse index? In this section, we point out several recent results in this direction of research.

8.2.1 Foliated Schwarz symmetry

When working with solutions of positive Morse index on the unit ball, radial symmetry may fail. For example, if $f(u) = \lambda_2 u$, where $\lambda_2 = \lambda_2(-\Delta; B)$, all solutions have index one. Furthermore, in this case, any solution to (1.3) takes the form
$$u(x) = \phi(r)\cos(\theta).$$
In the above formula, $r = |x|$, $\cos(\theta) = \frac{x}{|x|} \cdot p$, where $p \in \mathscr{S}^{N-1}$ is arbitrary, and the function ϕ is explicit, namely,

$$\phi(r) = Ar^{\frac{2-N}{2}} J(jr), \qquad r \in (0,1),$$

where A is an arbitrary constant, J is the Bessel function of the first kind of order $\frac{N-2}{2}$, and j is its first zero. In particular, although u is not radial, u has the following partial symmetry property.

Definition 8.2.1 *Let $N \geq 2$ and let $B \subset \mathbb{R}^N$ be either a ball or an annulus centered at the origin. A function $u \in C(\overline{B})$ is foliated Schwarz symmetric if there exists a unit vector $p \in \mathbb{R}^N$ such that*

- $u(x) = u(r, \theta)$ *is a function of only two variables: the distance* $r = |x|$ *of the point* x *to the origin and the angle* $\theta = \arccos(\frac{x}{|x|} \cdot p)$ *formed by the vectors* x *and* p.

- u *is nonincreasing in* θ.

It turns out that foliated Schwarz symmetry is a general property of solutions of low Morse index, as the following two results demonstrate. For the proof, we refer the reader to [176].

Theorem 8.2.1 ([175]) *Let* $N \geq 2$, B *denote either a ball or an annulus in* $\mathbb{R}^N$. *Assume that* $f \in C^{1,\alpha}(\mathbb{R})$ *is convex. Then, every solution to (1.18) with Morse index* $\mathrm{ind}(u) = 1$ *is foliated Schwarz symmetric.*

Theorem 8.2.2 ([177]) *Let* $N \geq 2$, B *denote either a ball or an annulus in* $\mathbb{R}^N$. *Assume that* $f \in C^{1,\alpha}(\mathbb{R})$ *is such that* f' *is convex. Then, every solution to (1.18) with Morse index* $\mathrm{ind}(u) \leq N$ *is foliated Schwarz symmetric.*

Whether such symmetry results hold for solutions of higher Morse index is an open problem. Partial results of Bouchez and VanSchaftingen indicate that there should exist solutions of index $N + 2$ or higher, which are not foliated Schwarz symmetric, see [24]. In this direction, we indicate the following symmetry breaking theorem.

Theorem 8.2.3 ([23]) *For every* $p > 1$ *sufficiently close to 1, there exists a rectangle* $\Omega \subset \mathbb{R}^2$ *such that any least energy nodal solution to*

$$\begin{cases} -\Delta u = |u|^{p-1}u & \text{in } \Omega, \\ u = 0 & \text{on } \partial\Omega, \end{cases} \tag{8.6}$$

is neither symmetric nor antisymmetric with respect to the medians of Ω.

Remark 8.2.1 *A least energy nodal solution to (8.6) is a minimizer of the energy functional*

$$\mathcal{E}_\Omega(u) = \frac{1}{2}\int_\Omega |\nabla u|^2\, dx - \frac{1}{p+1}\int_\Omega |u|^{p+1}\, dx$$

over the nodal Nehari set $\mathcal{M}_p$ *defined by*

$$\mathcal{N}_p = \{u \in H_0^1(\Omega) \setminus \{0\} \,:\, D\mathcal{E}_\Omega(u).u = 0\}, \quad \mathcal{M}_p = \{u \in H_0^1(\Omega) \,:\, u^\pm \in \mathcal{N}_p\}.$$

It can be shown that there exists a least energy nodal solution to (8.6) (see [54]) and that it has Morse index $\mathrm{ind}(u) = 1$.

8.2.2 Convex domains

We have seen in Proposition 1.3.4 that a stable solution u to (1.18) in the ball is radial. In addition, $u = u(r)$ is either constant, radially decreasing, or radially increasing. In other words, the level sets of u are hyperspheres.

Suppose now that the domain Ω is convex. Is it true that the level sets of u are convex? Here is a result in this direction.

Theorem 8.2.4 ([37, 109]) *Let $\Omega \subset \mathbb{R}^2$ be a smoothly bounded strictly convex domain. Suppose that $f \geq 0$ is a smooth function and that u is a positive stable classical solution to (1.3). Then, u has a unique critical point and the level curves of u are convex in Ω.*

Note also that under more restrictive assumptions on the nonlinearity f, it is known that any positive solution (whether stable or not) of (1.3) in a convex domain has convex level sets. This is true, for example, if $f(u) = u^p$, $p \in (0, 1)$ and any dimension N. We refer the reader to [134, 135, 138] for more details.

8.3 Beyond the stable branch

8.3.1 Turning point

In Section 3.3, we studied the branch of stable solutions to semilinear elliptic equations of the form (3.1). If the extremal solution is bounded, it can be shown that (λ^*, u^*) is a turning point on the bifurcation diagram, that is, solutions (λ, u) to (3.1) close to the extremal solution belong to a curve of solutions bending back at λ^*, in which the lower part is the stable branch and the upper part consists of solutions of higher Morse index (compare to Figure P.1).

Theorem 8.3.1 ([63]) *Assume $N \geq 1$. Let $\Omega \subset \mathbb{R}^N$ denote a smoothly bounded domain. Assume that $f \in C^3(\mathbb{R})$, $f > 0$. In addition, assume that f is nondecreasing and that f is superlinear in the sense of (3.26). In addition, assume that the extremal solution u^* to (3.1) is bounded. Then, there exists $\delta > 0$ such that the solutions to (3.1) near (λ^*, u^*) form a curve $\{(\lambda(s), u(s)) : s \in (-\delta, \delta)\}$, where*

- *the map $s \mapsto (\lambda(s), u(s))$ is twice continuously differentiable from $(-\delta, \delta)$ to $\mathbb{R} \times C_0^{2,\alpha}(\overline{\Omega})$,*

- *for $s = 0$, $\lambda(0) = \lambda^*, u(0) = u^*$,*

186

- *for $s \in (-\delta, 0)$, $\lambda(s) < \lambda^*$ and $u(s)$ is the minimal solution associated to $\lambda(s)$,*

- *for $s \in (0, \delta)$, $\lambda(s) < \lambda^*$ and $u(s)$ is an unstable solution to (3.1), and*

- *$\lambda'(0) = 0$ and $\lambda''(0) < 0$.*

Let us also mention that, at least for the Gelfand problem in an arbitrary domain, there are infinitely many bifurcation points on the solution curve whenever $3 \leq N \leq 9$. In the case of the ball, Figure P.1 shows that these are in fact turning points.

Theorem 8.3.2 ([71]) *Assume $3 \leq N \leq 9$. Let $\Omega \subset \mathbb{R}^N$ denote a smoothly bounded domain. Assume that $f : \mathbb{R} \to \mathbb{R}$ is analytic, $f > 0$, and $f'(u) \sim ce^u$, as $u \to +\infty$, for some $c > 0$. Then, there exists an unbounded connected curve $\hat{T} = \{(\lambda(s), u(s)) : s \geq 0\}$ starting from $(0,0)$ such that $\|u(s)\|_{C^1(\overline{\Omega})} + \lambda(s) \to +\infty$, as $s \to +\infty$. Moreover, $-\Delta - \lambda(s)f'(u(s)) : C_0^{2,\alpha}(\overline{\Omega}) \to C^\alpha(\overline{\Omega})$ is invertible except at isolated points, called bifurcation points, and given any bounded set $S \subset C([0,1]) \times \mathbb{R}$, the set $\hat{T} \setminus S$ contains infinitely many bifurcation points.*

Remark 8.3.1 *The bifurcation points can be turning points, as is the case for the Gelfand problem in the ball, see Figure P.1, but they can also be secondary bifurcation points, for example, if Ω is an annulus (see [165]).*

8.3.2 Mountain-pass solutions

When the problem is subcritical (with respect to the usual Sobolev exponent $p_S(N) = \frac{N+2}{N-2}$), it is possible to continue the bifurcation branch beyond the neighborhood of the turning point given in Theorem 8.3.1.

Theorem 8.3.3 ([31,63]) *Let Ω denote a smoothly bounded domain of $\mathbb{R}^N$, $N \geq 1$ and $p \in (1, p_S(N)]$. Then, for every $\lambda \in (0, \lambda^*)$, there exists a mountain-pass solution U_λ to*

$$\begin{cases} -\Delta u = \lambda(1+u)^p & \text{in } \Omega, \\ \quad u = 0 & \text{on } \partial\Omega. \end{cases}$$

Remark 8.3.2 *In nonconvex domains, unlike the case where Ω is a ball (recall Figure P.1), there can be more than one mountain-pass solution for a fixed value of λ (see, for example, [118]). However, in a left neighborhood of λ^*, the mountain-pass solution is necessarily unique (see [154]). $\square$*

8.3.3   Uniqueness for small λ

Take another look at Figure P.1 and observe that for $N \geq 3$ and $\lambda > 0$ small enough, the Gelfand problem has a unique solution: the stable solution. In this section, we show that this uniqueness result can be extended to the case of quite general supercritical nonlinearities.

Theorem 8.3.4 ([199]) *Let $N \geq 3$. Assume $\Omega \subset \mathbb{R}^N$ is smoothly bounded and star-shaped. Assume that $f \in C^2(\mathbb{R})$, $f > 0$, is supercritical in the following sense:*

$$\limsup_{t \to +\infty} \frac{F(t)}{tf(t)} < \frac{1}{2^*} = \frac{1}{2} - \frac{1}{N}, \tag{8.7}$$

where $F(t) = \int_0^t f(s)\, ds$.

Then, there exists $\lambda_u > 0$ such that for all $0 < \lambda < \lambda_u$, there exists at most one classical solution to (3.1).

In order to prove Theorem 8.3.4, we first establish two auxiliary results.

Lemma 8.3.1 *Let $N \geq 2$. Let $\Omega \subset \mathbb{R}^N$ denote a domain with C^1 boundary. Assume that Ω is star-shaped with respect to the origin, that is, for each $x \in \overline{\Omega}$, the line segment*

$$\{tx \ : \ t \in [0,1]\}$$

lies in $\overline{\Omega}$. Then,

$$x \cdot n(x) \geq 0 \qquad \text{for all } x \in \partial\Omega,$$

where n is the outward unit normal to the boundary of Ω.

Proof. We follow [93]. Since $\partial\Omega$ is C^1, for every $\epsilon > 0$, there exists $\delta > 0$ such that $|y - x| < \delta$ and $y \in \Omega$ imply $n(x) \cdot \frac{y-x}{|y-x|} \leq \epsilon$. In particular,

$$\limsup_{\substack{y \to x \\ y \in \overline{\Omega}}} n(x) \cdot \frac{y - x}{|y - x|} \leq 0.$$

Let $y = tx$, $t \in [0,1]$. Then, $y \in \overline{\Omega}$, since Ω is star-shaped. Thus,

$$n(x) \cdot \frac{x}{|x|} = -\lim_{t \to 1^-} n(x) \cdot \frac{tx - x}{|tx - x|} \geq 0.$$

$\square$

188

Lemma 8.3.2 ([181]) *Let $N \geq 2$. Let $\Omega \subset \mathbb{R}^N$ denote a smoothly bounded domain and let $w \in C^2(\Omega) \cap C^1(\overline{\Omega})$. Then,*

$$\int_\Omega \Delta w \, (x \cdot \nabla w) \, dx =$$
$$\frac{N-2}{2} \int_\Omega |\nabla w|^2 \, dx - \frac{1}{2} \int_{\partial\Omega} |\nabla w|^2 x \cdot n \, d\sigma + \int_{\partial\Omega} (\nabla w \cdot n)(\nabla w \cdot x) d\sigma, \quad (8.8)$$

where n denotes the outward unit normal to $\partial\Omega$. If in addition, w is constant on the boundary of Ω, then

$$\int_\Omega \Delta w \, (x \cdot \nabla w) \, dx = \frac{N-2}{2} \int_\Omega |\nabla w|^2 \, dx + \frac{1}{2} \int_{\partial\Omega} |\nabla w|^2 x \cdot n \, d\sigma. \quad (8.9)$$

Proof. Using integration by parts, we have

$$\int_\Omega \Delta w \, (x \cdot \nabla w) \, dx = \sum_{i,j=1...N} \int_\Omega w_{ii} x_j w_j \, dx =$$
$$\sum_{i,j=1...N} \left(-\int_\Omega w_i (x_j w_j)_i \, dx + \int_{\partial\Omega} w_i n_i x_j w_j \, d\sigma \right)$$
$$= \sum_{i,j=1...N} \left(-\int_\Omega w_i \delta_{ij} w_j \, dx - \int_\Omega w_i w_{ij} x_j \, dx + \int_{\partial\Omega} w_i n_i x_j w_j \, d\sigma \right)$$
$$= -\int_\Omega |\nabla w|^2 \, dx - \frac{1}{2} \int_\Omega \nabla(|\nabla w|^2) \cdot x \, dx + \int_{\partial\Omega} (\nabla w \cdot n)(\nabla w \cdot x) \, d\sigma$$
$$= \frac{N-2}{2} \int_\Omega |\nabla w|^2 \, dx - \frac{1}{2} \int_{\partial\Omega} |\nabla w|^2 x \cdot n \, d\sigma + \int_{\partial\Omega} (\nabla w \cdot n)(\nabla w \cdot x) \, d\sigma.$$

If in addition, w is constant on $\partial\Omega$, then $\nabla w = |\nabla w| n$ on $\partial\Omega$ and so (8.9) follows. □

Proof of Theorem 8.3.4. We follow [87]. Up to a translation of space, we may assume that Ω is star-shaped with respect to the origin. Assume that (3.1) has two solutions, u and w, for a given $\lambda > 0$. Without loss of generality, we may assume that $u = u_\lambda$ is the minimal solution to (3.1). In particular, $w = u + v$, for some $v \geq 0$ and

$$\lim_{\lambda \to 0^+} \|u_\lambda\|_{L^\infty(\Omega)} = 0. \quad (8.10)$$

The above equality follows, for example, from the fact that $\varepsilon\zeta_0$, where ζ_0 solves (3.8), is a supersolution to (3.1) for $\lambda > 0$ sufficiently small.

Hence, $v = w - u$ solves

$$-\Delta v = \lambda \left[f(u+v) - f(u) \right], \qquad \text{in } \Omega. \tag{8.11}$$

Multiply (8.11) by v and integrate. Then,

$$\int_\Omega |\nabla v|^2 \, dx = \lambda \int_\Omega v \left[f(u+v) - f(u) \right] \, dx. \tag{8.12}$$

Multiply (8.11) by $x \cdot \nabla v$ and integrate. Using Pohozaev's identity (Lemma 8.3.2), it follows that

$$\frac{N-2}{2} \int_\Omega |\nabla v|^2 \, dx + \frac{1}{2} \int_{\partial\Omega} |\nabla v|^2 (x \cdot n) \, d\sigma = -\lambda \int_\Omega [f(u+v) - f(u)] x \cdot \nabla v \, dx =$$

$$= -\lambda \int_\Omega [f(u+v)\nabla(u+v) - f(u)\nabla u - (f(u)\nabla v + vf'(u)\nabla u)] \cdot x \, dx +$$

$$+ \lambda \int_\Omega \left[f(u+v) - f(u) - f'(u)v \right] x \cdot \nabla u \, dx$$

$$= -\lambda \int_\Omega \nabla[F(u+v) - F(u) - f(u)v] \cdot x \, dx +$$

$$+ \lambda \int_\Omega \left[f(u+v) - f(u) - f'(u)v \right] x \cdot \nabla u \, dx$$

$$= N\lambda \int_\Omega \left[F(u+v) - F(u) - f(u)v \right] \, dx +$$

$$+ \lambda \int_\Omega (x \cdot \nabla u) \left[f(u+v) - f(u) - f'(u)v \right] \, dx. \tag{8.13}$$

Set $\eta = \limsup_{t \to +\infty} \frac{F(t)}{tf(t)}$. By (8.7), there exists $\eta_1 \in (\eta, 1/2^*)$. By (8.10) and elliptic regularity, given $\varepsilon > 0$, we can choose $\lambda > 0$ sufficiently small so that $|x \cdot \nabla u| \leq \varepsilon$ in Ω. Define

$$h_\varepsilon(u, v) := N \left[F(u+v) - F(u) - f(u)v \right] + \varepsilon |f(u+v) - f(u) - f'(u)v|$$
$$- N\eta_1 v \left[f(u+v) - f(u) \right].$$

Since f is C^2 and since (8.7) holds, the function $h_\varepsilon(u, v)/v^2$ is bounded above by some constant K, uniformly in ε, provided $\varepsilon > 0$ is sufficiently small. Since Ω is star-shaped, we also have $x \cdot n \geq 0$ on $\partial\Omega$, by Lemma 8.3.1. Using (8.12) and (8.13), we deduce that

$$\frac{N-2}{2} \int_\Omega |\nabla v|^2 \, dx \leq N\lambda K \int_\Omega v^2 \, dx + N\eta_1 \int_\Omega |\nabla v|^2 \, dx.$$

Applying Poincaré's inequality, the condition

$$\lambda < \frac{1}{\lambda_1(-\Delta;\Omega)K}\left(\frac{1}{2^*}-\eta_1\right)$$

implies $v = 0$ and the desired result follows. $\qquad\qquad\qquad\qquad\square$

8.3.4 Regularity of solutions of bounded Morse index

We saw in Chapter 4 that in low dimensions, stable solutions of (1.3) are bounded for a quite general class of nonlinearities. We also saw that for the specific nonlinearities $f(u) = e^u$ and $f(u) = (1 + u)^p$ solutions of bounded Morse index can be uniformly controlled whenever $N \leq 9$. For general nonlinearities, the regularity theory of solutions of bounded Morse index is still to be established (see [76] for partial results). We mention the following result, applying to quite general *subcritical* nonlinearities.

Theorem 8.3.5 ([224]) *Let Ω denote a smoothly bounded domain of $\mathbb{R}^N$, $N \geq 1$. Let $f \in C^1(\mathbb{R})$ be such that*

- *(superlinearity) there exists $\mu > 0$ such that*

$$f'(u)u^2 \geq (1+\mu)f(u)u > 0 \qquad \text{for } |u| > M, \text{ and}$$

- *(subcritical growth) there exists $\theta \in (0,1)$ such that*

$$\frac{2N}{N-2}F(u) \geq (1+\theta)f(u)u, \qquad \text{for } |u| > M$$

where $F(u) = \int_0^u f(s)\,ds$.

Then, given any solution $u \in C^2(\overline{\Omega})$ to (1.3),

$$\|u\|_{L^\infty(\Omega)} \leq C(ind(u)+1)^\beta,$$

where β depends on μ, N, θ only.

8.4 The parabolic equation

As discussed in Section 1.4, solutions to (1.3) can be seen as stationary states to the corresponding nonlinear heat equation (1.22). In addition, a solution must at least be stable, in the sense of Equation (1.5), in order to be asymptotically stable (as defined in Section 1.4).

Next, we describe deeper connections between the elliptic equation and its parabolic counterpart. For $f \in C^1(\mathbb{R})$, $u_0 \in C_0^2(\overline{\Omega})$, it is well known that (1.22) is well posed: there exists a maximal time $T \in (0, +\infty]$ and a unique solution $v \in C^1([0, T); C^2(\overline{\Omega}))$, see, for example, [184]. The first natural question is to determine whether the solution is global ($T = +\infty$) or whether it blows up in finite time. This is very much related to the size of the initial datum u_0 and to the existence of a stationary solution.

Theorem 8.4.1 ([29]) *Let $N \geq 1$ and let Ω be a smoothly bounded domain of $\mathbb{R}^N$. Assume that $f \in C^1(\mathbb{R})$ is a nondecreasing convex function such that $f(0) > 0$. Assume that there exists an L^1-weak solution u to (1.3). Then, for any initial condition $u_0 \in C_0^2(\overline{\Omega})$ such that $0 \leq u_0 \leq u$, the solution to (1.22) is global. Conversely, if in addition, we have*

$$\int_{t_0}^{+\infty} \frac{dt}{f(t)} < +\infty,$$

and if (1.22) has a global solution for some $u_0 \in C_0^2(\overline{\Omega})$, $u_0 \geq 0$, then there exists a weak solution to (1.3).

See also [14, 111, 112, 140] for earlier results. As an immediate corollary, see the following.

Corollary 8.4.1 ([29]) *Let $N \geq 1$, Ω be a smoothly bounded domain of $\mathbb{R}^N$, and $\lambda > 0$. Assume that $f \in C^1(\mathbb{R})$ is a nondecreasing convex function such that $f(0) > 0$. In addition, assume that f is superlinear in the sense of (4.17). Then, the solution to*

$$\begin{cases} \dfrac{\partial v}{\partial t} - \Delta v = \lambda f(v) & \text{in } \Omega \times (0, T), \\ v = 0 & \text{on } \partial\Omega \times (0, T), \\ v(x, 0) = 0 & \text{for } x \in \Omega, \end{cases} \tag{8.14}$$

is global if and only if $\lambda \leq \lambda^*$, where λ^* is the extremal parameter associated to (3.1).

Solutions may or may not continue to exist in the weak sense for $t \geq T^*$.

Theorem 8.4.2 ([150]) *Make the same assumptions as in Theorem 8.4.1. In addition, assume that $u_0 \geq 0$ and $-\Delta u_0 \geq f(u_0)$ and that the solution v to (1.22) blows up in finite time T. Then, v blows up completely after T, that is,* *for any nondecreasing sequence (f_n) of bounded continuous functions such that $f_n \nearrow f$ pointwise, the solution v_n to (1.22) with nonlinearity f_n satisfies*

$$\frac{v_n(t,x)}{d_\Omega(x)} \to +\infty, \qquad \text{uniformly for } t \in [T+\varepsilon, +\infty),$$

where $\varepsilon > 0$ is arbitrary and d_Ω is the distance to the boundary of Ω.

The hypothesis on the initial condition u_0 implies that v is a monotone nondecreasing function of time. This turns out to be crucial:

Theorem 8.4.3 ([107, 108]) *Let $3 \leq N \leq 9$ and B be the unit ball of $\mathbb{R}^N$. Fix $\lambda \in (0, \lambda^*)$, where λ^* is the extremal parameter associated to the Gelfand problem, Equation (2.1). Denote by u_λ^k the k-th solution to (2.1) associated to the parameter λ (ordered by its L^∞ norm, see Figure P.1).*
   Then, for any $k \geq 2$, there exists a radial function $u_0 \in C_0^2(\overline{B})$ such that the solution to

$$\begin{cases} \dfrac{\partial v}{\partial t} - \Delta v = \lambda e^v & \text{in } B \times (0,T), \\ v = 0 & \text{on } \partial B \times (0,T), \\ v(x,0) = u_0(x) & \text{for } x \in B, \end{cases} \tag{8.15}$$

satisfies

- *v blows up in finite time.*

- *v can be extended to a global L^1-weak solution.*

- *$v(t,\cdot) \to u_\lambda^0$ as $t \to +\infty$ in $C_{loc}^1((0,1])$.*

- *$v(t,\cdot)$ is well defined and smooth for all $t \in (-\infty, T)$ and $v(t,\cdot) \to u_\lambda^k$, as $t \to -\infty$.*

v is called an L^1 connection between u_λ^0 and u_λ^k.

In the above theorem, we used the following notion: $v \in C([0,T]; L^1(\Omega))$ is an L^1-weak solution to (1.22) if $f(u) \in L^1((0,T) \times \Omega)$ and

$$\int_\Omega [u\varphi]_{s=\tau}^{s=t} \, dx - \int_\tau^t \int_\Omega u\varphi_t \, dxds = \int_\tau^t \int_\Omega (u\Delta\varphi + f(u)\varphi) \, dxds,$$

for all $0 \leq \tau < t < T$ and $\varphi \in C^2([0,T] \times \overline{\Omega})$ with $\varphi = 0$ on $[0,T] \times \partial\Omega$.
   For more results on the parabolic problem, see, for example, [184].

8.5 Other energy functionals

8.5.1 The *p*-Laplacian

A natural generalization of the energy functional (1.1) considered throughout this book is the following

$$\mathcal{E}_\Omega(u) = \frac{1}{p} \int_\Omega |\nabla u|^p \, dx - \int_\Omega F(u) \, dx, \tag{8.16}$$

where $p \in (1, +\infty)$ and $F \in C^2(\mathbb{R})$. Working as in Chapter 1, one easily proves that $\mathcal{E}_\Omega$ is well defined on the space $X = W^{1,p}(\Omega) \cap L^\infty(\Omega)$ and that its critical points satisfy the Euler-Lagrange equation

$$-\Delta_p u := -\nabla \cdot (|\nabla u|^{p-2} \nabla u) = f(u) \qquad \text{in } \Omega$$

in the following weak sense: $u \in W^{1,p}(\Omega)$ and

$$\int_\Omega |\nabla u|^{p-2} \nabla u \cdot \nabla \varphi \, dx = \int_\Omega f(u) \varphi \, dx, \qquad \text{for all } \varphi \in C_c^1(\Omega). \tag{8.17}$$

In general, solutions to (8.17) need not be classical. Take for example $u = |x|^{p'}$, where p' is the conjugate exponent of p. Then, u solves (8.17) with constant right-hand side, yet u is not C^2 if $p > 2$. This lack of regularity can be understood through the loss of ellipticity of the p-Laplace operator near critical points of u. Still, letting

$$Z := \{x \in \Omega : \nabla u(x) = 0\}, \tag{8.18}$$

it can be proven that any solution u to (8.17) belongs to $C_{loc}^{1,\alpha}(\Omega) \cap C^2(\Omega \setminus Z)$ (see [85, 143, 211]). Taking a test function $\varphi \in C_c^1(\Omega \setminus Z)$ supported away from the singular set Z, the function $E(t) = \mathcal{E}_\Omega(u + t\varphi)$ is twice differentiable and $d^2E/dt^2|_{t=0}$ is given by

$$Q_u(\varphi) := \int_\Omega \rho |\nabla \varphi|^2 \, dx + (p-2) \int_\Omega \rho \left(\frac{\nabla u}{|\nabla u|} \cdot \nabla \varphi \right)^2 dx - \int_\Omega f'(u) \varphi^2 \, dx,$$

$$\text{where } \rho = |\nabla u|^{p-2}. \tag{8.19}$$

This leads us to the following natural definition.

Definition 8.5.1 *Let $p \in (1, +\infty)$, $\alpha \in (0,1)$, Ω an open set of $\mathbb{R}^N$, $N \geq 1$, and $f \in C^1(\mathbb{R})$. A function $u \in C_{loc}^{1,\alpha}(\Omega)$ solving (8.17) is stable away from its singular set if $Q_u(\varphi) \geq 0$, for all $\varphi \in C_c^1(\Omega \setminus Z)$.*

The notion of stability near the singular set is more delicate.

When $p \geq 2$, $\rho = |\nabla u|^{p-2} \in C(\Omega)$, since $u \in C^1(\Omega)$. In particular, the function $E(t) = \mathscr{E}_\Omega(u + t\varphi)$ is still twice differentiable, for any $\varphi \in C_c^1(\Omega)$. So, a stable solution to (8.17) should at least satisfy $Q_u(\varphi) \geq 0$, for all $\varphi \in C_c^1(\Omega)$.

When $p \in (1, 2)$, it is not clear in general what the natural notion of stability should be. However, if $f > 0$, then the singular set has zero Lebesgue measure: $|Z| = 0$, see [148]. So, $\rho = |\nabla u|^{p-2}$ is well defined almost everywhere and measurable. This leads us to the following definition.

Definition 8.5.2 *Let* $p \in (1, +\infty)$, $\alpha \in (0, 1)$, Ω *an open set of* $\mathbb{R}^N$, $N \geq 1$, *and* $f \in C^1(\mathbb{R})$. *If* $p < 2$, *assume that* $f > 0$. *A function* $u \in C_{loc}^{1,\alpha}(\Omega)$ *solving* (8.17) *is stable if* $Q_u(\varphi) \geq 0$, *for all* $\varphi \in C_c^1(\Omega)$ *such that*

$$\int_\Omega \rho |\nabla \varphi|^2 \, dx < +\infty, \qquad \text{where } \rho = |\nabla u|^{p-2}.$$

We note that in both cases $p \in (1, 2)$ and $p \geq 2$, it is not yet clear that the above definition includes enough test functions φ for practical purposes. For example, in order to derive the geometric Poincaré formula (Theorem 4.4.1), we used test functions of the form $\varphi = |\nabla u|\psi$. For positive solutions to (8.17), this is indeed a licit choice, thanks to the regularity results given in [67].

Bifurcation diagrams similar to Figure P.1 are given in [59, 131]. Regularity results for stable solutions to (8.17) are addressed in [40, 42, 52, 113, 114, 190, 191]. Liouville results are discussed in [53, 66], the geometric Poincaré formula and the generalization of the De Giorgi conjecture in [99, 100, 103].

8.5.2 The biharmonic operator

Another possible generalization of the energy functional (1.1) is

$$\mathscr{E}_\Omega(u) = \frac{1}{2} \int_\Omega |\Delta u|^2 \, dx - \int_\Omega F(u) \, dx. \qquad (8.20)$$

This energy arises, for example, when describing the deformations of an elastic thin plate, see [117]. Again, working as in Chapter 1, one easily proves that $\mathscr{E}_\Omega$ is well defined on the space $X = H^2(\Omega) \cap L^\infty(\Omega)$ and that its critical points satisfy the Euler-Lagrange equation

$$\Delta^2 u := -\Delta(-\Delta u) = f(u) \qquad \text{in } \Omega. \qquad (8.21)$$

The second variation of energy is given by

$$Q_u(\varphi) := \int_\Omega |\Delta \varphi|^2 \, dx - \int_\Omega f'(u)\varphi^2 \, dx, \qquad \text{for all } \varphi \in C_c^2(\Omega). \qquad (8.22)$$

Hence, the following definition.

Definition 8.5.3 *Let Ω be an open set of $\mathbb{R}^N$, $N \geq 1$, and $f \in C^1(\mathbb{R})$. A function $u \in C^4(\Omega)$ solving (8.21) is stable if $Q_u(\varphi) \geq 0$, for all $\varphi \in C_c^2(\Omega)$.*

When working on bounded domains Ω, equation (8.21) must be complemented with boundary conditions. Among the most studied are: the Dirichlet boundary condition $u = |\nabla u| = 0$ on $\partial\Omega$, the Navier boundary condition $u = \Delta u = 0$ on $\partial\Omega$, and the Steklov boundary condition $u = \Delta u - a\frac{\partial u}{\partial n} = 0$ on $\partial\Omega$.

Throughout this book we have used extensively the maximum principle (see Appendix A). Unfortunately, the maximum principle fails to be true for the biharmonic operator with Dirichlet boundary conditions (see, for example, [202]), unless stringent assumptions are made on the geometry of the domain, for example, Ω is a ball (see [21]). Additional difficulties arise with the spectral theory of the operator, even when one is solely interested in the principal eigenvalue. See the monograph [117].

Equation (8.21) with Navier boundary conditions can be rewritten as a system of equations in the unknown (u, v), where $v = -\Delta u$. If f is non-decreasing, the system is cooperative and so maximum principle tools are again available, at least for smooth domains. Still, other difficulties arise. For example, when proving regularity of stable solutions to the Gelfand problem (see Theorem 4.2.1), we used crucially the elementary calculus identity $\nabla u \cdot \nabla e^{2\alpha u} = \frac{2}{\alpha}|\nabla e^{\alpha u}|^2$, which has no counterpart when the nabla operator ∇ is replaced by the Laplace operator Δ.

Nevertheless, stable solutions (in particular regularity theory) have been actively studied, perhaps because they appear naturally in a physical context: the study of micro-electro-mechanical devices (MEMS), see the monographs [91, 178]. Full bifurcation diagrams (like Figure P.1) have been established in [79, 80] for the exponential and for power-type nonlinearities. It should be noted that in the latter case and for negative exponents only, the diagrams are qualitatively different than for the Laplace operator. The question of regularity of the stable branch is addressed by [10, 15, 55, 60, 61, 77, 105, 106, 160, 161, 218]. Classification results in entire space appear in [9, 105, 116, 219, 223].

8.5.3 The fractional Laplacian

Consider again the energy functional (1.1) in the special case of $\Omega = \mathbb{R}^N$. Using the Fourier transform, the energy can be rewritten as

$$\frac{1}{2}\int_{\mathbb{R}^N} |\xi|^2 |\hat{u}|^2 \, d\xi - \int_{\mathbb{R}^N} F(u) \, dx,$$

where $\hat{u}$ is the Fourier transform of u. Yet another generalization of (1.1) that is being currently investigated consists of using a lower order Sobolev norm:

$$\mathcal{E}(u) = \frac{1}{2} \int_{\mathbb{R}^N} |\xi|^{2s} |\hat{u}|^2 \, d\xi - \int_{\mathbb{R}^N} F(u) \, dx, \qquad (8.23)$$

where $s \in (0,1)$. The Euler-Lagrange equation associated to (8.23) reads

$$(-\Delta)^s u = f(u) \qquad \text{in } \mathbb{R}^N, \qquad (8.24)$$

where, letting $\mathscr{F}^{-1}$ denote the inverse Fourier transform,

$$(-\Delta)^s u := \mathscr{F}^{-1}(|\xi|^{2s} \hat{u}).$$

Up to a constant multiplicative factor, the fractional Laplacian is also expressed through second-order difference quotients of u by

$$(-\Delta)^s u = -\int_{\mathbb{R}^N} \frac{u(x+y) - 2u(x) + u(x-y)}{|y|^{N+2s}} \, dy.$$

Indeed,

$$\mathscr{F}\left(\int_{\mathbb{R}^N} \frac{u(x+y) - 2u(x) + u(x-y)}{|y|^{N+2s}} \, dy \right) = \hat{u}(\xi) \int_{\mathbb{R}^N} \frac{e^{iy\cdot\xi} - 2 + e^{-iy\cdot\xi}}{|y|^{N+2s}} \, dy$$

$$= 2\hat{u}(\xi) \int_{\mathbb{R}^N} \frac{\cos(y\cdot\xi) - 1}{|y|^{N+2s}} \, dy$$

$$= 2\hat{u}(\xi) \int_{\mathbb{R}^N} \frac{\cos(y|\xi| \cdot \frac{\xi}{|\xi|}) - 1}{|y|^{N+2s}} \, dy$$

$$= 2|\xi|^{2s} \hat{u}(\xi) \int_{\mathbb{R}^N} \frac{\cos(z \cdot e_1) - 1}{|z|^{N+2s}} \, dz$$

$$= -c_{N,s} |\xi|^{2s} \hat{u}(\xi).$$

Note in particular, that the fractional Laplacian is a nonlocal operator. There exists at least two other ways to define this operator: as the generator of a random walk with long jumps (see, for example, [213]), or as the Dirichlet-to-Neumann operator of a degenerate elliptic equation in $N+1$ variables (see, for example, [47]). To illustrate the latter point of view, we present here the case $s = 1/2$. Given a function u defined over $\mathbb{R}^N$, let $v = v(x,t) =: Eu$ denote its harmonic extension in $\mathbb{R}^N \times (0,+\infty)$, that is, v solves

$$\begin{cases} \Delta v = 0 & \text{in } \mathbb{R}^N \times (0,+\infty), \\ v(x,0) = u(x) & \text{for all } x \in \mathbb{R}^N. \end{cases}$$

Note that v is well defined and unique in the class of bounded functions if, say, u is smooth and bounded. Now, define the Dirichlet-to-Neumann operator

$$Tu := -\partial_t(Eu)\big|_{t=0}.$$

We claim that $Tu = (-\Delta)^{1/2}u$. To see this, let $w = -\partial_t Eu$. Then,

$$\Delta w = -\partial_t \Delta Eu = 0 \qquad \text{in } \mathbb{R}^N \times (0,+\infty)$$

and $w|_{t=0} = Tu$. Hence, $w = E(Tu)$. But then,

$$T^2u = -\partial_t E(Tu)\big|_{t=0} = -\partial_t w\big|_{t=0} = \partial_{tt} Eu\big|_{t=0}.$$

Since Eu is harmonic, $0 = \Delta Eu = \Delta_x Eu + \partial_{tt}Eu$ and so

$$T^2u = \partial_{tt} Eu\big|_{t=0} = -\Delta_x Eu\big|_{t=0} = -\Delta u,$$

that is, T is a square root of the Laplace operator. With the previous interpretation in mind, we can reformulate (8.24) as a boundary reaction problem for its harmonic extension $v = Eu$, in the case $s = 1/2$:

$$\begin{cases} \Delta v = 0 & \text{in } \mathbb{R}^N \times (0,+\infty), \\ -\partial_t v(x,0) = f(v(x,0)) & \text{for all } x \in \mathbb{R}^N. \end{cases}$$

Working as in Chapter 1, we define stability as follows.

Definition 8.5.4 *A bounded solution $u \in C^2(\mathbb{R}^N)$ to (8.24) is stable if*

$$\int_{\mathbb{R}^N} |(-\Delta)^{s/2}\varphi|^2 \, dx \geq \int_{\mathbb{R}^N} f'(u)\varphi^2 \, dx, \qquad \text{for all } \varphi \in C_c^\infty(\mathbb{R}^N).$$

Seeing again the fractional Laplacian as a Dirichlet-to-Neumann operator over $\mathbb{R}^N \times (0,+\infty)$, stability is equivalently defined through

$$\int_{\mathbb{R}^N \times (0,+\infty)} t^{1-2s}|\nabla\psi|^2 \, dxdt \geq \int_{\mathbb{R}^N} f'(u)\psi^2(x,0) \, dx,$$

for all $\psi \in C_c^\infty(\mathbb{R}^N \times [0,+\infty))$, see, for example, [90]. Regularity of stable solutions of boundary-reaction problems have been investigated by [78], and for the fractional Laplacian in [50]. Liouville theorems are studied by [41, 43, 90, 205]. See also [8, 45, 46, 195, 196] for the theory of nonlocal minimal surfaces.

8.5.4 The area functional

A classical problem in geometry is the following: Determine the graph of a smooth function $u = u(x, y)$ over a two-dimensional bounded open domain Ω, having the least area among all graphs that assume given values at the boundary of Ω. The area of a graph $u : \Omega \subset \mathbb{R}^N \to \mathbb{R}$ can be computed as

$$\mathscr{E}_\Omega(u) = \int_\Omega \sqrt{1 + |\nabla u|^2} \, dx,$$

and the corresponding (Euler-)Lagrange equation is the famous minimal surface equation:

$$-\nabla \cdot \left(\frac{\nabla u}{\sqrt{1 + |\nabla u|^2}} \right) = 0, \qquad \text{in } \Omega. \qquad (8.25)$$

From geometry's point of view, this is equivalent to requesting that the mean curvature of the graph of u, seen as a nonparametric surface in $\mathbb{R}^{N+1}$, vanishes identically. Solutions to (8.25) have been actively investigated. We mention in particular the following Bernstein theorem.

Theorem 8.5.1 ([6, 19, 22, 82, 204]) *Let $u \in C^2(\mathbb{R}^N, \mathbb{R})$ be a solution to the minimal surface equation (Equation (8.25)) on the entire space $\mathbb{R}^N$. Then, u is an affine function, that is, the graph of u is a plane in $\mathbb{R}^{N+1}$, if and only if $N \leq 7$.*

The proof of the above theorem when $N = 2$ follows from a Liouville-type theorem for elliptic operators (see, for example, [96]), while in higher dimensions one uses a connection between minimal graphs defined over $\mathbb{R}^N$ and minimal hypercones in $\mathbb{R}^N$ (see, for instance, [127, 203]). In fact, the stability of such cones (that is, the assumption that the second variation of area is nonnegative) is used crucially, see, for example, [39].

For more on minimal surfaces, see the monograph by [127]. For recent developments on the regularity of stable solutions to equations involving the minimal surface operator, see [151].

8.5.5 Stable solutions on manifolds

What is known when the ambient space $\mathbb{R}^N$ is replaced by a manifold M in the definition of the energy functional (1.1)? More precisely, consider a complete, connected, smooth, m-dimensional manifold M without boundary, endowed with a smooth Riemannian metric $g = (g_{ij})$. Given a bounded open set Ω of M, we are interested in the energy functional defined for $u \in C^2(\overline{\Omega})$ by

$$\mathscr{E}_\Omega(u) = \frac{1}{2} \int_\Omega |\nabla_g u|^2 \, d\sigma - \int_\Omega F(u) \, d\sigma,$$

where ∇_g is the Riemannian gradient on M, $d\sigma$ its volume element, and $|X|^2 = g(X,X)$ for $X \in TM$ the norm induced by the metric g. Using the notations of Section C.2, when M is a submanifold of the Euclidean space $\mathbb{R}^N$, $|\nabla_g u|^2 = \nabla_T u \cdot \nabla_T u$, where $\nabla_T u$ is the tangential gradient of u and $X \cdot Y$ the canonical dot product of two vectors $X, Y \in \mathbb{R}^N$. As in the Euclidean case, a critical point of $\mathscr{E}_\Omega$ is a function $u \in C^2(\overline{\Omega})$ solving

$$-\Delta_M u = f(u) \qquad \text{in } \Omega,$$

where Δ_M is the Laplace-Beltrami operator of M endowed with the metric g. Stable solutions verify

$$\int_\Omega |\nabla_g \varphi|^2 \, d\sigma \geq \int_\Omega f'(u)\varphi^2 \, d\sigma, \qquad \text{for all } \varphi \in C_c^1(\Omega).$$

We shall also need the following definition.

Definition 8.5.5 *A Riemannian manifold M is parabolic if for any $x \in M$, there exists a compact neighborhood Ω of x, such that for any $\varepsilon > 0$, there exists $\varphi_\varepsilon \in C_c^\infty(M)$ such that $\varphi_\varepsilon \equiv 1$ in Ω, and*

$$\int_M |\nabla_g \varphi_\varepsilon|^2 \, d\sigma < \varepsilon.$$

For example, $M = \mathbb{R}^2$ endowed with the standard Euclidean metric is parabolic, while this is not the case of $\mathbb{R}^N$, $N \geq 3$.

We have the following Liouville-type theorems.

Theorem 8.5.2 ([101]) *Let M be a connected manifold and let $u \in C^2(M)$ be a stable solution to*

$$-\Delta_M u = f(u) \qquad \text{in } M. \tag{8.26}$$

Assume that the Ricci curvature of M is nonnegative. In addition, assume that

- *either M is compact,*

- *or M is complete, parabolic, $|\nabla_g u| \in L^\infty(M)$, and Ric_g does not vanish identically.*

Then, u is constant.

The above theorem need not hold on a (noncompact) manifold with zero Ricci curvature. Indeed, $M = \mathbb{R}^2$ endowed with the standard Euclidean metric is parabolic, with identically zero Ricci tensor. However, the function $u(x_1, x_2) = \tanh(x_1/\sqrt{2})$ is a stable nonconstant solution to the two-dimensional Allen-Cahn equation, see Chapter 7. The previous example motivates the following.

Theorem 8.5.3 ([101]) *Let M be a complete, connected Riemannian surface (that is, a complete, connected Riemannian manifold of dimension 2). Assume that the Ricci curvature of M vanishes identically. Let u be a stable solution of (8.26), with $|\nabla_g u| \in L^\infty(M)$.*

Then, any connected component of the level set of u on which $\nabla_g u$ does not vanish, is a geodesic.

Appendix A

Maximum principles

This section is devoted to the various versions of the maximum principle used in this book. We begin by reviewing the model operator Δ.

A.1 Elementary properties of the Laplace operator

Given an open set $\Omega \subset \mathbb{R}^N$, a function $u \in C^2(\Omega)$, and a point $x_0 \in \Omega$, the Taylor expansion of u at x_0 is given by

$$u(x_0 + h) = u(x_0) + Du(x_0).h + \frac{1}{2}D^2u(x_0 + th)(h, h), \qquad \text{(A.1)}$$

where $|h|$ is small, t is some number in the interval $(0, 1)$ and where $Du \in \mathscr{L}(\mathbb{R}^N, \mathbb{R}), D^2u \in \mathscr{B}(\mathbb{R}^n \times \mathbb{R}^N, \mathbb{R})$ denote the first- and second-order differentials of u. The gradient of u at x_0 is defined as the unique vector in $\mathbb{R}^N$ such that $\nabla u(x_0) \cdot h = Du(x_0).h$ for all $h \in \mathbb{R}^N$ (where $a \cdot b$ is the canonical inner product of $a, b \in \mathbb{R}^N$). The Hessian matrix of u at x_0 is the unique matrix in $\mathbb{R}^N \times \mathbb{R}^N$ such that $(Hu(x_0).h) \cdot h = D^2u(x_0)(h, h)$ for all $h \in \mathbb{R}^N$. In particular, if $x_1, x_2, \ldots, x_N$ denote coordinates in an orthonormal basis of $\mathbb{R}^N$, (A.1) can be rewritten in the familiar form

$$u(x_0 + h) = u(x_0) + \nabla u(x_0) \cdot h + \frac{1}{2}(Hu(x_0 + th).h) \cdot h$$

$$= u(x_0) + \sum_{i=1}^{N} \frac{\partial u}{\partial x_i}(x_0)h_i + \frac{1}{2}\sum_{i,j=1}^{N} \frac{\partial^2 u}{\partial x_i \partial x_j}(x_0 + th)h_i h_j.$$

Definition A.1.1 *The Laplacian of $u \in C^2(\Omega)$ is defined as the trace of the Hessian matrix of u. In particular, if $x_1, x_2, \ldots, x_N$ denote coordinates in an orthonormal basis of $\mathbb{R}^N$, then*

$$\Delta u = \frac{\partial^2 u}{\partial x_1^2} + \cdots + \frac{\partial^2 u}{\partial x_N^2}.$$

We begin by listing the fundamental invariance properties of the Laplace operator:

Proposition A.1.1 *Consider a function $u \in C^2(\mathbb{R}^N)$, $N \geq 1$.*

The Laplace operator commutes with translations. More precisely, given a vector $\tau \in \mathbb{R}^N$,

$$\Delta(u(\cdot + \tau)) = (\Delta u)(\cdot + \tau). \tag{A.2}$$

The Laplace operator commutes with orthogonal transformations. More precisely, if $T \in \mathscr{L}(\mathbb{R}^n)$ is such that $|Tx| = |x|$ for all $x \in \mathbb{R}^N$, then

$$\Delta(u \circ T) = (\Delta u) \circ T. \tag{A.3}$$

The Laplace operator scales like a homogeneous polynomial of degree 2, that is, given $\lambda \in \mathbb{R}$,

$$\Delta(u(\lambda \cdot)) = \lambda^2 (\Delta u)(\lambda \cdot). \tag{A.4}$$

Proof. Equations (A.2) and (A.4) are straightforward. For (A.3), we simply observe that orthonormal bases are preserved by an orthogonal transformation and then apply Definition A.1.1. Let $\{e_i\}_{i=1,\ldots,N}$ denote an orthonormal basis of $\mathbb{R}^N$ and $e_i' = Te_i$ for $i = 1, \ldots, N$. Then, $|e_i'| = |Te_i| = |e_i| = 1$. In addition, we have for $i \neq j$,

$$|e_i' + e_j'|^2 = \left|e_i'\right|^2 + \left|e_j'\right|^2 + 2e_i' \cdot e_j' = \left|e_i\right|^2 + \left|e_j\right|^2 + 2e_i' \cdot e_j'.$$

Now, we also have

$$|e_i' + e_j'|^2 = |T(e_i + e_j)|^2 = |e_i + e_j|^2 = \left|e_i\right|^2 + \left|e_j\right|^2,$$

since $e_i \cdot e_j = 0$. Hence, $e_i' \cdot e_j' = 0$ and $\{e_i'\}_{i=1,\ldots,N}$ is orthonormal. $\square$

The simplest equation involving the Laplace operator is the one satisfied by harmonic functions.

Definition A.1.2 *Let $u \in C^2(\Omega)$. u is harmonic in Ω if $\Delta u = 0$ in Ω.*

Let us look for harmonic functions that have the same invariance properties as the Laplace operator. Translation invariant functions are simply the constant functions. A function u that is invariant under all orthogonal transformations is radial, that is, u is of the form $u(x) = u(r)$, where $r = |x|$. Now, $r^2 = x_1^2 + \cdots + x_N^2$. So, for $i = 1, \ldots, N$, $r \partial r / \partial x_i = x_i$. So,

$$\frac{\partial u}{\partial x_i} = \frac{du}{dr}\frac{\partial r}{\partial x_i} = \frac{du}{dr}\frac{x_i}{r}.$$

Differentiating once more,

$$\frac{\partial^2 u}{\partial x_i^2} = \frac{d^2 u}{dr^2}\left(\frac{x_i}{r}\right)^2 + \frac{du}{dr}\left(\frac{1}{r} - \frac{x_i^2}{r^3}\right).$$

Summing over i, we finally obtain

$$\Delta u = \frac{d^2 u}{dr^2} + \frac{N-1}{r}\frac{du}{dr}. \tag{A.5}$$

If u is harmonic, it follows that $v = du/dr$ satisfies a first-order differential equation, which can be solved explicitly, using separation of variables. We obtain that any radial harmonic function u must take the form $u(r) = a + br^{2-N}$, if $N \neq 2$ and $u(r) = a + b\ln r$, if $N = 2$. Note that such functions are harmonic in $\Omega = \mathbb{R}^N \setminus \{0\}$ but *not* in $\mathbb{R}^N$, if $b \neq 0$. We shall be particularly interested in the so-called fundamental solution of the Laplace operator given for $N \neq 2$ by

$$\Gamma(x) = \Gamma(r) = \frac{1}{N(N-2)|B|}r^{2-N}, \tag{A.6}$$

where $|B|$ denotes the volume of the unit ball in $\mathbb{R}^N$, and by

$$\Gamma(x) = \Gamma(r) = -\frac{1}{2\pi}\ln r, \tag{A.7}$$

if $N = 2$. Using Green's formula in $\Omega = B(0,R) \setminus B(0,\varepsilon)$, with $R > 0$ large and $\varepsilon > 0$ converging to 0, one can easily check (see, for example, Section 2.2 in [93]) that Γ solves

$$-\Delta\Gamma = \delta_0 \quad \text{in } \mathscr{D}'(\mathbb{R}^N), \tag{A.8}$$

where δ_0 denotes the Dirac mass at the origin.

Exercise A.1.1 Find harmonic functions that remain invariant under the transformation $u \to \lambda^{-2}u(\lambda\cdot)$.

Exercise A.1.2 Given $N \geq 1$, find a function Γ_2 solving

$$\Delta^2 \Gamma_2 := \Delta(\Delta \Gamma_2) = \delta_0 \qquad \text{in } \mathcal{D}'(\mathbb{R}^N).$$

We continue our discussion with the celebrated mean-value formula. Recall that Taylor's formula (A.1) expresses the value $u(x_0 + h)$ at a point $x_0 + h$ close to x_0 in terms of values of u and its differentials at x_0. Suppose now that instead of the *value* of u at x, one is interested in the *mean* of u over a given ball $B(x_0, R) \subset\subset \Omega$ or its boundary $\partial B(x_0, R)$.

Proposition A.1.2 ([115]) *Consider an open set $\Omega \subset \mathbb{R}^N$, $N \geq 2$ and a function $u \in C^2(\Omega)$. For any ball $B(x_0, R) \subset \Omega$, there exists $y_0, y_1 \in B(x_0, R)$ such that*

$$\fint_{\partial B(x_0,R)} u \, d\sigma = u(x_0) + \frac{1}{2N} R^2 \Delta u(y_0) \tag{A.9}$$

and

$$\fint_{B(x_0,R)} u \, dx = u(x_0) + \frac{1}{2(N+2)} R^2 \Delta u(y_1). \tag{A.10}$$

Remark A.1.1 *Much like Taylor expansions, higher-order expansions of the form*

$$\fint_{\partial B(x_0,R)} u \, d\sigma = u(x_0) + \sum_{j=1}^{k-1} a_j R^{2j} \Delta^j u(x_0) + a_k R^{2k} \Delta^k u(y_0)$$

hold true for $u \in C^{2k}(\Omega)$. A formula with an integral remainder term is also available. For all these results, see [168].

Proof. Working with v defined for $x \in B = B(0,1)$ by $v(x) = u(Rx + x_0)$ if necessary, we may always assume that $x_0 = 0$ and $R = 1$. We also restrict to the case $N \geq 3$, the case $N = 2$ being similar. Consider the fundamental solution of the Laplace operator, that is, the radial function Γ given by (A.6). For $\varepsilon > 0$, integrate $(-\Delta\Gamma)u$ on $B \setminus B(0,\varepsilon)$, apply Green's identity and let $\varepsilon \to 0$. Then,

$$u(0) = \int_B (-\Delta u)\Gamma \, dx + \int_{\partial B} \left(-\frac{\partial \Gamma}{\partial \nu} u + \Gamma \frac{\partial u}{\partial \nu} \right) d\sigma.$$

By (A.6), $-\partial\Gamma/\partial v|_{\partial B} = 1/|\partial B|$. It follows that

$$
\begin{aligned}
u(0) &= \int_B (-\Delta u)\Gamma \, dx + \fint_{\partial B} u \, d\sigma + \Gamma(1)\int_{\partial B} \frac{\partial u}{\partial v} \, d\sigma \\
&= \int_B (\Gamma(x) - \Gamma(1))(-\Delta u) \, dx + \fint_{\partial B} u \, d\sigma \\
&= \int_0^1 (\Gamma(r) - \Gamma(1))\left(\int_{\partial B_r} (-\Delta u) \, d\sigma\right) dr + \fint_{\partial B} u \, d\sigma \\
&= \int_0^1 \frac{1}{N-2}\left(r - r^{N-1}\right)\left(\fint_{\partial B_r} (-\Delta u) \, d\sigma\right) dr + \fint_{\partial B} u \, d\sigma \\
&= \fint_{\partial B_{r_0}} (-\Delta u) \, d\sigma \int_0^1 \frac{1}{N-2}\left(r - r^{N-1}\right) dr + \fint_{\partial B} u \, d\sigma \\
&= \frac{1}{2N}(-\Delta u)(y_0) + \fint_{\partial B} u \, d\sigma,
\end{aligned}
$$

where $r_0 \in (0,1), y_0 \in B$ have been obtained using the first mean-value theorem for integration. We have just proved (A.9). Equation (A.10) follows by observing that

$$
\fint_B u \, dx = \frac{1}{|B|}\int_0^1 \left(|\partial B_r|\fint_{\partial B_r} u \, d\sigma\right) dr
$$

and integrating (A.9) accordingly. $\qquad\square$

The mean-value formulae are of particular interest when dealing with sub- and superharmonic functions.

Definition A.1.3 *A function $u \in C^2(\Omega)$ is superharmonic in Ω if*

$$
-\Delta u \geq 0 \qquad \text{in } \Omega.
$$

Similarly, a function $u \in C^2(\Omega)$ is subharmonic in Ω if $-\Delta u \leq 0$ in Ω.

Exercise A.1.3 Prove that a function $u \in C^2(\Omega)$ is superharmonic in Ω if and only if for all balls $B(x_0, R) \subset \Omega$,

$$
\fint_{B(x_0,R)} u \, dx \leq u(x_0). \tag{A.11}
$$

The notion of superharmonic function (respectively subharmonic) can be extended to solutions having weaker regularity.

- Prove that $u \in L^1_{loc}(\Omega)$ satisfies $-\Delta u \geq 0$ in $\mathscr{D}'(\Omega)$ if and only if (A.11) holds for almost all $x_0 \in \Omega$ and all $R > 0$ such that $B(x_0, R) \subset \Omega$.

- Let $u \in L^1_{loc}(\Omega)$ solve $-\Delta u = 0$ in $\mathscr{D}'(\Omega)$. Prove that $u \in C^2(\Omega)$.

- Let $u \in C^1(\Omega)$ solve $-\Delta u = 0$ in $\mathscr{D}'(\Omega \setminus \{0\})$. Prove that $u \in C^2(\Omega)$ and u is harmonic in Ω.

Exercise A.1.4 Let B be the unit ball of $\mathbb{R}^N$ and $B^+ = \{x \in B \ : \ x_N > 0\}$. Assume that $u \in C^2(B^+) \cap C^1(\overline{B}^+)$ solves

$$\begin{cases} -\Delta u = 0 & \text{in } B^+ \\ \quad u = 0 & \text{on } [x_N = 0]. \end{cases}$$

Let $\tilde{u}$ be the odd extension of u through $[x_N = 0]$, that is,

$$\tilde{u}(x) = \begin{cases} u(x) & \text{if } x \in B^+, \\ -u(x', -x_N) & \text{if } x \in B \setminus B^+. \end{cases}$$

Prove that $\tilde{u}$ is harmonic in B.

Exercise A.1.5 Let $\lambda > 0$ and $u \in C^2(\Omega)$, $u \geq 0$, satisfy

$$-\Delta u \leq \lambda u \qquad \text{in } \Omega.$$

Prove that the function v defined for $(x, t) \in \Omega \times \mathbb{R}$ by $v(x, t) = e^{\sqrt{\lambda} t} u(x)$ is subharmonic. Deduce that there exists a constant $C = C(N) > 0$, such that

$$u(x_0) \leq C e^{\sqrt{\lambda} R} \fint_{B(x_0, R)} u \, dx,$$

for every ball $B(x_0, R) \subset \Omega$.

A.2 The maximum principle

A crucial corollary of the mean-value formulae is the following strong maximum principle.

Proposition A.2.1 *Let Ω denote a domain of $\mathbb{R}^N$, $N \geq 1$ and $u \in C^2(\Omega)$ a superharmonic function. Then, u cannot achieve an interior point of minimum, unless u is constant.*

Proof. Let $m = \inf_\Omega u$ and assume that there exists a point $x_0 \in \Omega$ such that $u(x_0) = m$. Then, $F = \{x \in \Omega : u(x) = m\}$ is nonempty. Since u is continuous, F is relatively closed in Ω. Now take $x_1 \in F$ and $R > 0$ so small that $B(x_1, R) \subset \Omega$. Apply the mean-value formula (A.11). Then,

$$m \leq \fint_{B(x_1,R)} u \, dx \leq u(x_1) = m.$$

In particular, $B(x_1, R) \subset F$ so F is open and closed in Ω. Since Ω is connected, we deduce that $F = \Omega$, that is, $u \equiv m$. $\qquad\square$

The strong maximum principle immediately implies the following strong comparison principle.

Proposition A.2.2 *Let Ω denote a bounded domain of $\mathbb{R}^N$, $N \geq 1$ and $u \in C^2(\Omega) \cap C^0(\overline{\Omega})$ a function satisfying*

$$\begin{cases} -\Delta u \geq 0 & \text{in } \Omega, \\ u \geq 0 & \text{on } \partial\Omega. \end{cases} \tag{A.12}$$

Then, either $u \equiv 0$, either $u > 0$ in Ω.

Proof. Let $m = \min_{\overline{\Omega}} u$ and assume that u is nonconstant (the remaining case being straightforward). By the strong maximum principle, u may not achieve m at an interior point, hence $m = \min_{\partial\Omega} u \geq 0$ and $u > 0$ in Ω. $\qquad\square$

A.3 Harnack's inequality

Another useful consequence of the mean-value formulae is the following inequality.

Proposition A.3.1 *Let $\Omega \subset \mathbb{R}^N$, $N \geq 1$ denote a domain. Take a point $x_0 \in \Omega$ and $r > 0$ such that $B(x_0, 4r) \subset \Omega$. Assume that u is nonnegative and harmonic in Ω. Then,*

$$\sup_{B(x_0,r)} u \leq 3^N \inf_{B(x_0,r)} u.$$

Proof. Take any two points $y, z \in B(x_0, r)$ and apply the mean-value theorem (Proposition A.1.2). Then,

$$u(y) = \frac{1}{|B_r|} \int_{B(y,r)} u(x) \, dx \leq \frac{1}{|B_r|} \int_{B(z,3r)} u(x) \, dx =$$

$$= 3^N \frac{1}{|B_{3r}|} \int_{B(z,3r)} u(x) \, dx = 3^N u(z).$$

The desired estimate follows. □

Harnack's inequality can be extended to a much wider class of equations (see, for example, [126]). For our purposes, the following generalization will suffice.

Proposition A.3.2 *Let* $\Omega \subset \mathbb{R}^N$, $N \geq 1$, *denote a domain. Assume that* $V(x)$ *is a bounded function on* Ω *and* $u \in C^2(\Omega)$ *is a nonnegative solution to*

$$-\Delta u = V(x)u \qquad in \ \Omega.$$

Take a point $x_0 \in \Omega$ *and* $r > 0$ *such that* $B(x_0, 6r) \subset \Omega$. *Then, there exists a constant* $C = C(N, \|V\|_{L^\infty(\Omega)}, r)$ *such that*

$$\sup_{B(x_0,r)} u \leq C \inf_{B(x_0,r)} u.$$

Proof. Let $\lambda = \|V\|_{L^\infty(\Omega)}$. It suffices to prove the proposition for $r < \pi/(10\sqrt{\lambda})$. Take any two points $y, z \in B(x_0, r)$. By Exercise A.1.5, we have

$$u(y) \leq C \fint_{B(y,r)} u \, dx \leq C \fint_{B(z,3r)} u(x) \, dx.$$

Consider the function $v(t,x) = \cos(\sqrt{\lambda}t)u(x)$ defined for $(t,x) \in \mathbb{R} \times \Omega$. Then, v is superharmonic, $v \geq 0$ in $(-5r, 5r) \times \Omega$, and so

$$u(z) = v(0,z) \geq \fint_{B((0,z),5r)} v(t,x) \, dx dt$$

$$\geq \frac{1}{|B((0,z),5r)|} \int_{-4r}^{4r} \cos(\sqrt{\lambda}t) \, dt \int_{B(z,3r)} u(x) \, dx \geq c \fint_{B(z,3r)} u(x) \, dx.$$

Collecting the above two inequalities, the proposition follows. □

A.4 The boundary-point lemma

We have just seen that a superharmonic function u must achieve its minimum on the boundary of the domain. In particular, at any such point of minimum, u must be nonincreasing in the direction of the exterior normal to the boundary. The boundary point lemma states that in fact more can be said, provided $\partial\Omega$ is sufficiently smooth: u must be *strictly* decreasing at such boundary points.

Definition A.4.1 *Let $N \geq 1$. A domain $\Omega \subset \mathbb{R}^N$ satisfies an interior sphere condition at a point $x_0 \in \partial\Omega$, if there exists a ball $B = B(y_0, r_0) \subset \Omega$ such that $x_0 \in \partial\Omega \cap \partial B$.*

Definition A.4.2 *Let $N \geq 1$. A vector $v \in \mathbb{R}^N$ is an interior vector to a ball B if there exists $t_0 > 0$ and $x_0 \in \partial B$ such that $x_0 + tv \in B$ for all t in the interval $(0, t_0)$.*

Proposition A.4.1 *([225]) Let $N \geq 1$ and let $\Omega \subset \mathbb{R}^N$ denote a domain satisfying an interior sphere condition at some point $x_0 \in \partial\Omega$. Then, given any function $u \in C^2(\Omega) \cap C(\overline{\Omega})$ satisfying*

$$\begin{cases} -\Delta u \geq 0 & \text{in } \Omega, \\ u > 0 & \text{in } \Omega, \\ u(x_0) = 0, \end{cases} \tag{A.13}$$

there holds

$$\liminf_{t \to 0^+} \frac{u(x_0 + tv)}{t} > 0, \tag{A.14}$$

where v is any interior vector to B at the point x_0 (and where B is an interior sphere tangent to $\partial\Omega$ at x_0).

Proof. We follow [188]. First, note that the function $v(x) = |x|^{-\lambda}$, $\lambda > N - 2$, satisfies

$$-\Delta v = \lambda(-\lambda + N - 2)|x|^{-\lambda-2} \leq 0 \quad \text{in } \mathbb{R}^N \setminus \{0\}.$$

Next, observe that $u \geq c > 0$ on $\partial B(y_0, r_0/2)$. Now consider

$$v_1(x) = c_1\left(|x - y_0|^{-\lambda} - r_0^{-\lambda}\right), \qquad x \in A = B(y_0, r_0) \setminus B(y_0, r_0/2),$$

where $c_1 > 0$ is chosen so small that $v_1 \leq c \leq u$ on $\partial B(y_0, r_0/2)$. Since, $v_1 = 0$ on $\partial B(y_0, r_0)$, we conclude that $v_1 \leq u$ on ∂A. By the maximum principle, since $-\Delta(u - v_1) \geq 0$ in A, we obtain that $u \geq v_1$ in A. Equation (A.14) holds for v_1 by direct inspection and so (A.14) also holds for u. $\square$

Corollary A.4.1 *Let $N \geq 1$, let $\Omega \subset \mathbb{R}^N$ denote a smoothly bounded domain, and let n denote the exterior normal unit vector to $\partial\Omega$. Then, given any function $u \in C^2(\Omega) \cap C^1(\overline{\Omega})$ solving*

$$\begin{cases} -\Delta u \geq 0 & \text{in } \Omega, \\ u = 0 & \text{on } \partial\Omega, \end{cases} \tag{A.15}$$

there holds at every $x_0 \in \partial\Omega$,

$$-\frac{\partial u}{\partial n}(x_0) > 0. \tag{A.16}$$

In particular, there exists a constant $c > 0$, depending on u and Ω only, such that

$$u(x) \geq c \, d_\Omega(x) \qquad \forall x \in \Omega, \tag{A.17}$$

where $d_\Omega(x) = \text{dist}(x, \partial\Omega)$ is the distance to the boundary.

Proof. For (A.16), we simply note that if B denotes an interior sphere tangent to $\partial\Omega$ at x_0, then $-n(x_0)$ is an interior vector to B at x_0. We then apply (A.14):

$$-\frac{\partial u}{\partial n}(x_0) = \liminf_{t \to 0^+} \frac{u(x_0 - tn(x_0))}{t} > 0.$$

Since $\partial\Omega$ is compact, it follows that $-\frac{\partial u}{\partial n} \geq c > 0$ on $\partial\Omega$. Letting $c > 0$ smaller if necessary, the inequality remains valid on some given neighborhood $\omega \subset \Omega$ of $\partial\Omega$. Now, every $x \in \Omega$ sufficiently close to the boundary has a unique projection $x_0 \in \partial\Omega$. Let $\omega \subset \Omega$ denote a neighborhood of $\partial\Omega$ containing all such points. It clearly suffices to establish (A.17) in ω. Given $x \in \omega$, the segment $[x, x_0]$ lies in $\overline{\omega}$ and by the fundamental theorem of calculus,

$$u(x) = u(x) - u(x_0) =$$

$$= \int_0^1 \frac{d}{dt} u(x_0 + t(x - x_0)) \, dt = \int_0^1 \nabla u(x_0 + t(x - x_0)) \cdot (x - x_0) \, dt =$$

$$= \left(\int_0^1 -\frac{\partial u}{\partial n}(x_0 + t(x - x_0)) \, dt \right) |x - x_0| \geq c \, d_\Omega(x).$$

(A.17) follows. $\qquad\square$

We have seen in (A.17), that superharmonic functions vanishing on $\partial\Omega$, are bounded below by a constant $c > 0$ times the distance to the boundary. The constant c can be further quantified, as the following proposition shows.

Proposition A.4.2 ([163, 226]) *Let $N \geq 1$, let $\Omega \subset \mathbb{R}^N$ denote a smoothly bounded domain and let $d_\Omega(x) = \text{dist}(x, \partial\Omega)$ denote the distance to the boundary of Ω. Assume that $f \geq 0$ belongs to $L^\infty(\Omega)$ and let u denote the solution to*

$$\begin{cases} -\Delta u = f & \text{in } \Omega, \\ \quad\;\; u = 0 & \text{on } \partial\Omega. \end{cases} \tag{A.18}$$

Then,

$$u(x) \geq c \left(\int_\Omega f \, d_\Omega(x) \, dx \right) d_\Omega(x), \qquad \text{for all } x \in \Omega, \qquad (A.19)$$

where $c > 0$ is a constant depending on Ω only.

Proof. We follow [27].
Step 1. For any compact set $K \subset \Omega$, we show that

$$u(x) \geq c \int_\Omega f \, d_\Omega(x) \, dx, \qquad \forall x \in K, \qquad (A.20)$$

where $c > 0$ depends only on K and Ω. To prove (A.20), let $\rho = \text{dist}(K, \partial\Omega)/2$ and cover K by m balls of radius ρ:

$$K \subset B_\rho(x_1) \cup \cdots \cup B_\rho(x_m) \subset \Omega.$$

For $i = 1, \ldots, m$, let ζ_i denote the solution to

$$\begin{cases} -\Delta\zeta_i = \chi_{B_\rho(x_i)} & \text{in } \Omega, \\ u = 0 & \text{on } \partial\Omega, \end{cases}$$

where χ_A denotes the characteristic function of A. The boundary point lemma (Corollary A.4.1) implies that there exists a constant $c > 0$ such that

$$\zeta_i(x) \geq c \, d_\Omega(x) \qquad \forall x \in \Omega \qquad \forall i = 1, \ldots, m.$$

Take $x \in K$ and a ball $B_\rho(x_i)$ containing x. Then, $B_\rho(x_i) \subset B_{2\rho}(x) \subset \Omega$ and since $-\Delta u \geq 0$ in Ω, we conclude that

$$u(x) \geq \fint_{B_{2\rho}(x)} u \, dx = c \int_{B_{2\rho}(x)} u \, dx \geq c \int_{B_\rho(x_i)} u \, dx =$$

$$= c \int_\Omega u(-\Delta\zeta_i) \, dx = c \int_\Omega f \zeta_i \, dx \geq c \int_\Omega f d_\Omega \, dx.$$

Step 2. Fix a smooth compact set $K \subset \Omega$. By (A.20), $u \geq c \int_\Omega f d_\Omega \, dx$ in K, so that it suffices to prove (A.19) for $x \in \Omega \setminus K$. Let w be the solution to

$$\begin{cases} -\Delta w = 0 & \text{in } \Omega \setminus K, \\ w = 0 & \text{on } \partial\Omega, \\ w = 1 & \text{on } \partial K. \end{cases}$$

The boundary point lemma (Corollary A.4.1) gives again

$$w(x) \geq c\, d_\Omega(x) \qquad \forall x \in \Omega \setminus K.$$

Since u is superharmonic and $u \geq c \int_\Omega f\, d_\Omega\, dx$ on ∂K, the maximum principle implies that

$$u(x) \geq c \left(\int_\Omega f\, d_\Omega\, dx \right) w(x) \geq c \left(\int_\Omega f\, d_\Omega\, dx \right) d_\Omega(x) \qquad \forall x \in \Omega \setminus K.$$

This completes the proof. $\qquad\qquad\qquad\qquad\qquad\qquad\qquad\qquad\qquad\qquad$ $\square$

A.5 Elliptic operators

In this section and the next, we shall discuss the generalization of the maximum principle to more general elliptic operators. To this end, observe that the conclusion of Proposition A.2.2 can be divided in two parts:

- (Weak comparison principle) If u is superharmonic in Ω and $u \geq 0$ on $\partial\Omega$, then $u \geq 0$ in Ω.

- (Strong comparison principle) If u is superharmonic in Ω and $u \geq 0$ in all of Ω, then in fact $u > 0$ in Ω, unless $u \equiv 0$.

This two-step procedure will be used in what follows. We consider elliptic operators of the form

$$-Lu = -A(x).Hu + B(x) \cdot \nabla u + V(x)u$$

$$= -\sum_{i,j=1}^{N} a_{ij}(x)\frac{\partial^2 u}{\partial x_i \partial x_j} + \sum_{i=1}^{N} b_i(x)\frac{\partial u}{\partial x_i} + V(x)u, \qquad (A.21)$$

where $a_{ij}, b_i, V \in C(\Omega)$ for all $i,j = 1,\ldots,N$. In addition, we assume that

$$V \geq 0 \qquad \text{in } \Omega. \qquad\qquad (A.22)$$

Definition A.5.1 *An operator of the form (A.21) is said to be elliptic at a point $x \in \Omega$ if the matrix $A(x) = [a_{ij}(x)]$ is positive; that is, if $\lambda(x)$ and $\Lambda(x)$ denote respectively the minimum and maximum eigenvalues of $A(x)$, then*

$$0 < \lambda(x)|\xi|^2 \leq (A(x).\xi) \cdot \xi \leq \Lambda(x)|\xi|^2, \qquad \text{for all } \xi \in \mathbb{R}^N \setminus \{0\}. \quad (A.23)$$

The operator is uniformly elliptic if Λ/λ is bounded in Ω.

Proposition A.5.1 *Let Ω denote a bounded domain of $\mathbb{R}^N$, $N \geq 1$, let L denote an elliptic operator in Ω such that (A.22) holds. Let $u \in C^2(\overline{\Omega})$ denote a function satisfying*

$$\begin{cases} -Lu \geq 0 & \text{in } \Omega, \\ u \geq 0 & \text{on } \partial\Omega. \end{cases}$$

Then, $u \geq 0$ in Ω.

Proof. We follow [126]. Assume first that $-Lu > 0$ in Ω. Assume by contradiction that $u(x_0) < 0$, for some $x_0 \in \Omega$. We may always assume that u achieves its minimum at x_0. Then, $u(x_0) \leq 0$, $\nabla u(x_0) = 0$ and $Hu(x_0)$ is a nonnegative matrix. Since L is elliptic, $A(x_0)$ is positive, hence $-Lu(x_0) = -A(x_0)Hu(x_0) + V(x_0)u(x_0) \leq 0$, contradicting $-Lu > 0$ in Ω.

Now take a subdomain $\Omega' \subset\subset \Omega$. Since $a_{i,j}, b_i, V$ are continuous and since L is elliptic, the quotients $|b_1|/\lambda$ and $|V|/\lambda$ are bounded by some constant b_0 on Ω'. Then, since $a_{11} \geq \lambda$, there exists a constant γ sufficiently large such that

$$L e^{\gamma x_1} = \left(\gamma^2 a_{11} - \gamma b_1 - V\right) e^{\gamma x_1} \geq \lambda \left(\gamma^2 - \gamma b_0 - b_0\right) e^{\gamma x_1} > 0 \qquad \text{in } \Omega'.$$

Take a constant $c > 0$ so large that $e^{\gamma x_1} \leq c$ in Ω. Then, for any $\varepsilon > 0$, $-L(u + \varepsilon(c - e^{\gamma x_1})) > 0$ in Ω' and $u + \varepsilon(c - e^{\gamma x_1}) \geq 0$ on $\partial\Omega'$. Hence, $u + \varepsilon(c - e^{\gamma x_1}) \geq 0$ in Ω'. This being true for all $\varepsilon > 0$, $\Omega' \subset\subset \Omega$, the proposition follows. $\qquad \square$

To prove the strong maximum principle for general elliptic operators, we use the following boundary point lemma.

Lemma A.5.1 ([129, 169]) *Let Ω denote a smoothly bounded domain of $\mathbb{R}^N$, $N \geq 1$. Assume that L is a uniformly elliptic operator such that (A.22) holds. Let $u \in C^2(\overline{\Omega})$ satisfy*

$$-Lu \geq 0 \qquad \text{in } \Omega.$$

Assume that for some $x_0 \in \partial\Omega$, $u(x_0) \leq 0$ and $u(x_0) < u(x)$ for all $x \in \Omega$. Then, the outer normal derivative of u at x_0 satisfies

$$-\frac{\partial u}{\partial n}(x_0) > 0. \tag{A.24}$$

Exercise A.5.1 Adapt the proof of Proposition A.4.1 to establish (A.24).

Corollary A.5.1 ([129, 169]) *Let Ω denote **any** domain of $\mathbb{R}^N$, $N \geq 1$. Assume that L is a uniformly elliptic operator. Let $u \in C^2(\Omega)$ satisfy*

$$\begin{cases} -Lu \geq 0 & \text{in } \Omega, \\ u \geq 0 & \text{in } \Omega. \end{cases}$$

Then, $u > 0$ in Ω, unless $u \equiv 0$.

Proof. Assume by contradiction that u vanishes at some point in Ω, while the set $\Omega^+ = \{x \in \Omega \; : \; u(x) > 0\}$ is nonempty. We first note that $\partial \Omega^+ \cap \Omega \neq \emptyset$. Otherwise, writing $\Omega^0 = \{x \in \Omega \; : \; u(x) = 0\}$, we would have $\Omega = \Omega^+ \sqcup \Omega^0 = \overline{\Omega^+} \sqcup \Omega^0$, contradicting the fact that Ω is connected. So, there exists a point $x_1 \in \partial \Omega^+ \cap \Omega$. Let $d = d(x_1, \partial \Omega) > 0$ and take $a \in \Omega^+$ such that $|a - x_1| < d/3$. In particular, $d(a, \partial \Omega) \geq 2d/3$. Now set $R = d(a, \Omega^0) > 0$, so that $B(a, R) \subset \Omega^+$. Since $R \leq |a - x_1| < d/3$, we also have $\overline{B(a, R)} \subset \Omega$. Then, $u(x_0) = 0$ for some $x_0 \in \partial B(a, R)$ (hence x_0 is a point of minimum of u), while $u > 0$ in $B(a, R)$. Applying Lemma A.5.1, we deduce that $\nabla u(x_0) \neq 0$. This contradicts the fact that x_0 is an interior point of minimum of u.

$\square$

A.6   The Laplace operator with a potential

In this section, we work with elliptic operators of the form

$$-Lu = -\Delta u - V(x)u,$$

where $V \in C^{0,\alpha}(\overline{\Omega})$. This time, we do not make any assumption on the sign of V. The validity of the maximum principle for $-L$ is very much related to its *spectrum*.

Definition A.6.1 *Let Ω denote a smoothly bounded domain of $\mathbb{R}^N$, $N \geq 1$ and let $V \in C(\overline{\Omega})$. Assume that there exists $\lambda \in \mathbb{R}$ and $\varphi \in C^2(\overline{\Omega})$, $\varphi \not\equiv 0$ such that*

$$\begin{cases} -\Delta\varphi - V(x)\varphi = \lambda\varphi & \text{in } \Omega, \\ \varphi = 0 & \text{on } \partial\Omega. \end{cases} \tag{A.25}$$

Then, λ is called an eigenvalue of $-L = -\Delta - V(x)$ (with Dirichlet boundary conditions) and φ an eigenvector associated to λ.

Definition A.6.2 *Let Ω denote a smoothly bounded domain of $\mathbb{R}^N$, $N \geq 1$ and let $V \in C(\overline{\Omega})$. The principal eigenvalue of the operator $-L = -\Delta - V(x)$ is denoted by $\lambda_1 = \lambda_1(-\Delta - V(x); \Omega)$ and defined by*

$$\lambda_1 = \inf_{\substack{\varphi \in C_c^1(\Omega), \\ \|\varphi\|_{L^2(\Omega)} = 1}} \left(\int_\Omega |\nabla\varphi|^2 \, dx - \int_\Omega V(x)\varphi^2 \, dx \right). \tag{A.26}$$

We shall soon verify that λ_1 is indeed an eigenvalue of $-L$. We start by proving that the maximum principle for $-L$ holds if $\lambda_1 > 0$.

216

Appendix A. Maximum principles

Proposition A.6.1 *Let Ω denote a smoothly bounded domain of $\mathbb{R}^N$, $N \geq 1$ and let $V \in C^{0,\alpha}(\overline{\Omega})$. Assume that the principal eigenvalue of the operator $-L = -\Delta - V(x)$ satisfies*

$$\lambda_1(-\Delta - V(x); \Omega) > 0.$$

If $u \in C^2(\overline{\Omega})$ satisfies

$$\begin{cases} -\Delta u - V(x)u \geq 0 & \text{in } \Omega, \\ u \geq 0 & \text{on } \partial\Omega, \end{cases} \tag{A.27}$$

then, $u \geq 0$ in Ω.

Proof. Let u denote a solution to (A.27). Clearly, $u^- \in H_0^1(\Omega)$ and so we may multiply (A.27) by u^- and integrate by parts.

$$0 \leq \int_\Omega \nabla u \nabla u^- \, dx - \int_\Omega V(x) u \, u^- \, dx$$
$$= -\left(\int_\Omega |\nabla u^-|^2 \, dx - \int_\Omega V(x) \left(u^-\right)^2 \, dx \right).$$

Since $\lambda_1(-\Delta - V(x); \Omega) > 0$ is given by (A.26), we deduce that $u^- \equiv 0$. □

We now return to the characterization of the principal eigenvalue λ_1. It turns out that λ_1 is an eigenvalue of $-L$. More precisely, λ_1 is the smallest eigenvalue of $-L$.

Theorem A.6.1 *Let Ω denote a smoothly bounded domain of $\mathbb{R}^N$, $N \geq 1$ and let $V \in C^1(\overline{\Omega})$. Then, $-L = -\Delta - V(x)$ has a smallest eigenvalue, called the principal eigenvalue of $-L$ and denoted by*

$$\lambda_1 = \lambda_1(-\Delta - V(x); \Omega).$$

Furthermore, λ_1 is characterized by either of the following statements.

1. *λ_1 is given by (A.26).*

2. *λ_1 is a simple eigenvalue and there exists an eigenvector φ_1 associated to λ_1 such that $\varphi_1 > 0$ in Ω. Furthermore, if $\varphi > 0$ is an eigenvector associated to an eigenvalue λ, then in fact $\lambda = \lambda_1$.*

3. *λ_1 is the supremum of all $\lambda \in \mathbb{R}$ such that there exists $v \in C^2(\Omega)$ such that $v > 0$ in Ω and*

$$-\Delta v - V(x)v \geq \lambda v \quad \text{in } \Omega. \tag{A.28}$$

217

Proof. Let $\mu_1 = \inf_{\varphi \in C_c^1(\Omega), \|\varphi\|_{L^2(\Omega)}=1} Q(\varphi)$, where

$$Q(\varphi) = \int_\Omega |\nabla \varphi|^2 \, dx - \int_\Omega V(x)\varphi^2 \, dx. \qquad (A.29)$$

Since V is bounded, $Q(\varphi) \geq -\|V\|_{L^\infty(\Omega)}$ whenever $\|\varphi\|_{L^2(\Omega)} = 1$. We deduce that $\mu_1 > -\infty$. Let (φ_n) denote a minimizing sequence for μ_1, that is,

$$\mu_1 \leq Q(\varphi_n) \leq \mu_1 + \frac{1}{n}$$

and $\|\varphi_n\|_{L^2(\Omega)} = 1$. Then, (φ_n) is bounded in $H_0^1(\Omega)$, since

$$\int_\Omega |\nabla \varphi_n|^2 \, dx = Q(\varphi_n) + \int_\Omega V(x)\varphi_n^2 \, dx \leq \mu_1 + \frac{1}{n} + \|V\|_{L^\infty(\Omega)} \leq C.$$

In particular, a subsequence $\left(\varphi_{k_n}\right)$ converges weakly in $H_0^1(\Omega)$ and strongly in $L^2(\Omega)$ (see Theorem IX.16 in [25]) to a function $\varphi_1 \in H_0^1(\Omega)$ such that $\|\varphi_1\|_{L^2(\Omega)} = 1$ and $Q(\varphi_1) \leq \mu_1$. We also deduce that $Q(\varphi_1) = \mu_1 = \min_{\varphi \in H_0^1(\Omega), \|\varphi\|_{L^2(\Omega)}=1} Q(\varphi)$, since $C_c^1(\Omega)$ is dense in $H_0^1(\Omega)$. We claim that φ_1 is a weak solution to (A.25) with $\lambda = \mu_1$, that is,

$$B(\varphi_1, \varphi) := \int_\Omega \nabla\varphi_1 \nabla\varphi \, dx - \int_\Omega V(x)\varphi_1\varphi \, dx - \mu_1 \int_\Omega \varphi_1\varphi \, dx = 0, \quad (A.30)$$

for all $\varphi \in H_0^1(\Omega)$. Let $Q_2(\varphi, \varphi) := B(\varphi, \varphi)$. Then, for any $t \in \mathbb{R}$ and $\varphi \in H_0^1(\Omega)$,

$$0 \leq Q_2(\varphi_1 + t\varphi) = B(\varphi_1 + t\varphi, \varphi_1 + t\varphi)$$
$$= Q_2(\varphi_1) + t^2 Q_2(\varphi) + 2tB(\varphi_1, \varphi) = t^2 Q_2(\varphi) + 2tB(\varphi_1, \varphi).$$

So that,

$$-2tB(\varphi_1, \varphi) \leq t^2 Q_2(\varphi).$$

Dividing by $|t|$ and letting $t \to 0^\pm$, it follows that $B(\varphi_1, \varphi) = 0$, that is, (A.30) holds. Using elliptic regularity theory (see Appendix B.1), we deduce that $\varphi_1 \in C^2(\overline{\Omega})$ is an eigenvector of (A.25) associated to the eigenvalue $\lambda = \mu_1$.

Let $\lambda \in \mathbb{R}$ denote another eigenvalue with eigenvector φ normalized by $\|\varphi\|_{L^2(\Omega)} = 1$. Multiply (A.25) by φ and integrate. Then, $\lambda = Q(\varphi) \geq \mu_1$, since μ_1 minimizes Q among all functions $\varphi \in H_0^1(\Omega)$ such that $\|\varphi\|_{L^2(\Omega)} = 1$. Hence, μ_1 is the smallest eigenvalue of $-L = -\Delta - V(x)$ and $\mu_1 =: \lambda_1$ is given by (A.26).

Now let φ_1 denote an eigenvector associated to λ_1, normalized by $\|\varphi_1\|_{L^2(\Omega)} = 1$. We claim that φ_1 is of constant sign throughout Ω. Indeed, with Q given by (A.29), there holds

$$\lambda_1 = Q(\varphi_1) = Q(\varphi_1^+) + Q(\varphi_1^-) \geq \lambda_1 \int_\Omega \left(\varphi_1^+\right)^2 dx + \lambda_1 \int_\Omega \left(\varphi_1^-\right)^2 dx = \lambda_1.$$

In particular, $Q(\varphi_1^\pm) = \lambda_1 \|\varphi_1^\pm\|_{L^2(\Omega)}^2$. At least one of the functions φ_1^+, φ_1^- cannot be identical to zero. Say $\varphi_1^+ \not\equiv 0$. Then, $\varphi_1^+/\|\varphi_1^+\|_{L^2(\Omega)}$ is a minimizer for (1.10). Working as previously, we deduce that φ_1^+ is itself an eigenvector associated to λ_1. By the strong maximum principle, we deduce that $\varphi_1^+ > 0$ in Ω, hence $\varphi_1 = \varphi_1^+ > 0$. So, the sign of any eigenvector associated to λ_1 is constant. Now take two eigenvectors $\varphi_1, \tilde{\varphi}_1$ and let

$$t = \frac{\int_\Omega \tilde{\varphi}_1\, dx}{\int_\Omega \varphi_1\, dx}. \tag{A.31}$$

Then, $\tilde{\varphi}_1 = t\varphi_1$, that is, λ_1 is simple. If this were not the case, then $\psi := \tilde{\varphi}_1 - t\varphi_1$ would be an eigenvector, hence it would be of constant sign. But then $\int_\Omega \psi\, dx = 0$ by (A.31), a contradiction.

Next, we show that if $\varphi > 0$ is an eigenvector associated to λ, then $\lambda = \lambda_1$. Let $\varphi_1 > 0$ denote an eigenvector associated to λ_1, multiply (A.25) by φ_1, and integrate. We obtain

$$\lambda \int_\Omega \varphi\varphi_1\, dx = \int_\Omega \left(\nabla\varphi\nabla\varphi_1 - V(x)\varphi\varphi_1\right)\, dx = \lambda_1 \int_\Omega \varphi\varphi_1\, dx$$

and thus, $\lambda = \lambda_1$. This proves Point 2 of the theorem.

Let μ_1 denote the supremum of all $\lambda \in \mathbb{R}$ such that for some $v \in C^2(\Omega)$, $v > 0$ in Ω, (A.28) holds. Fix $\lambda < \mu_1$. Take $\varphi \in C_c^1(\Omega)$ and multiply (A.28) by φ^2/v:

$$\lambda \int_\Omega \varphi^2\, dx \leq \int_\Omega (-\Delta v - V(x)v)\frac{\varphi^2}{v}\, dx$$

$$\leq \int_\Omega \left(\nabla v \cdot \nabla\frac{\varphi^2}{v} - V(x)\varphi^2\right)\, dx$$

$$\leq -\int_\Omega \frac{\varphi^2}{v^2}|\nabla v|^2\, dx + 2\int_\Omega \frac{\varphi}{v}\nabla v \cdot \nabla\varphi\, dx - \int_\Omega V(x)\varphi^2\, dx$$

$$\leq \int_\Omega |\nabla\varphi|^2\, dx - \int_\Omega V(x)\varphi^2\, dx,$$

where we used Young's inequality in the last inequality. By (A.26), we deduce that $\lambda \leq \lambda_1$, hence $\mu_1 \leq \lambda_1$. Now, by Point 2 of the theorem, for $\lambda = \lambda_1$, there exists $v = \varphi_1 > 0$ satisfying (A.28). So, $\lambda_1 \leq \mu_1$, hence $\lambda_1 = \mu_1$ and Point 3 of the theorem is proven. $\qquad\square$

A.7 Thin domains and unbounded domains

In this section, we collect two useful versions of the maximum principle. The first result applies to bounded domains of small measure.

Proposition A.7.1 *([18]) Let Ω denote a bounded open set of $\mathbb{R}^N$, $N \geq 1$, $V(x) \in L^p(\Omega)$, $p > N/2$. There exists $\varepsilon > 0$ such that if $|\Omega| < \varepsilon$, then for any function $u \in C^2(\Omega) \cap C^1(\overline{\Omega})$ satisfying*

$$\begin{cases} -\Delta u + V(x)u \geq 0 & \text{in } \Omega, \\ u \geq 0 & \text{on } \partial\Omega, \end{cases}$$

we have $u \geq 0$ in Ω.

Proof. Multiply the equation by u^- and integrate by parts. We obtain

$$\int_\Omega |\nabla u^-|^2 \, dx + \int_\Omega V(x)|u^-|^2 \, dx \leq 0.$$

By Sobolev and Hölder's inequalities, there exists a constant $C_N > 0$ such that

$$C_N \left(\int_\Omega |u^-|^{2N/(N-2)} \, dx \right)^{\frac{N-2}{N}} \leq \int_\Omega |\nabla u^-|^2 \, dx \leq - \int_\Omega V(x)|u^-|^2 \, dx$$

$$\leq \|V(x)\|_{L^p(\Omega)} \left(\int_\Omega |u^-|^{2N/(N-2)} \, dx \right)^{\frac{N-2}{N}} |\Omega|^{\frac{2p-N}{2p}},$$

hence

$$(C_N - \|V(x)\|_{L^p(\Omega)} |\Omega|^{\frac{2p-N}{2p}}) \left(\int_\Omega |u^-|^{2N/(N-2)} \, dx \right)^{\frac{N-2}{N}} \leq 0.$$

If $|\Omega| > 0$ is sufficiently small, $C_N - \|V(x)\|_{L^p(\Omega)} |\Omega|^{\frac{2p-N}{2p}} > 0$ and we deduce that $u^- \equiv 0$, that is, $u \geq 0$ in Ω. $\qquad\square$

Our next result applies in any domain, provided there exists a suitable barrier function.

Proposition A.7.2 *Let Ω denote an open set of $\mathbb{R}^N$, $N \geq 1$, $V(x) \in C(\Omega)$. Assume that there exists a function $w \in C^2(\overline{\Omega})$ such that*

$$\begin{cases} -\Delta w - V(x)w \geq 0 & \text{in } \Omega \\ w > 0 & \text{in } \overline{\Omega}. \end{cases} \tag{A.32}$$

Assume that $u \in C^2(\overline{\Omega})$ satisfies

$$\begin{cases} -\Delta u - V(x)u \geq 0 & \text{in } \Omega \\ u \geq 0 & \text{on } \partial\Omega, \end{cases} \tag{A.33}$$

and if Ω is unbounded, also assume that

$$\lim_{\substack{|x|\to+\infty,\\ x\in\Omega}} \frac{u(x)}{w(x)} = 0. \tag{A.34}$$

Then, $u \geq 0$ in Ω.

Proof. Assume by contradiction that $\omega = \{x \in \Omega \,:\, u(x) < 0\}$ is nonempty. Then, $\sigma = \frac{u}{w}$ solves

$$-\nabla \cdot (w^2 \nabla \sigma) = w(-\Delta u) - u(-\Delta w) =$$
$$w(-\Delta u - V(x)u) - u(-\Delta w - V(x)w) \geq 0 \qquad \text{in } \omega.$$

In addition, $\sigma \geq 0$ on $\partial\omega$ and $\lim_{|x|\to+\infty} \sigma(x) = 0$. So, given $\varepsilon > 0$, there exists $R_0 > 0$ such that for every $R > R_0$,

$$\begin{cases} -\nabla \cdot (w^2 \nabla \sigma) \geq 0 & \text{in } \omega \cap B_R, \\ \sigma \geq -\varepsilon & \text{on } \partial(\omega \cap B_R). \end{cases} \tag{A.35}$$

By the maximum principle (Proposition A.5.1), we deduce that $\sigma \geq -\varepsilon$ in $\omega \cap B_R$ for every $R > R_0$. Letting $R \to +\infty$ and then $\varepsilon \to 0$, we deduce that $\sigma \geq 0$ in ω. This is a contradiction with the definition of ω. $\qquad\square$

A.8   Nonlinear comparison principle

Proposition A.8.1 *Let Ω denote a bounded domain of $\mathbb{R}^N$, $N \geq 1$, $f : \mathbb{R} \to \mathbb{R}$ a locally Lipschitz function and $\underline{u}, \overline{u} \in C^2(\overline{\Omega})$ a sub- and a supersolution to (1.3). Then, if $\underline{u} \leq \overline{u}$ in Ω, either $\underline{u} \equiv \overline{u}$ or $\underline{u} < \overline{u}$.*

Proof. Let $a = \|\underline{u}\|_\infty + \|\overline{u}\|_\infty$ and let K denote the Lipschitz constant of f on $[-a, a]$. Then,

$$f(\overline{u}) - f(\underline{u}) \geq -K(\overline{u} - \underline{u}).$$

In particular, $w := \overline{u} - \underline{u}$ solves

$$\begin{cases} -\Delta w + Kw = f(\overline{u}) - f(\underline{u}) + Kw \geq 0 & \text{in } \Omega, \\ w \geq 0 & \text{on } \partial\Omega. \end{cases}$$

By the strong maximum principle (Corollary A.5.1), $w \equiv 0$ or $w > 0$ in Ω. $\quad\square$

A.9 L^1 theory for the Laplace operator

A.9.1 Linear theory and weak comparison principle

Lemma A.9.1 ([29]) *Let $N \geq 1$ and $\Omega \subset \mathbb{R}^N$ denote a smoothly bounded domain. Let d_Ω denote the distance to the boundary of Ω, as defined in (3.4). Given $f \in L^1(\Omega, d_\Omega(x)dx)$, there exists a unique solution $u \in L^1(\Omega)$ of*

$$\begin{cases} -\Delta u = f & \text{in } \Omega, \\ u = 0 & \text{on } \partial\Omega, \end{cases}$$

in the sense that

$$\int_\Omega u(-\Delta\varphi)\,dx = \int_\Omega f\varphi\,dx,$$

for every $\varphi \in C_0^2(\overline{\Omega})$. In addition, there exists a constant $C = C(\Omega, N) > 0$ such that

$$\|u\|_{L^1(\Omega)} \leq C\|f\|_{L^1(\Omega, d_\Omega(x)dx)} \tag{A.36}$$

and the following comparison principle holds:

$$f \geq 0 \quad \text{a.e.} \quad \Longrightarrow \quad u \geq 0 \quad \text{a.e.}$$

Proof.
Step 1. Existence of a solution
  By splitting f into its positive and negative parts $f = f^+ - f^-$, it suffices to consider the case $f \geq 0$ a.e. Given $n \in \mathbb{N}$, let $f_n = \min(f, n)$. Since f_n is bounded, there exists a unique u_n solving

$$\begin{cases} -\Delta u_n = f_n & \text{in } \Omega, \\ u_n = 0 & \text{on } \partial\Omega. \end{cases} \tag{A.37}$$

Note that by the standard maximum principle, the sequence (u_n) is nonde-creasing. Now take $\psi \in C_c^\infty(\Omega)$ and let $\varphi \in C_0^2(\overline{\Omega})$ denote the solution to

$$\begin{cases} -\Delta\varphi = \psi & \text{in } \Omega, \\ \varphi = 0 & \text{on } \partial\Omega. \end{cases} \tag{A.38}$$

Multiplying the above equation by u_n and integrating, we obtain

$$\int_\Omega u_n \psi \, dx = \int_\Omega f_n \varphi \, dx.$$

It follows that

$$\left| \int_\Omega u_n \psi \, dx \right| \leq C \|f_n\|_{L^1(\Omega, d_\Omega(x)dx)} \|\varphi\|_{C^1(\overline{\Omega})} \leq C' \|f\|_{L^1(\Omega, d_\Omega(x)dx)} \|\psi\|_{L^\infty(\Omega)}.$$

Taking the supremum over all functions ψ such that $\|\psi\|_{L^\infty(\Omega)} \leq 1$, we deduce that

$$\|u_n\|_{L^1(\Omega)} \leq C' \|f\|_{L^1(\Omega, d_\Omega(x)dx)}.$$

By monotone convergence, the sequence (u_n) tends to $u \in L^1(\Omega)$ solving the equation in the weak sense and satisfying (A.36).

Step 2. Comparison principle and uniqueness

   Clearly, uniqueness is a direct consequence of the comparison principle. So, it suffices to prove the latter. If $\psi \geq 0$, we have, as in Step 1,

$$\int_\Omega u\psi \, dx = \int_\Omega f\varphi \, dx \geq 0.$$

This being true for every $\psi \geq 0$, we deduce that $u \geq 0$ a.e. $\qquad \square$

   Lemma A.9.1 can be extended to the setting of measures, as follows.

Corollary A.9.1 *Let $N \geq 1$ and $\Omega \subset \mathbb{R}^N$ denote a smoothly bounded domain. Given a Radon measure μ such that $d_\Omega \in L^1(\Omega, d|\mu|)$, there exists a unique solution $u \in L^1(\Omega)$ of*

$$\begin{cases} -\Delta u = \mu & \text{in } \Omega, \\ u = 0 & \text{on } \partial\Omega, \end{cases}$$

in the sense that

$$\int_\Omega u(-\Delta\varphi) \, dx = \int_\Omega \varphi \, d\mu,$$

for every $\varphi \in C_0^2(\overline{\Omega})$. In addition,

(i) *For every* $p \in [1, \frac{N}{N-1})$, *there exists a constant* $C = C(\Omega, N, p) > 0$ *such that*

$$\|u\|_{L^p(\Omega)} \le C \|d_\Omega \mu\|_{\mathcal{M}(\Omega)}, \qquad (A.39)$$

where $\| \cdot \|_{\mathcal{M}(\Omega)}$ *denotes that total variation of a measure.*

(ii) *If a sequence* $(d_\Omega \mu_n)$ *is bounded in* $\mathcal{M}(\Omega)$, *then the corresponding sequence of solutions* (u_n) *is relatively compact in* $L^p(\Omega)$ *for all* $p \in [1, \frac{N}{N-1})$.

(iii) *The following comparison principle holds:*

$$\mu \ge 0 \quad \text{as a measure} \quad \Longrightarrow \quad u \ge 0 \quad \text{a.e.}$$

Proof. The comparison principle and uniqueness are proved exactly as in Step 2 of the previous lemma. For the existence of a solution, choose a sequence (f_n) in $L^1(\Omega, d_\Omega(x)dx)$ such that $d_\Omega f_n \rightharpoonup d_\Omega \mu$ in $\mathcal{M}(\Omega)$. Let $u_n \in L^1(\Omega)$ denote the solution to (A.37) and φ the solution to (A.38) for a given $\psi \in C_c^\infty(\Omega)$.

Multiplying (A.38) by u_n and integrating, we obtain

$$\int_\Omega u_n \psi \, dx = \int_\Omega f_n \varphi \, dx.$$

It follows that

$$\left| \int_\Omega u_n \psi \, dx \right| \le C \|f_n\|_{L^1(\Omega, d_\Omega(x)dx)} \|\varphi\|_{C^1(\overline{\Omega})} \le C' \|d_\Omega \mu\|_{\mathcal{M}(\Omega)} \|\psi\|_{L^q(\Omega)},$$

where $q > N/2$ (so that $\|\varphi\|_{C^1(\overline{\Omega})} \le C \|\psi\|_{L^q}$, by elliptic regularity). Letting p denote the conjugate exponent of q and taking the supremum over all functions ψ such that $\|\psi\|_{L^q(\Omega)} \le 1$, we deduce that

$$\|u_n\|_{L^p(\Omega)} \le C' \|d_\Omega \mu\|_{\mathcal{M}(\Omega)}.$$

In particular, a subsequence of (u_n) converges weakly in $L^p(\Omega)$ to a solution u of the equation satisfying (A.39).

It remains to be proven the compactness result *(ii)*. Take a sequence of measures (μ_n) such that $\|d_\Omega \mu_n\|_{\mathcal{M}(\Omega)} \le C$ and a smooth domain $\omega \subset\subset \Omega$. Let $v_n \in L^1(\omega)$ denote the solution to

$$\begin{cases} -\Delta v_n = \mu_n |_\omega & \text{in } \omega, \\ v_n = 0 & \text{on } \partial \omega. \end{cases}$$

We claim that (v_n) is bounded in $W^{1,p}(\omega)$ for every $p < \frac{N}{N-1}$. To see this, take any $\psi \in C_c^1(\omega)$ and let $\varphi \in C_0^2(\overline{\omega})$ denote the solution to

$$\begin{cases} -\Delta\varphi = \psi & \text{in } \omega, \\ \varphi = 0 & \text{on } \partial\omega. \end{cases}$$

By Theorem B.4.1,

$$\|\varphi\|_{L^\infty(\omega)} \le C\|\psi\|_{W^{-1,p'}(\omega)}.$$

So,

$$\left| \int_\omega v_n \psi \, dx \right| = \left| \int_\omega \varphi \, d\mu_n \right| \le \|\varphi\|_{L^\infty(\omega)} \|\mu_n\|_{\mathscr{M}(\omega)} \le C\|\varphi\|_{L^\infty(\omega)} \le C\|\psi\|_{W^{-1,p'}(\omega)}.$$

This being true for all ψ, the claim follows.

In addition to this, since $u_n - v_n$ is harmonic in ω, we also have for $\omega' \subset\subset \omega$,

$$\|u_n - v_n\|_{C^1(\overline{\omega'})} \le C_{\omega'}\|u_n - v_n\|_{L^1(\omega)} \le C'\|\mu_n d_\Omega\|_{\mathscr{M}(\Omega)} \le C''.$$

So, the sequence (u_n) is bounded in $W^{1,p}(\omega')$ for any $p < \frac{N}{N-1}$ and any $\omega' \subset\subset \Omega$. By a standard diagonalization argument, there exists a subsequence $u_{k_n} \to u$ a.e. in Ω. Since (u_{k_n}) is bounded in $L^p(\Omega)$ for $p < \frac{N}{N-1}$, we conclude using Egorov's theorem. $\qquad\square$

Exercise A.9.1 Assume that $\mu \in M(\Omega)$, that is, μ is integrable up to the boundary. Prove that for every $p \in [1, \frac{N}{N-2})$, there exists a constant $C = C(\Omega, N, p)$ such that

$$\|u\|_{L^p(\Omega)} \le C\|\mu\|_{\mathscr{M}(\Omega)}.$$

A.9.2 The boundary-point lemma

The refined boundary point lemma (see Proposition A.4.2) also holds in the general L^1 setting. This is the content of the next corollary.

Corollary A.9.2 *Let $N \ge 1$, let $\Omega \subset \mathbb{R}^N$ denote a smoothly bounded domain and let d_Ω given by (3.4). Assume that $\mu \ge 0$ is a Radon measure such that d_Ω is μ-integrable and let $u \in L^1(\Omega)$ denote the solution to*

$$\begin{cases} -\Delta u = \mu & \text{in } \Omega, \\ u = 0 & \text{on } \partial\Omega. \end{cases} \tag{A.40}$$

Then,

$$u(x) \geq c \left(\int_\Omega d_\Omega \, d\mu \right) d_\Omega(x), \qquad \textit{for a.e. } x \in \Omega, \qquad (\text{A.41})$$

where $c > 0$ is a constant depending on Ω only.

Proof. Take a sequence of functions $f_n \in C_c^\infty(\Omega)$ such that $d_\Omega f_n \rightharpoonup d_\Omega \mu$ in $\mathscr{M}(\Omega)$. Let u_n denote the associated solution to (A.37). By (A.39), $u_n \to u$ in $L^p(\Omega)$ for $p \in (1, \frac{N}{N-1})$. By Lemma A.4.2, we also have for $x \in \Omega$,

$$u_n(x) \geq c \left(\int_\Omega f_n \, d_\Omega \, dy \right) d_\Omega(x).$$

Take any $\psi \in C_c^\infty(\Omega)$ such that $\psi \geq 0$ and integrate. Then,

$$\int_\Omega \left(u_n(x) - c \left(\int_\Omega f_n \, d_\Omega \, dy \right) d_\Omega(x) \right) \psi \, dx \geq 0.$$

Passing to the (weak) limit, it follows that

$$\int_\Omega \left(u(x) - c \left(\int_\Omega d_\Omega \, d\mu \right) d_\Omega(x) \right) \psi \, dx \geq 0.$$

The above inequality being true for all $\psi \geq 0$, (A.41) follows. $\qquad\qquad\square$

A.9.3   Sub- and supersolutions in the L^1 setting

We establish the following extension of the method of sub- and supersolutions.

Theorem A.9.1 ([159]) *Let $N \geq 1$ and $\Omega \subset \mathbb{R}^N$ denote a smoothly bounded domain. Let d_Ω denote the distance to the boundary of Ω, as defined in (3.4). Let $f \in C^1(\mathbb{R})$. Assume that there exists $\underline{u}, \overline{u} \in L^1(\Omega)$ a weak sub- and supersolution of (1.3), that is, $f(\underline{u})d_\Omega, f(\overline{u})d_\Omega \in L^1(\Omega)$ and*

$$- \int_\Omega \underline{u} \Delta \varphi \, dx \leq \int_\Omega f(\underline{u}) \varphi \, dx, \qquad \textit{for all } \varphi \in C_0^2(\overline{\Omega}), \ \varphi \geq 0$$

and the reverse inequality holds for $\overline{u}$. Assume that $\underline{u} \leq \overline{u}$ a.e. and

$$f(v)d_\Omega \in L^1(\Omega), \qquad \textit{for every } v \in L^1(\Omega) \textit{ such that } \underline{u} \leq v \leq \overline{u} \textit{ a.e.} \qquad (\text{A.42})$$

Then, there exists the minimal solution $u \in L^1(\Omega)$ of (1.3) relative to $\underline{u}$, that is, u solves (1.3) and $\underline{u} \leq u \leq \overline{v}$ a.e., for every weak supersolution $\overline{v}$ such that $\overline{v} \geq \underline{u}$ a.e.

Exercise A.9.2 Using the monotone iteration scheme (see Exercise 1.2.1), prove Theorem A.9.1 under the additional assumption that f is nondecreasing. Note that in this case, (A.42) is automatically satisfied.

To prove the theorem, we first establish a series of intermediate results.

Lemma A.9.2 *Let* (w_n) *denote a sequence of functions in* $L^1(\Omega)$ *and let* (E_n) *a sequence of measurable subsets of* Ω *such that*

$$|E_n| \to 0 \quad and \quad \int_{E_n} |w_n|\, dx \geq 1 \quad \forall n \geq 1. \tag{A.43}$$

Then, there exists a subsequence (w_{n_k}) *and a sequence of disjoint measurable sets* (F_k) *such that*

$$F_k \subset E_{n_k} \quad and \quad \int_{F_k} |w_{n_k}|\, dx \geq \frac{1}{2} \quad \forall k \geq 1. \tag{A.44}$$

Proof. Set $n_1 = 1$, $n_2 = 2$, $A_1 = E_1$, and $A_2 = E_2$. By induction, we construct an increasing sequence of integers (n_k) and measurable sets (A_k) as follows. Assume that $k \geq 3$, $n_1, \ldots, n_{k-1}$ and $A_1, \ldots, A_{k-1}$ are such that $A_j \subset E_j$ and

$$\int_{A_{j+1} \cup \cdots \cup A_{k-1}} |w_{n_j}|\, dx \leq \frac{1}{2} - \frac{1}{2^{k-1}} \quad \forall j = 1, \ldots, k-2.$$

Since $|E_n| \to 0$, then for $n_k > n_{k-1}$ sufficiently large we have

$$\int_{E_{n_k}} |w_{n_j}|\, dx \leq \frac{1}{2^k} \quad \forall j = 1, \ldots, k-1.$$

Let $A_k = E_{n_k}$. Then,

$$\int_{A_{j+1} \cup \cdots \cup A_k} |w_{n_j}|\, dx \leq \frac{1}{2} - \frac{1}{2^k} \quad \forall j = 1, \ldots, k-1.$$

Now set

$$F_k = A_k \setminus \cup_{i=k+1}^{\infty} A_i.$$

Then the sets F_k are disjoint and

$$\int_{F_k} |w_{n_k}|\, dx = \lim_{i \to +\infty} \int_{A_k \setminus (A_{k+1} \cup \cdots \cup A_i)} |w_{n_k}|\, dx \geq \frac{1}{2}.$$

$\square$

Proposition A.9.1 *Let* $g : \Omega \times \mathbb{R} \to \mathbb{R}$ *be a Carathéodory function and* $\underline{u}, \overline{u} \in L^1(\Omega)$ *be such that* $\underline{u} \leq \overline{u}$ *a.e. Assume that*

$$g(\cdot, v)d_\Omega \in L^1(\Omega) \quad \text{for every } v \in L^1(\Omega) \text{ such that } \underline{u} \leq v \leq \overline{u} \text{ a.e.} \qquad (A.45)$$

Then, the set

$$\mathscr{B} = \left\{ g(\cdot, v) \in L^1(\Omega, d_\Omega dx) : v \in L^1(\Omega) \text{ and } \underline{u} \leq v \leq \overline{u} \text{ a.e.} \right\} \qquad (A.46)$$

is bounded and equi-integrable in $L^1(\Omega, d_\Omega \, dx)$.

Proof. Recall that a set $\mathscr{B} \subset L^1(\Omega; d_\Omega dx)$ is equi-integrable if for every $\varepsilon > 0$, there exists $\delta > 0$ such that

$$E \subset \Omega \quad \text{and} \quad |E| < \delta \Longrightarrow \int_E |g| d_\Omega \, dx < \varepsilon \quad \forall g \in \mathscr{B}.$$

Since Ω is bounded, it suffices to show that $\mathscr{B}$ is equi-integrable. Assume by contradiction that $\mathscr{B}$ is not equi-integrable. Then, there exists $\varepsilon > 0$, a sequence (u_n) in $L^1(\Omega)$ with $\underline{u} \leq u_n \leq \overline{u}$ a.e., and a sequence of measurable sets (E_n) such that

$$|E_n| \to 0 \quad \text{and} \quad \int_{E_n} g(x, u_n) d_\Omega \, dx \geq \varepsilon \quad \forall n \geq 1.$$

Applying Lemma A.9.2 with $w_n = g(\cdot, u_n)d_\Omega / \varepsilon$, we find a subsequence (u_{n_k}) and a sequence of disjoint measurable sets (F_k) such that

$$\int_{F_k} |g(x, u_{n_k})| d_\Omega \, dx \geq \frac{\varepsilon}{2} \quad \forall k \geq 1. \qquad (A.47)$$

Set

$$v(x) = \begin{cases} u_{n_k}(x) & \text{if } x \in F_k \text{ for some } k \geq 1, \\ \underline{u}(x) & \text{otherwise.} \end{cases}$$

Then, $\underline{u} \leq v \leq \overline{u}$ a.e. Hence, $v \in L^1(\Omega)$. Moreover,

$$\int_\Omega |g(x, v)| d_\Omega \, dx \geq \sum_{k=1}^{+\infty} \int_{F_k} |g(x, u_{n_k})| d_\Omega \, dx = +\infty.$$

This contradicts (A.46). Therefore $\mathscr{B}$ is equi-integrable in $L^1(\Omega; d_\Omega dx)$. $\qquad \square$

Proposition A.9.2 *Let $g : \Omega \times \mathbb{R} \to \mathbb{R}$ be a Carathéodory function such that*

$$g(\cdot, v)d_\Omega \in L^1(\Omega) \qquad \text{for every } v \in L^1(\Omega). \tag{A.48}$$

Then, the Nemytskii operator

$$G : \begin{cases} L^1(\Omega) \to L^1(\Omega; d_\Omega dx) \\ \qquad v \mapsto g(\cdot, v) \end{cases}$$

is continuous.

Proof. Assume that $v_n \to v$ in $L^1(\Omega)$. Extract a sequence (v_{n_k}) such that $v_{n_k} \to v$ a.e. and $|v_{n_k}| \le V$ a.e., for some $V \in L^1(\Omega)$. In particular,

$$g(\cdot, v_{n_k}) \to g(\cdot, v) \qquad \text{a.e.}$$

Moreover, by Proposition A.9.1 (applied with $\underline{u} = -V$, $\overline{U} = V$), the sequence $(g(\cdot, v_{n_k}))$ is equi-integrable in $L^1(\Omega; d_\Omega dx)$. By Egorov's theorem,

$$g(\cdot, v_{n_k}) \to g(\cdot, v) \qquad \text{in } L^1(\Omega; d_\Omega dx).$$

Since the limit is independent of the subsequence (v_{n_k}), we deduce that

$$G(v_n) \to G(v) \quad \text{in } L^1(\Omega; d_\Omega dx).$$

$\square$

Proof of Theorem A.9.1.
Step 1. There exists a solution u of (1.3) such that $\underline{u} \le u \le \overline{u}$ a.e.
   For $t \ge 0, x \in \Omega$, set

$$g(x, t) = \begin{cases} f(\underline{u}(x)) & \text{if } t < \underline{u}(x), \\ \qquad f(t) & \text{if } \underline{u}(x) \le t \le \underline{u}(x), \\ f(\overline{u}(x)) & \text{if } t > \overline{u}(x). \end{cases}$$

Then g is a Carathéodory function and by (A.42),

$$g(\cdot, v)d_\Omega \in L^1(\Omega) \qquad \text{for every } v \in L^1(\Omega).$$

Consider the operators

$$G : \begin{cases} L^1(\Omega) \to L^1(\Omega; d_\Omega dx), \\ \qquad v \mapsto g(\cdot, v), \end{cases}$$

229

and

$$K : \begin{cases} L^1(\Omega; d_\Omega dx) \to L^1(\Omega) \\ \qquad\qquad h \mapsto w, \end{cases}$$

where w is the unique solution to

$$\begin{cases} -\Delta w = h & \text{in } \Omega, \\ \quad\; w = 0 & \text{on } \partial\Omega. \end{cases}$$

Proposition A.9.2 shows that G is continuous, while Corollary A.9.1 implies that K is compact. Hence, $KG : L^1(\Omega) \to L^1(\Omega)$ is compact. By Proposition A.9.1, $G(L^1(\Omega))$ is bounded in $L^1(\Omega; d_\Omega dx)$. So,

$$\|K(G(v))\|_{L^1(\Omega)} \le C_1 \|G(v) d_\Omega\|_{L^1(\Omega)} \le C \quad \forall v \in L^1(\Omega).$$

By Schauder's fixed-point theorem, KG has a fixed point $u \in L^1(\Omega)$. In other words, u solves

$$\begin{cases} -\Delta u = g(x, u) & \text{in } \Omega, \\ \quad\; u = 0 & \text{on } \partial\Omega. \end{cases}$$

We claim that

$$\underline{u} \le u \le \overline{u} \quad \text{a.e.} \tag{A.49}$$

This implies in particular that u solves (1.3). To prove (A.49), we note that

$$g(\cdot, u) = g(\cdot, \overline{u}) \quad \text{a.e. on the set } [u \ge \overline{u}].$$

Thus, applying Kato's inequality (see (3.19)) to $w = u - \overline{u}$, we get

$$-\int_\Omega w^+ \Delta \varphi \, dx \le \int_{[u \ge \overline{u}]} (g(x, u) - g(x, \overline{u})) \varphi \, dx = 0 \quad \forall \varphi \in C_0^2(\overline{\Omega}), \varphi \ge 0.$$

By the maximum principle (see Corollary A.9.1), $w^+ \le 0$ a.e. This implies that $u \le \overline{u}$ a.e. The inequality $u \ge \underline{u}$ is obtained similarly. So, we have obtained a solution u of (1.3) such that $\underline{u} \le u \le \overline{u}$.

Step 2. There exists the smallest solution u in the interval $[\underline{u}, \overline{u}]$.

Claim. If u_1, u_2 are two solutions of (1.3) such that $\underline{u} \le u_1, u_2 \le \overline{u}$ a.e., then there exists a solution w such that

$$\underline{u} \le w \le \min(u_1, u_2) \le \overline{u} \quad \text{a.e.} \tag{A.50}$$

Indeed, by Kato's inequality, $\min(u_1, u_2) = u_2 - (u_1 - u_2)^-$ is a supersolution to (1.3). We then apply Step 1 to find the desired solution w. Now set

$$A = \inf \left\{ \int_\Omega u \, dx \; : \; \underline{u} \le u \le \overline{u} \text{ a.e. and } u \text{ is a solution to (1.3)} \right\}.$$

It follows from the above claim that one can find a *nonincreasing* sequence of solutions (w_n) such that

$$\underline{u} \le w_n \le \overline{u} \quad \text{and} \quad \int_\Omega w_n \, dx \to A. \tag{A.51}$$

By monotone convergence, there exists $u \in L^1(\Omega)$ such that $w_n \to u$ a.e.,

$$\underline{u} \le u \le \overline{u} \quad \text{and} \quad \int_\Omega w_n \, dx \to \int_\Omega u \, dx = A.$$

By Proposition A.9.1, $(f(w_n))$ is equi-integrable and bounded in $L^1(\Omega; d_\Omega dx)$, hence

$$f(w_n) \to f(u) \quad \text{in } L^1(\Omega; d_\Omega dx).$$

So, u is a solution to (1.3) and

$$\int_\Omega u \, dx = A.$$

According to the claim, u is the smallest solution to (1.3) lying in the interval $[\underline{u}, \overline{u}]$.

Step 3. $u \le \overline{v}$ a.e., for every supersolution $\overline{v}$ such that $\overline{v} \ge \underline{u}$ a.e.

Using Kato's inequality, $w = \min(u, \overline{v}) = u - (\overline{v} - u)^-$ is a supersolution to (1.3) and $\underline{u} \le w$. By Step 1, there exists a solution $\tilde{u}$ lying in the interval $[\underline{u}, w]$. By definition of w, $\tilde{u} \le w \le u$. Since u is the smallest solution in the interval $[\underline{u}, \overline{u}]$ and since $\tilde{u}$ lies in that interval, we also have $u \le \tilde{u}$. Hence, $u = \tilde{u} \le w = \min(u, \overline{v})$ and so $u \le \overline{v}$. $\qquad\square$

Appendix B

Regularity theory for elliptic operators

The following two sections follow [11, 126, 222].

B.1 Harmonic functions

In this section, we establish basic elliptic regularity estimates for harmonic functions and solve the Dirichlet problem.

B.1.1 Interior regularity

Proposition B.1.1 *Let Ω be an open set of $\mathbb{R}^N$, $N \geq 1$. Let $u \in L^1_{loc}(\Omega)$ be a harmonic function. Then, $u \in C^\infty(\Omega)$, and given any $k \in \mathbb{N}^N$, there exists a constant $C_{N,k}$, depending on N and k only, such that*

$$|D^k u(x_0)| \leq \frac{C_{N,k}}{r^{|k|}} \|u\|_{L^\infty(B(x_0,r))}, \tag{B.1}$$

for every ball $B(x_0, r) \subset \Omega$.

Proof. Let $u \in L^1_{loc}(\Omega)$ be a harmonic function. By Exercise A.1.3, $u \in C^2(\Omega)$. We establish (B.1) by induction on $|k|$. The case $|k| = 0$ is a straightforward consequence of the mean-value formula (A.10). Now differentiate the mean-value formula (A.10) with respect to x_i. Then, $v = \partial u / \partial x_i$ also satisfies the mean-value formula. By Exercise A.1.3, v is harmonic. In particular, $u \in C^3(\Omega)$. Iterating this argument, we deduce that $u \in C^\infty(\Omega)$. In addition,

$$
\begin{aligned}
\left| \frac{\partial u}{\partial x_i}(x_0) \right| &= \left| \fint_{B(x_0, r/2)} \frac{\partial u}{\partial x_i} \, dy \right| \\
&= \frac{1}{|B_1|} \left(\frac{2}{r} \right)^N \left| \fint_{\partial B(x_0, r/2)} u n_i \, d\sigma \right| \\
&\le \frac{C}{r} \|u\|_{L^\infty(\partial B(x_0, r/2))}.
\end{aligned}
\tag{B.2}
$$

This implies (B.1) for $|k| = 1$. Now take $n \ge 1$ and assume that (B.1.1) holds for all $|k| \le n$. Take $k' \in \mathbb{N}^N$ such that $|k'| = n + 1$. Then,

$$
D^{k'} u = \frac{\partial}{\partial x_i} D^k u,
$$

for some $i \in \{1, \dots, N\}$ and $k \in \mathbb{N}^N$ such that $|k| = n$. By (B.2) applied to $v = D^k u$, we have

$$
|D^{k'} u(x_0)| \le \frac{C}{r} \|D^k u\|_{L^\infty(\partial B(x_0, r/2))}.
\tag{B.3}
$$

Now, if $x \in \partial B(x_0, r/2)$, then $B(x, r/2) \subset B(x_0, r) \subset \Omega$. Using the induction hypothesis, we deduce that

$$
\|D^k u\|_{L^\infty(\partial B(x_0, r/2))} \le \frac{C}{r^{|k|}} \|u\|_{L^\infty(B(x_0, r))}.
$$

This together with (B.3) yields the desired result. $\qquad\square$

Exercise B.1.1 Let u be a harmonic function in $\Omega \subset \mathbb{R}^N$, $N \ge 1$. Prove that for every ball $B(x_0, r) \subset \Omega$ and every $k \in \mathbb{N}^N$,

$$
|D^k u(x_0)| \le \frac{C_{N,k}}{r^{N+|k|}} \|u\|_{L^1(B(x_0, r))},
$$

where $C_{N,0} = 1/|B_1|$ and for $|k| \ge 1$,

$$
C_{N,k} \le \frac{\left(2^{N+1} N |k| \right)^{|k|}}{|B_1|}.
$$

Deduce that u is analytic.

B.1.2 Solving the Dirichlet problem on the unit ball

In this section, we show that given any function $g \in C(\partial B)$, there exists a unique function $u \in C^2(B) \cap C(\overline{B})$ solving

$$\begin{cases} -\Delta u = 0 & \text{in } B, \\ \quad u = g & \text{on } \partial B. \end{cases} \tag{B.4}$$

Suppose for the moment that $u \in C^2(\overline{B})$ is a solution to (B.4). Fix a point $x \in B$, multiply by $\Gamma(y-x)$, where Γ is the fundamental solution of the Laplace operator, given by (A.6) and (A.7), and integrate over $B \setminus B(x,\varepsilon)$, $\varepsilon > 0$. Then,

$$0 = \int_{B \setminus B(x,\varepsilon)} -\Delta u(y)\Gamma(y-x)\, dy =$$

$$\int_{\partial B} \left(-\frac{\partial u}{\partial n}(y)\Gamma(y-x) + u(y)\frac{\partial \Gamma}{\partial n}(y-x) \right) d\sigma(y) -$$

$$\int_{\partial B(x,\varepsilon)} \left(-\frac{\partial u}{\partial n}(y)\Gamma(y-x) + u(y)\frac{\partial \Gamma}{\partial n}(y-x) \right) d\sigma(y).$$

Letting $\varepsilon \to 0$, it is not hard to deduce that

$$u(x) = \int_{\partial B} \left(-\frac{\partial u}{\partial n}(y)\Gamma(y-x) + u(y)\frac{\partial \Gamma}{\partial n}(y-x) \right) d\sigma(y).$$

More generally, if $h \in C^2(\overline{B})$ is harmonic and $G(x,y) = \Gamma(y-x) - h(y)$, the same computation leads to

$$u(x) = \int_{\partial B} \left(-\frac{\partial u}{\partial n}(y)G(x,y) + u(y)\frac{\partial G}{\partial n_y}(x,y) \right) d\sigma(y).$$

If h can be chosen so that $G(x,y) = 0$ for all $y \in \partial B$, then the above equation simplifies to

$$u(x) = \int_{\partial B} g(y)\frac{\partial G}{\partial n_y}(x,y)\, d\sigma(y). \tag{B.5}$$

Note that (B.5) provides a representation formula for u in terms of its boundary value g. G is called the Green's function for the Laplace operator on the ball, and h the associated corrector function. It remains to construct such a corrector function. To this end, we shall use the following simple geometric identity.

235

Lemma B.1.1 *Take two nonzero vectors $x, y \in \mathbb{R}^N$, $N \geq 1$. Then,*

$$\left| \frac{y}{|y|} - |y|x \right| = \left| \frac{x}{|x|} - |x|y \right|.$$

Proof. Simply square both sides and expand using the inner product. $\quad\square$

Recall that we are looking for a harmonic function $h(y) = h_x(y)$ that agrees with $\Gamma(x - y)$ for $y \in \partial B$. Let

$$x^* = x/|x|^2 \tag{B.6}$$

be the image of x under the inversion in the unit sphere ∂B. In particular, since $x \in B$, $x^* \in \mathbb{R}^N \setminus \overline{B}$ and so, given any constant $k \in \mathbb{R}$, $h_k(y) = \Gamma(k(y - x^*))$ is harmonic in B. By the symmetry lemma (Lemma B.1.1), we have for all $y \in \partial B$,

$$|y - x| = \left| \frac{x}{|x|} - |x|y \right| = |x| \cdot |y - x^*|.$$

Hence, choosing $k = |x|$, we see that $h(y) = \Gamma(|x|(y - x^*))$ is the desired corrector function. Plugging this in the definition of the Green's function $G = G(x, y)\Gamma(y - x) - h(y)$, we obtain after simplification

$$P(x, y) = \frac{\partial G}{\partial n_y}(x, y) = \frac{1 - |x|^2}{|x - y|^N}, \quad \text{for } x \in B, y \in \partial B. \tag{B.7}$$

P is called the Poisson kernel for the ball.

Theorem B.1.1 *Let $g \in C(\partial B)$. There exists a unique solution $u \in C^2(B) \cap C(\overline{B})$ to* (B.4)*, given by*

$$u(x) = \begin{cases} \displaystyle\int_{\partial B} P(x, y)g(y)\, d\sigma(y) & \text{if } x \in B, \\ g(x) & \text{if } x \in \partial B, \end{cases} \tag{B.8}$$

where P is the Poisson kernel given by (B.7)*.*

Proof. The uniqueness of u follows from the maximum principle. To prove that u is indeed a solution, we proceed in steps.
Step 1. Let $y \in \partial B$. Then, $P(\cdot, y)$ is harmonic in $\mathbb{R}^N \setminus \{y\}$.

By construction, $y \mapsto G(x,y) = \Gamma(x,y) - h(y)$ is a harmonic function of y in $\mathbb{R}^N \setminus \{x\}$. In addition, using the symmetry lemma (Lemma B.1.1),

$$G(x,y) = \Gamma(x-y) - \Gamma(|x| \cdot |y - x^*|) = G(y,x),$$

for all $x \neq y$. Hence, $x \mapsto G(x,y)$ is harmonic in $\mathbb{R}^N \setminus \{y\}$, and so must be $P(\cdot, y) = \frac{\partial G}{\partial n_y}(\cdot, y)$.

Step 2. The Poisson kernel verifies

(a) $P(x,y) > 0$, for all $x \in B$, $y \in \partial B$,

(b) $\int_{\partial B} P(x,y)\, d\sigma(y) = 1$, for all $x \in B$, and

(c) for every $y_0 \in \partial B$ and every $\delta > 0$, $\displaystyle \lim_{x \to y_0} \int_{|y-y_0|>\delta} P(x,y)\, d\sigma(y) = 0$.

Properties (a) and (c) follow directly from (B.7), while (b) is a consequence of (B.5) and (B.7) , applied to $u \equiv 1$.

Step 3. The function u given by (B.8) is harmonic in B.

By Step 1, $P(\cdot, y)$ is harmonic in B. We may then safely differentiate under the integral sign to conclude that u is harmonic in B.

Step 4. u is continuous on $\overline{B}$.

Fix $y_0 \in \partial B$ and $\varepsilon > 0$. Choose $\delta > 0$ so small that $|g(y) - g(y_0)| < \varepsilon$, for $|y - y_0| < \delta$. Using Step 2, we deduce that

$$|u(x) - u(y_0)| = \left| \int_{\partial B} (g(y) - g(y_0)) P(x,y)\, d\sigma(y) \right|$$

$$\leq \int_{|y-y_0| \leq \delta} |g(y) - g(y_0)| P(x,y)\, d\sigma(y) + \int_{|y-y_0|>\delta} |g(y) - g(y_0)| P(x,y)\, d\sigma(y)$$

$$\leq \varepsilon + 2\|g\|_{L^\infty(\partial B)} \int_{|y-y_0|>\delta} P(x,y)\, d\sigma(y).$$

The last term above is less than ε for x sufficiently close to y_0, hence, u is continuous at y_0. $\qquad \square$

B.1.3   Solving the Dirichlet problem on smooth domains

In this section, we describe the Perron method for solving the Dirichlet problem in a smoothly bounded domain $\Omega \subset \mathbb{R}^N$, that is, we prove the following theorem.

Theorem B.1.2 *Assume that* $\Omega \subset \mathbb{R}^N$ *is a smoothly bounded domain and let* $g \in C(\partial\Omega)$. *There exists a unique solution* $u \in C^2(\Omega) \cap C(\overline{\Omega})$ *to*

$$\begin{cases} -\Delta u = 0 & \text{in } \Omega, \\ \quad u = g & \text{on } \partial\Omega. \end{cases} \tag{B.9}$$

To prepare the proof for the above theorem, we shall use the following basic properties of subharmonic functions.

Proposition B.1.2 *If* u_1 *and* u_2 *are subharmonic functions on* Ω, *then so is* $\max(u_1, u_2)$.

Proof. This simply follows from the fact that a function is subharmonic if and only if it satisfies the mean-value inequality (see Exercise A.1.3). $\quad\square$

Proposition B.1.3 *Let* u *be subharmonic in* Ω *and take a ball* $\overline{B} \subset \Omega$. *Let* w *be the subharmonic lift of* u, *that is, take* $v \in C^2(B) \cap C(\overline{B})$ *the solution to*

$$\begin{cases} -\Delta v = 0 & \text{in } B, \\ \quad v = u & \text{on } \partial B. \end{cases}$$

given by Theorem B.1.1 and set

$$\begin{cases} w = v & \text{in } B, \\ w = u & \text{in } \Omega \setminus B. \end{cases}$$

Then, w is subharmonic, and $u \leq w$.

Proof. Apply the maximum principle in B. It follows that $u \leq w$ in Ω. Now fix a point $x \in \Omega$. If $x \in B$, then w is harmonic in any ball $B(x,r) \subset B$. In particular,

$$w(x) = \fint_{\partial B(x,r)} w \, d\sigma.$$

If $x \notin B$, then, $w(x) = u(x)$. Since u is subharmonic and $u \leq w$, we have for all the balls $B(x,r) \subset \Omega$,

$$w(x) = u(x) \leq \fint_{\partial B(x,r)} u \, d\sigma \leq \fint_{\partial B(x,r)} w \, d\sigma.$$

So, w satisfies the mean-value inequality for every $x \in \Omega$ and for every sufficiently small ball $B(x,r)$, hence w is subharmonic in Ω. $\quad\square$

Proof of Theorem B.1.2.

Take $g \in C(\partial\Omega)$, and let $m = \min_{\partial\Omega} g$, $M = \max_{\partial\Omega} g$. We write $v \in S_g$ whenever $v \in C(\overline{\Omega})$ is subharmonic in Ω and $v \leq g$ on $\partial\Omega$. Note that S_g is nonempty, since $v \equiv m \in S_g$. We want to prove that the function u defined for $x \in \overline{\Omega}$ by

$$u(x) = \sup\{v(x) : v \in S_g\} \tag{B.10}$$

is the solution to (B.9). Note that u is well defined and that $m \leq u \leq M$ on $\overline{\Omega}$.

Step 1. u is harmonic on Ω.

Fix a point $x_0 \in \Omega$ and a ball $\overline{B} \subset \Omega$ centered at x_0. Choose a sequence (v_k) in S_g, such that $v_k(x_0) \to u(x_0)$. Replace v_k by the fonction w_k obtained as the subharmonic lift of $\max\{m, v_1, \ldots, v_k\}$ in B. Then, $w_k \in S_g$, w_k is harmonic in B, and (w_k) is a nondecreasing sequence. In particular, $w_k(x_0) \to u(x_0)$. Also, (w_k) is uniformly bounded. By Proposition B.1.1, (w_k) converges uniformly on compact subsets of B to a smooth function w. Passing to the limit in the mean-value inequality, we deduce that w is harmonic in B.

It remains to be proven that $u = w$ in B. Clearly, $w \leq u$ in B. To prove the reverse inequality, take $v \in S_g$, and let z_k be the subharmonic lift of $\max\{w_k, v\}$ in B. Since $u(x_0) = w(x_0)$ and $z_k \in S_g$, we have $z_k(x_0) \leq w(x_0)$ for all k. In addition, z_k is harmonic in B and $\max\{w_k, v\} \leq z_k$ in B by the maximum principle. Thus,

$$w(x_0) \geq z_k(x_0) = \fint_B z_k \, dx \geq \fint_B \max\{w_k, v\} \, dx.$$

Applying the mean-value equality on the left-hand side and letting $k \to +\infty$ on the right-hand side, we obtain

$$\fint_B w \, dx \geq \fint_B \max(w, v) \, dx.$$

It follows that $v \leq w$ in B. This being true for all $v \in S_g$, we deduce that $u \leq w$ in B, as desired.

Step 2. Given any $x_0 \in \partial\Omega$, there exists a barrier function ζ at x_0, that is, $\zeta \in C(\overline{\Omega})$ is superharmonic in Ω, $\zeta > 0$ in $\overline{\Omega} \setminus \{x_0\}$, and $\zeta(x_0) = 0$.

Since Ω is bounded and smooth, there exists an exterior ball $\overline{B} = \overline{B}(y_0, r) \subset \mathbb{R}^N \setminus \Omega$ such that $\partial\Omega \cap \partial B = \{x_0\}$. Define ζ in $\mathbb{R}^N \setminus \{y_0\}$ by

$$\zeta(x) = \Gamma(x - y_0) - \Gamma(r),$$

where Γ is the fundamental solution to the Laplace equation given by (A.6) and (A.7). Then, ζ is the desired barrier.

Step 3. End of proof. Let ζ be the barrier at $x_0 \in \partial\Omega$, constructed in Step 2. Let $\varepsilon > 0$. By continuity of g on $\partial\Omega$, if $r > 0$ is sufficiently small, $g(x_0) - \varepsilon < g < g(x_0) + \varepsilon$ in $\partial\Omega \cap B(x_0, r)$. Since ζ is positive and continuous on the compact set $\overline{\Omega} \setminus B(x_0, r)$, there exists a positive constant $C > 0$ such that

$$g(x_0) - \varepsilon - C\zeta < g < g(x_0) + \varepsilon + C\zeta, \qquad \text{on } \partial\Omega \setminus B(x_0, r). \tag{B.11}$$

Note that $g(x_0) - \varepsilon - C\zeta \in S_g$, hence $g(x_0) - \varepsilon - C\zeta \leq u$, where u is given by (B.10). Now, for any $v \in S_g$, $v \leq g$ on $\partial\Omega$, and therefore $v - C\zeta < g(x_0) + \varepsilon$ on $\partial\Omega$, by (B.11). By the maximum principle, $v - C\zeta < g(x_0) + \varepsilon$ on $\overline{\Omega}$. This being true for all $v \in S_g$, we obtain at last

$$g(x_0) - \varepsilon - C\zeta \leq u \leq g(x_0) + \varepsilon + C\zeta, \qquad \text{on } \overline{\Omega}.$$

Since ζ is continuous at x_0, $\zeta(x_0) = 0$, and $\varepsilon > 0$ is arbitrary, we deduce that $\lim_{x \to x_0} u(x) = g(x_0)$, as desired. $\qquad\square$

B.2 Schauder estimates

B.2.1 Poisson's equation on the unit ball

We consider next the Poisson equation

$$-\Delta u = f \qquad \text{in } B, \tag{B.12}$$

posed on the unit ball $B \subset \mathbb{R}^N$, $N \geq 1$. By the fundamental theorem of calculus, in dimension $N = 1$, every solution to $-u'' = f$, with $f \in C(-1, 1)$, belongs to the space $C^2(-1, 1)$. Unfortunately, such a regularity result fails in higher dimensions.

Example B.2.1 ([126]) Let $N \geq 2$ and let B be the unit ball of $\mathbb{R}^N$. There exists a function $f \in C(B)$ such that no function $u \in C^2(B)$ solves (B.12).

Proof. Take a harmonic homogeneous polynomial of degree two, for example, $P(x) = x_1 x_2$, and a cutoff function $\eta \in C_c^2(\mathbb{R}^N)$, such that $\eta \equiv 1$ in B_1 and $\eta \equiv 0$ in $\mathbb{R}^N \setminus B_2$. Now let

$$-f(x) = \sum_{k=1}^{+\infty} \frac{1}{k} \Delta(\eta P)(2^k x), \qquad \text{for } x \in B.$$

Since P is harmonic, $\Delta(\eta P) = P\Delta\eta + 2\nabla\eta \cdot \nabla P$. So, $\Delta(\eta P)(2^k x)$ is supported in $B_{2^{-k+1}} \setminus B_{2^{-k}}$. It follows that $f(0) = 0$ and for $x \in B \setminus \{0\}$,

$$-f(x) = \frac{1}{k_x}\Delta(\eta P)(2^{k_x} x),$$

where $k = k_x$ is the unique integer such that $2^{-k} < |x| \leq 2^{-k+1}$. Hence, $f \in C(B)$. Now, let

$$u(x) = \sum_{k=1}^{+\infty} \frac{1}{k4^k}(\eta P)(2^k x), \qquad \text{for } x \in B.$$

Then, $u \in C^1(B) \cap C^2(B \setminus \{0\})$, $-\Delta u = f$ in $B \setminus \{0\}$, and

$$\frac{\partial^2 u}{\partial x_1 \partial x_2}(x) = \sum_{k=1}^{+\infty} \frac{1}{k}\frac{\partial^2(\eta P)}{\partial x_1 \partial x_2}(2^k x) = \sum_{k=1}^{+\infty} \frac{1}{k}\left(\eta\frac{\partial^2 P}{\partial x_1 \partial x_2}\right)(2^k x) + v(x),$$

for some $v \in C(B)$. Now, for $P(x) = x_1 x_2$,

$$\sum_{k=1}^{+\infty} \frac{1}{k}\left(\eta\frac{\partial^2 P}{\partial x_1 \partial x_2}\right)(2^k x) \geq \sum_{k=1}^{k_x-1} \frac{1}{k} \to +\infty, \qquad \text{as } |x| \to 0.$$

So, $u \notin C^2(B)$. Assume by contradiction that some function $\tilde{u} \in C^2(B)$ solves $-\Delta\tilde{u} = f$ in B. Then, $z = u - \tilde{u}$ is harmonic in $B \setminus \{0\}$, and in fact, also in B, by Exercise A.1.3. This is not possible, by Proposition B.1.1. □

Example B.2.1 shows that some quantitative information on the modulus of continuity of f is needed in order to gain C^2 regularity of solutions. This is what we discuss next.

Definition B.2.1 *Let Ω be an open set of $\mathbb{R}^N$, $N \geq 1$. We say that f is Dini continuous in Ω and write $f \in C_{\mathrm{Dini}}(\Omega)$ if*

$$\int_0^d \frac{\omega(r)}{r}\,dr < +\infty,$$

where d is the diameter of Ω, and where

$$\omega(r) = \sup_{\{x,y\in\Omega\,:\,|x-y|<r\}} |f(x) - f(y)|, \qquad \text{for } r \in (0,d).$$

241

The following theorem gives an interior *a priori* estimate on the modulus of continuity of D^2u, for any $u \in C^2$ solving the Poisson equation (Equation (B.12)), with Dini continuous right-hand side f.

Theorem B.2.1 ([33]) *Let B denote a ball of radius 1 in $\mathbb{R}^N$, $N \geq 1$. Let $f \in C_{\mathrm{Dini}}(B)$. Assume that $u \in C^2(B)$ solves (B.12). Then, there exists a constant C_N depending on N only, such that for all $x, y \in B_{1/2}$,*

$$|D^2u(x) - D^2u(y)| \leq C_N \left[d \sup_{B_1} |u| + \int_0^d \frac{\omega(r)}{r} \, dr + d \int_d^1 \frac{\omega(r)}{r^2} \, dr \right], \quad \text{(B.13)}$$

where $d = |x - y|$.

Proof. We follow [222]. It suffices to establish (B.13) for $x = 0$ and $|y| < 1/8$. For $k \geq 1$, let $B_k = B_{\rho^k}(0)$, where $\rho = 1/2$. Let u_k be the solution to

$$\begin{cases} -\Delta u_k = f(0) & \text{in } B_k, \\ u_k = u & \text{on } \partial B_k. \end{cases} \quad \text{(B.14)}$$

We claim that

$$\|u_k - u\|_{L^\infty(B_k)} \leq C\rho^{2k}\omega(\rho^k). \quad \text{(B.15)}$$

Indeed, let

$$\zeta_k(x) = \frac{1}{2N} \left(\rho^{2k} - |x|^2 \right)$$

be the solution to

$$\begin{cases} -\Delta\zeta = 1 & \text{in } B_k, \\ \zeta = 0 & \text{on } \partial B_k. \end{cases}$$

Since $v = u_k - u$ solves

$$\begin{cases} -\Delta v = f(0) - f & \text{in } B_k, \\ v = 0 & \text{on } \partial B_k, \end{cases}$$

it follows from the maximum principle that

$$-\omega(\rho^k)\zeta_k \leq v \leq \omega(\rho^k)\zeta_k.$$

Equation (B.15) follows. We also deduce that

$$\|u_k - u_{k+1}\|_{L^\infty(B_{k+1})} \leq C\rho^{2k}\omega(\rho^k). \quad \text{(B.16)}$$

Since $u_{k+1} - u_k$ is harmonic, it follows from (B.1) that

$$\|\nabla(u_{k+1} - u_k)\|_{L^\infty(B_{k+2})} \leq C\rho^k \omega(\rho^k),$$
$$\|D^2(u_{k+1} - u_k)\|_{L^\infty(B_{k+2})} \leq C\omega(\rho^k). \tag{B.17}$$

Let v_k denote the difference of u_k and the quadratic part of u at 0, that is,

$$v_k(x) = u_k(x) - \left(u(0) + \nabla u(0) \cdot x + \frac{1}{2}\left(D^2 u(0).x\right) \cdot x\right).$$

Since $u \in C^2(B)$, we deduce from (B.15) that $v_k = o(\rho^{2k})$ in B_k. By definition of u_k, we also have that v_k is harmonic in B_k. Hence, by (B.1),

$$\nabla u(0) = \lim_{k \to +\infty} \nabla u_k(0), \quad D^2 u(0) = \lim_{k \to +\infty} D^2 u_k(0). \tag{B.18}$$

Fix a point $y \in B_{1/8}(0)$. We have

$$|D^2 u(y) - D^2 u(0)| \leq |D^2 u(y) - D^2 u_k(y)| + |D^2 u_k(y) - D^2 u_k(0)| +$$
$$+ |D^2 u_k(0) - D^2 u(0)| =: I_1 + I_2 + I_3.$$

Then, by (B.17) and (B.18),

$$I_3 \leq \sum_{j=k}^{+\infty} |D^2 \tilde{u}_j(y) - D^2 \tilde{u}_{j+1}(y)| \leq C \sum_{j=k}^{+\infty} \omega(\rho^j) \leq C \int_0^{\rho^{k-1}} \frac{\omega(r)}{r} \, dr. \tag{B.19}$$

Next, we estimate I_1. For $j \geq k + 1$, let $B_j(y) = B(y, \rho^j)$ and let $\tilde{u}_j$ be the solution to

$$\begin{cases} -\Delta \tilde{u}_j = f(y) & \text{in } B_j(y), \\ \tilde{u}_j = u & \text{on } \partial B_j(y). \end{cases}$$

Then,

$$I_1 \leq |D^2 u(y) - D^2 \tilde{u}_{k+1}(y)| + |D^2 \tilde{u}_{k+1}(y) - D^2 u_k(y)|$$

$$\leq \sum_{j=k+1}^{+\infty} |D^2 \tilde{u}_j(y) - D^2 \tilde{u}_{j+1}(y)| + |D^2 \tilde{u}_{k+1}(y) - D^2 u_k(y)|$$

$$\leq C \int_0^{\rho^k} \frac{\omega(r)}{r} \, dr + |D^2 \tilde{u}_{k+1}(y) - D^2 u_k(y)|. \tag{B.20}$$

243

Let $k \geq 1$ be the unique integer such that

$$\rho^{k+3} \leq |y| < \rho^{k+2}. \tag{B.21}$$

Then, $B_{k+1}(y) = B(y, \rho^{k+1}) \subset B(0, \rho^k)$ and the function

$$w := \tilde{u}_{k+1} - u_k - (f(y) - f(0))\zeta_k$$

is harmonic in $B_{k+1}(y)$. Hence,

$$|D^2 w(y)| \leq \frac{C}{\rho^{2k}} \|w\|_{L^\infty(B_{k+1}(y))} \leq C\omega(\rho^k).$$

In addition,

$$|(f(y) - f(0))D^2\zeta_k(y)| \leq C\omega(\rho^k),$$

so that

$$|D^2\tilde{u}_{k+1}(y) - D^2 u_k(y)| \leq C\omega(\rho^k).$$

Using this in (B.20), we deduce that

$$I_1 \leq C \int_0^{\rho^{k-1}} \frac{\omega(r)}{r}\, dr. \tag{B.22}$$

We estimate at last I_2. For $j \leq k$, let $h_j = u_{j+1} - u_j$. It follows from (B.1) and (B.16) that

$$|D^2 h_j(y) - D^2 h_j(0)| \leq \|D^3 h_j\|_{L^\infty(B_{k+2})} |y|$$

$$\leq \|D^3 h_j\|_{L^\infty(B_{j+2})} |y| \leq C \frac{|y|}{\rho^j}\omega(\rho^j).$$

Hence,

$$I_2 \leq |D^2 u_{k-1}(y) - D^2 u_{k-1}(0)| + |D^2 h_{k-1}(y) - D^2 h_{k-1}(0)|$$

$$\leq |D^2 u_1(y) - D^2 u_1(0)| + \sum_{j=1}^{k-1} |D^2 h_j(y) - D^2 h_j(0)|$$

$$\leq C|y| \left(\|u_1\|_{L^\infty(B_1)} + \sum_{j=1}^{k-1} \rho^{-j}\omega(\rho^j) \right)$$

$$\leq C|y| \left(\|u\|_{L^\infty(B)} + \int_{|y|}^1 \frac{\omega(r)}{r^2}\, dr \right). \tag{B.23}$$

Collecting (B.19), (B.22), and (B.23), we obtain

$$|D^2u(y) - D^2u(0)| \leq C \left(\int_0^{\rho^{k-1}} \frac{\omega(r)}{r} \, dr + |y| \|u\|_{L^\infty(B)} + |y| \int_{|y|}^1 \frac{\omega(r)}{r^2} \, dr \right).$$

Recalling (B.21), (B.13) follows. $\qquad\qquad\qquad\qquad\qquad\qquad\qquad$ □

Corollary B.2.1 *Let B_1 be a ball of radius 1 in $\mathbb{R}^N$, $N \geq 1$, and $\alpha \in (0,1)$. Let $f \in C^\alpha(B)$. Assume that $u \in C^2(B)$ solves (B.12). Then, there exists a constant C_N depending on N only, such that*

$$\|u\|_{C^{2,\alpha}(B_{1/2})} \leq C_N \left[\|u\|_{L^\infty(B_1)} + \frac{1}{\alpha(1-\alpha)} \|f\|_{C^\alpha(B)} \right]. \tag{B.24}$$

If f is Lipschitz, then for all $x, y \in B_{1/2}$,

$$|D^2u(x) - D^2u(y)| \leq C_N d \left(\|u\|_{L^\infty(B_1)} + \|f\|_{C^{0,1}(B_1)} |\ln d| \right), \tag{B.25}$$

where $d = |x - y|$.

The previous corollary is an immediate consequence of Theorem B.2.1 and the following interpolation inequality (with $\varepsilon = 1$), the proof of which can be found in [126].

Lemma B.2.1 *Let $\alpha \in (0,1]$, $N \geq 1$ and Ω a smoothly bounded domain of $\mathbb{R}^N$. For every $\varepsilon > 0$, there exists a constant $C = C(N, \Omega, \varepsilon)$ such that for every $u \in C^{2,\alpha}(\Omega)$,*

$$\|u\|_{C^2(\overline{\Omega})} \leq C \|u\|_{L^\infty(\Omega)} + \varepsilon [D^2u]_\alpha,$$

where

$$[D^2u]_\alpha = \sup \left\{ \frac{|D^2u(x) - D^2u(y)|}{|x-y|^\alpha} : x, y \in \Omega, x \neq y \right\}.$$

Boundary estimates can be obtained similarly to Theorem B.2.1.

Theorem B.2.2 *Let $B^+ = \{x \in B : x_N > 0\}$ be a half-ball of radius 1 in $\mathbb{R}^N$, $N \geq 1$. Let $f \in C_{\mathrm{Dini}}(B^+)$. Assume that $u \in C^2(B) \cap C(\overline{B^+})$ solves*

$$\begin{cases} -\Delta u = f & \text{in } B^+, \\ \quad u = 0 & \text{on } [x_N = 0]. \end{cases} \tag{B.26}$$

Then, there exists a constant C depending on N only, such that for all $x, y \in B_{1/2}^+$, (B.13) holds. In addition, if $f \in C^\alpha(B^+)$ for some $\alpha \in (0, 1)$, then

$$\|u\|_{C^{2,\alpha}(B_{1/2}^+)} \le C \left[\|u\|_{L^\infty(B^+)} + \frac{1}{\alpha(1-\alpha)} \|f\|_{C^\alpha(B^+)} \right]. \tag{B.27}$$

and if f is Lipshitz continuous, then

$$|D^2 u(x) - D^2 u(y)| \le Cd \left(\|u\|_{L^\infty(B^+)} + \|f\|_{C^{0,1}(B^+)} |\ln d| \right), \tag{B.28}$$

where $d = |x - y|$.

Proof. The proof is the same as that for Theorem B.2.1, provided we replace B_k by $B_k \cap [x_N > 0]$ and note that if w is a harmonic function in B^+ such that $w = 0$ on $[x_N = 0]$, then w is harmonic in B after its odd extension in the x_N variable (see Exercise A.1.4). □

Thanks to the *a priori* estimates, we may now solve Poisson's equation on a ball.

Theorem B.2.3 *Let B be a ball of radius 1 in $\mathbb{R}^N$, $N \ge 1$. Let $f \in C^\alpha(B)$, $\alpha \in (0, 1)$. There exists a unique solution $u \in C^{2,\alpha}(B)$ to*

$$\begin{cases} -\Delta u = f & \text{in } B, \\ u = 0 & \text{on } \partial B. \end{cases} \tag{B.29}$$

Furthermore, there exists a constant C depending on N only, such that

$$\|u\|_{C^{2,\alpha}(B)} \le C \|f\|_{C^\alpha(B)}. \tag{B.30}$$

Proof. By a standard density argument, it suffices to consider the case where $f \in C_c^\infty(\mathbb{R}^N)$. Let $v = \Gamma * f$, where Γ is the fundamental solution of the Laplace operator, given by (A.6) and (A.7). Then, $v \in C^\infty(\mathbb{R}^N)$ and $-\Delta v = f$ in $\mathbb{R}^N$. Hence, any solution to (B.29) can be written as $u = v + w$, where w solves the Dirichlet problem (B.4) with boundary data $g = -v|_{\partial B}$. By Theorem B.1.1, such a solution is unique and belongs to $C^2(B) \cap C(\bar{B})$. Hence, the same holds true for the solution u to (B.29). It remains to be proven that $u \in C^{2,\alpha}(B)$. By Corollary B.2.1, for every $x_0 \in B$, there exists $r > 0$ and a constant C, which may depend on r such that

$$\|u\|_{C^{2,\alpha}(B(x_0,r))} \le C \left(\|u\|_{L^\infty(B)} + \|f\|_{C^\alpha(B)} \right) \le C' \|f\|_{C^\alpha(B)}, \tag{B.31}$$

where the last inequality is obtained by comparing u and $\pm\|f\|_{L^\infty(B)}\frac{1}{2N}(1 - |x|^2)$.

Now take a point $x_0 \in \partial B$. Without loss of generality, we may assume that $0 \in \partial B$ is the antipodal point of x_0. Note that the sphere inversion x^* given by (B.6) maps B onto a half-space H, ∂B onto ∂H, and x_0 to $x_0^* \neq 0$. Let u^* denote the Kelvin transform of u, that is, for $y \in H$, let

$$u^*(x) = |y|^{2-N}u(y^*).$$

Then, $u^* \in C^2(H) \cap C(\overline{H})$ solves

$$\begin{cases} -\Delta u^* = |y|^{-2-N}f(y^*) & \text{in } H, \\ \quad\quad u^* = 0 & \text{on } \partial H. \end{cases}$$

Using Theorem B.2.2 and the smoothness of the mapping $x \mapsto x^*$ near x_0, we deduce that for some $r > 0$, there exists a constant C such that

$$\|u\|_{C^{2,\alpha}(B(x_0,r)\cap\Omega)} \leq C\|f\|_{C^\alpha(B)}. \tag{B.32}$$

Since $\overline{B}$ is compact, it can be covered by finitely many balls where either (B.31) or (B.32) holds. Equation (B.30) follows. $\qquad\square$

B.2.2 *A priori* **estimates for** $C^{2,\alpha}$ **solutions**

In this section, we want to generalize Theorem B.2.3 to uniformly elliptic operators defined on a smoothly bounded domain. As a first step, we prove *a priori* estimates for $C^{2,\alpha}$ solutions.

Theorem B.2.4 ([198]) *Let $N \geq 1$, let Ω denote a smoothly bounded domain of $\mathbb{R}^N$ and given $\alpha \in (0,1)$, let $f \in C^\alpha(\Omega)$. Let L denote a uniformly elliptic operator, that is, (A.21) and (A.23) hold. In addition, assume that the coefficients of L are such that $A, B, V \in C^\alpha(\Omega)$. Assume that $u \in C^{2,\alpha}(\Omega)$ solves*

$$\begin{cases} -Lu = f & \text{in } \Omega, \\ \quad\, u = 0 & \text{on } \partial\Omega. \end{cases} \tag{B.33}$$

Then, there exists a constant C depending only on N, Ω, α, the ellipticity constant of L, and the norms $\|A\|_{C^\alpha(\Omega)}, \|B\|_{C^\alpha(\Omega)}, \|V\|_{C^\alpha(\Omega)}$, such that

$$\|u\|_{C^{2,\alpha}(\Omega)} \leq C\left(\|u\|_{L^\infty(\Omega)} + \|f\|_{C^\alpha(\Omega)}\right). \tag{B.34}$$

Remark B.2.1 *The above theorem does not say that there indeed exists a solution u to (B.33) belonging to the class $C^{2,\alpha}(\Omega)$. This will be proven later on under the extra requirement that $V \geq 0$ in Ω.*

Proof.

Changing coordinates if necessary, we may assume that $0 \in \Omega$ and $a_{ij}(0) = \delta_{ij}$. Let $u \in C^{2,\alpha}(\Omega)$ be a solution to (B.33). Then,

$$-\Delta u = f + \left(\sum_{i,j=1}^{N} (a_{ij}(x) - a_{ij}(0)) \partial_{ij} u + \sum_{i=1}^{N} b_i \partial_i u + V(x) u \right) =: R \qquad \text{in } \Omega.$$

Applying Corollary B.2.1 (to $v(x) = u(rx)$), there exists constants C and C' such that for any $r > 0$ small,

$$\|u\|_{C^{2,\alpha}(B_{r/2})} \leq C \left(\frac{1}{r^{2+\alpha}} \|u\|_{L^\infty(B_r)} + \|R\|_{C^\alpha(B_r)} \right)$$

$$\leq C' \left(\frac{1}{r^{2+\alpha}} \|u\|_{L^\infty(B_r)} + \|f\|_{C^\alpha(B_r)} + \|u\|_{C^2(B_r)} + \right.$$

$$\left. \sup_{i,j=1...N} \|a_{ij}(\cdot) - a_{ij}(0)\|_{L^\infty(B_r)} \|u\|_{C^{2,\alpha}(B_r)} \right). \qquad (B.35)$$

Since a_{ij} is continuous,

$$C' \sup_{i,j=1...N} \|a_{ij}(\cdot) - a_{ij}(0)\|_{L^\infty(B_r)} < 1/4, \qquad (B.36)$$

for $r > 0$ sufficiently small. Fix such a r. By Lemma B.2.1, there exists a constant C such that

$$C' \|u\|_{C^2(B_r)} \leq C \|u\|_{L^\infty(B_r)} + \frac{1}{4} \|u\|_{C^{2,\alpha}(B_r)}. \qquad (B.37)$$

Plugging (B.36) and (B.37) in (B.35), we obtain

$$\|u\|_{C^{2,\alpha}(B_{r/2})} \leq C \left(\|u\|_{L^\infty(\Omega)} + \|f\|_{C^\alpha(\Omega)} \right) + \frac{1}{2} \|u\|_{C^{2,\alpha}(\Omega)}. \qquad (B.38)$$

The above estimate applies to any sufficiently small ball $B(x_0, r) \subset \Omega$. Now take a point $x_0 \in \partial\Omega$. Since Ω has a smooth boundary, there exists $r > 0$ and a smooth coordinate chart $y = \phi(x)$ such that $\phi(\Omega \cap B(x_0, r)) = B^+$

248

and $\phi(\partial\Omega \cap B(x_0, r)) = \overline{B^+} \cap [x_N = 0]$. Writing Equation (B.33) in the new coordinates, we obtain

$$\begin{cases} -L'u = f & \text{in } B^+, \\ \quad u = 0 & \text{on } \overline{B^+} \cap [x_N = 0], \end{cases} \tag{B.39}$$

where $L'u = \sum_{i,j=1}^{N} a'_{ij}(y)\frac{\partial^2 u}{\partial y_i \partial y_j} + \sum_{i=1}^{N} b'_i(y)\frac{\partial u}{\partial y_i} + Vu$ is a uniformly elliptic operator, which coefficients depend on L and ϕ. We may further assume that $a'_{ij}(0) = \delta_{ij}$, so that (B.39) can be rewritten as

$$\begin{cases} -\Delta u = f + \left(\sum_{i,j=1}^{N} (a'_{ij}(y) - a'_{ij}(0))\partial_{ij}u + \sum_{i=1}^{N} b'_i \partial_i u + Vu \right) =: R' & \text{in } B^+, \\ \quad u = 0 \quad \text{on } \overline{B^+} \cap [x_N = 0]. \end{cases}$$

Applying Theorem B.2.2, we deduce that

$$\|u\|_{C^{2,\alpha}(B^+_{1/2})} \leq C \left(\|u\|_{L^\infty(B^+)} + \|R'\|_{C^\alpha(B^+)} \right).$$

By localization and interpolation, we obtain as previously that

$$\|u\|_{C^{2,\alpha}(B^+_{r'})} \leq C \left(\|u\|_{L^\infty(\Omega)} + \|f\|_{C^\alpha(\Omega)} \right) + \frac{1}{2}\|u\|_{C^{2,\alpha}(\Omega)},$$

for some $r' > 0$. Going back to the original variables, it follows that for some $r > 0$,

$$\|u\|_{C^{2,\alpha}(\Omega \cap B(x_0, r))} \leq C \left(\|u\|_{L^\infty(\Omega)} + \|f\|_{C^\alpha(\Omega)} \right) + \frac{1}{2}\|u\|_{C^{2,\alpha}(\Omega)}. \tag{B.40}$$

Since $\overline{\Omega}$ is compact, it can be covered by finitely many balls where either (B.38) or (B.40) holds, hence

$$\|u\|_{C^{2,\alpha}(\Omega)} \leq C \left(\|u\|_{L^\infty(\Omega)} + \|f\|_{C^\alpha(\Omega)} \right),$$

as desired. $\qquad \square$

B.2.3   Existence of $C^{2,\alpha}$ solutions

We now generalize Theorem B.2.3 to the case of a uniformly elliptic equation posed on the ball.

Theorem B.2.5 *Let B be a ball of radius 1 in $\mathbb{R}^N$, $N \geq 1$. Let $f \in C^\alpha(B)$, $g \in C^{2,\alpha}(\partial B)$, $\alpha \in (0,1)$. Let L be a uniformly elliptic operator, that is, (A.21) and (A.23) hold. In addition, assume that the coefficients of L are such that $A, B, V \in C^\alpha(B)$, and $V \geq 0$ in B. Then, there exists a unique solution $u \in C^{2,\alpha}(B)$ to*

$$\begin{cases} -Lu = f & \text{in } B, \\ \quad u = g & \text{on } \partial B. \end{cases} \tag{B.41}$$

Furthermore, there exists a constant C depending on N, α, the ellipticity constant of L, and the norms $\|A\|_{C^\alpha(B)}, \|B\|_{C^\alpha(B)}, \|V\|_{C^\alpha(B)}$ only, such that

$$\|u\|_{C^{2,\alpha}(B)} \leq C \left(\|f\|_{C^\alpha(B)} + \|g\|_{C^{2,\alpha}(\partial B)} \right). \tag{B.42}$$

To establish Theorem B.2.5, we shall use the following method of continuity.

Proposition B.2.1 *Let X, Y be two Banach spaces and L_0, L_1 two bounded linear operators from X to Y. For each $t \in [0,1]$, set*

$$L_t = (1-t)L_0 + tL_1$$

and suppose that there exists a constant C such that

$$\|u\|_X \leq C\|L_t u\|_Y, \qquad \text{for all } u \in X, \ t \in [0,1].$$

If L_0 is invertible, then so is L_1.

Proof. Suppose that L_s is invertible for some $s \in [0,1]$. For $t \in [0,1]$ and $f \in Y$, the equation $L_t u = f$ can be rewritten as

$$L_s u = f + (L_s - L_t)u = f + (t-s)(L_0 - L_1)u,$$

that is,

$$u = L_s^{-1}f + (t-s)L_s^{-1}(L_0 - L_1)u =: Tu.$$

The mapping T is a contraction in X if

$$|t-s| < \delta := \frac{1}{C(\|L_0\| + \|L_1\|)},$$

and hence L_s is invertible for all $t \in [0,1]$ such that $|t-s| < \delta$. By dividing the interval $[0,1]$ into subintervals of length less than δ, we see that the mapping L_t is invertible for all $t \in [0,1]$ provided L_0 is. $\qquad\square$

Proof of Theorem B.2.5. First, we note that we can assume that $g = 0$. Indeed, take $\eta \in C_c^\infty(0,2)$ such that $\eta(1) = 1$ and let $G(x) = \eta(|x|)g(x/|x|)$, for $x \in B$. Then, $G \in C^{2,\alpha}(B)$, $\|G\|_{C^{2,\alpha}(B)} \leq C\|g\|_{C^{2,\alpha}(\partial B)}$, and $G|_{\partial B} = g$. Then, $u = v + G$, where v vanishes on ∂B and solves $-\Delta u = f + \Delta G$ in B.

Now let X be the Banach space of functions $u \in C^{2,\alpha}(B)$ such that $u = 0$ on ∂B. Also let $Y = C^{\alpha}(B)$. For $t \in [0,1]$, let

$$L_t : \begin{cases} X \to Y \\ u \mapsto (1-t)\Delta u + tLu. \end{cases}$$

Given $f \in Y$, we want to find a solution to

$$-L_1 u = f. \tag{B.43}$$

By Proposition B.2.1 and Corollary B.2.3, it suffices to show that there exists a constant C such that for all $u \in X$ and $t \in [0,1]$,

$$\|u\|_{C^{2,\alpha}(B)} \leq C\|L_t u\|_{C^{\alpha}(B)}. \tag{B.44}$$

Noting that the ellipticity constant and C^{α} norms of the coefficients of L_t are uniformly controlled independently of $t \in [0,1]$, it follows from Theorem B.2.4 that

$$\|u\|_{C^{2,\alpha}(B)} \leq C\left(\|u\|_{L^{\infty}(B)} + \|L_t u\|_{C^{\alpha}(B)}\right). \tag{B.45}$$

For $\gamma > 0$ sufficiently large, the function $\zeta(x) = e^{\gamma} - e^{\gamma x_1}$ is positive in B and satisfies $-L\zeta > 1$ in B. Comparing u with $\pm\|L_t u\|_{L^{\infty}(B)}\zeta$, we deduce that $\|u\|_{L^{\infty}(B)} \leq C\|L_t u\|_{L^{\infty}(B)}$. (B.44) follows. $\qquad\square$

We turn at last to the case of a uniformly elliptic equation posed on a smoothly bounded domain.

Theorem B.2.6 *Let Ω be a smoothly bounded domain in $\mathbb{R}^N$, $N \geq 1$. Let $f \in C^{\alpha}(\Omega)$, $\alpha \in (0,1)$. Let L be a uniformly elliptic operator, that is, (A.21) and (A.23) hold. In addition, assume that the coefficients of L are such that $A, B, V \in C^{\alpha}(\Omega)$, and $V \geq 0$ in Ω. Then, there exists a unique solution $u \in C^{2,\alpha}(\Omega)$ to*

$$\begin{cases} -Lu = f & \text{in } \Omega, \\ u = 0 & \text{on } \partial\Omega. \end{cases} \tag{B.46}$$

Furthermore, there exists a constant C depending on N, α, Ω, the ellipticity constant of L, and the norms $\|A\|_{C^{\alpha}(\Omega)}, \|B\|_{C^{\alpha}(\Omega)}, \|V\|_{C^{\alpha}(\Omega)}$ only, such that

$$\|u\|_{C^{2,\alpha}(\Omega)} \leq C\|f\|_{C^{\alpha}(\Omega)}.$$

Proof. Thanks to the continuity method (Proposition B.2.1) and the *a priori* estimate of Theorem B.2.4, it suffices to prove that for $L = \Delta$, $f \in C_c^{\infty}(\Omega)$, there exists a solution $u \in C^{2,\alpha}(\Omega)$ to (B.46). Writing $u = \Gamma * f + v$, where Γ is the fundamental solution to the Laplace operator, and where v is the solution to the Dirichlet problem with boundary data $g = -\Gamma * f\big|_{\partial\Omega}$ given by Perron's method (Theorem B.1.2), we obtain a solution $u \in C^2(\Omega) \cap C(\overline{\Omega})$. By Corollary B.2.1, $u \in C^{2,\alpha}_{loc}(\Omega)$. It remains to be proven that u is $C^{2,\alpha}$ up to the boundary. Take a point $x_0 \in \partial\Omega$. Since Ω has a smooth boundary, there exists a neighborhood V of x_0 and a coordinate chart $y = \phi(x)$, such that $(0,\ldots,0,1) = \phi(x_0)$, $\phi(\Omega \cap V) = B$, and $\phi(\partial\Omega \cap V) = \{y \in \partial B : y_N > 0\}$. In the new coordinates, we obtain for some uniformly elliptic operator L',

$$\begin{cases} -L'u = f & \text{in } B, \\ u = 0 & \text{on } \partial B \cap [y_N > 0]. \end{cases}$$

Take $g_n \in C^{2,\alpha}(\partial B)$, such that g_n converges uniformly to $u|_{\partial B}$ on ∂B and g_n vanishes on $\partial B \cap [y_N > 1/2]$. Let $u_n \in C^{2,\alpha}(B)$ be the solution to

$$\begin{cases} -L'u_n = f & \text{in } B, \\ u_n = g_n & \text{on } \partial B, \end{cases}$$

given by Theorem B.2.5. Working with the proof of Theorem B.2.4, we obtain

$$\|u_n\|_{C^{2,\alpha}(B \cap [y_N > 3/4])} \le C \left(\|u_n\|_{L^\infty(B)} + \|f\|_{L^\infty(B)} \right).$$

Passing to the limit as $n \to +\infty$, we deduce that u is $C^{2,\alpha}$ in a neighborhood of x_0, as desired. $\qquad\square$

B.3 Calderon-Zygmund estimates

We state here without proof the following extension of Theorem B.2.4 to the setting of Sobolev spaces. For a proof in the case $p = 2$, see, for example, [25].

Theorem B.3.1 ([1,48]) *Let $N \ge 1$, let Ω denote a smoothly bounded domain of $\mathbb{R}^N$ and given $p \in (1, +\infty)$, let $f \in L^p(\Omega)$. Let L denote a uniformly elliptic operator, that is, (A.21) and (A.23) hold. In addition, assume that the coefficients of L are such that $A \in C(\overline{\Omega})$, $B, V \in L^\infty(\Omega)$, and $V \ge 0$ a.e. in Ω.*

Then, there exists a unique solution $u \in W^{2,p}(\Omega)$ of (B.33) in the sense that $-Lu = f$ a.e. in Ω and $u \in W^{1,p}_0(\Omega) \cap W^{2,p}(\Omega)$. Furthermore, there exists a constant $C = C(\Omega, L, N, p)$ such that

$$\|u\|_{W^{2,p}(\Omega)} \le C\|f\|_{L^p(\Omega)}. \tag{B.47}$$

Remark B.3.1 *As follows from Example B.2.1 (and its proof), the above theorem is false for $p = +\infty$. It is also false for $p = 1$ (work by contradiction with $f = \rho_n$, a standard mollifier converging to the Dirac mass δ_0).*

B.4 Moser iteration

In this section, we give two useful elliptic regularity results giving uniform *a priori* estimates via the Moser iteration method.

Theorem B.4.1 ([206]) *Let $N \geq 2$, let Ω denote a bounded domain of $\mathbb{R}^N$ and given $p > N$, let $f \in W^{-1,p}(\Omega)$. Then, there exists a unique solution $u \in H_0^1(\Omega) \cap L^\infty(\Omega)$ of*

$$\begin{cases} -\Delta u = f & \text{in } \Omega, \\ \quad\ u = 0 & \text{on } \partial\Omega. \end{cases} \tag{B.48}$$

Furthermore, there exists a constant $C = C(\Omega, n, p)$ such that

$$\|u\|_{L^\infty(\Omega)} \leq C \|f\|_{W^{-1,p}(\Omega)}. \tag{B.49}$$

Proof. Since $p > N$, $W^{-1,p}(\Omega) \subset H^{-1}(\Omega)$. It follows from the Lax-Milgram lemma that (B.33) is uniquely solvable in $H_0^1(\Omega)$. It remains to prove (B.49). By an obvious scaling argument, we may assume that $\|f\|_{W^{-1,p}(\Omega)} = 1$. Since $f \in W^{-1,p}(\Omega)$ and Ω is bounded, there exists $f_i \in L^p(\Omega)$, $i = 1, \ldots, N$, such that $f = \sum_{i=1}^N \frac{\partial f_i}{\partial x_i}$ and $\max_{i=1,\ldots,N} \|f_i\|_{L^p(\Omega)} = \|f\|_{W^{-1,p}(\Omega)} = 1$ (see Proposition IX.20 in [25]).

Given $k \geq 0$, consider the function $v = G(u) \in H_0^1(\Omega)$, where G is the Lipschitz function defined for $t \in \mathbb{R}$ by

$$G(t) = \begin{cases} t + k & \text{if } t \leq -k \\ \quad 0 & \text{if } -k < t < k \\ t - k & \text{if } t \geq k. \end{cases}$$

Multiply (B.48) by v and integrate. Then,

$$\int_\Omega |\nabla v|^2 \, dx = \int_\Omega \nabla u \cdot \nabla v \, dx = -\sum_{j=1}^N \int_{A_k} f_j \frac{\partial v}{\partial x_j} \, dx,$$

where A_k is the set $[|u| \geq k]$. Applying the Cauchy-Schwarz inequality, we deduce that

$$\int_\Omega |\nabla v|^2 \, dx \leq \sum_{j=1}^N \int_{A_k} f_j^2 \, dx.$$

Using Sobolev's inequality on the one hand, and Hölder's inequality on the other hand, it follows that

$$\left(\int_\Omega |v|^{2^*} \, dx \right)^{2/2^*} \le C\|f\|^2_{W^{-1,p}(\Omega)} |A_k|^{1-2/p} \le C|A_k|^{1-2/p}.$$

In other words,

$$\left(\int_{A_k} (|u| - k)^{2^*} \, dx \right)^{2/2^*} \le C|A_k|^{1-2/p}.$$

Now take $h > k \ge 0$. In particular, $A_h \subset A_k$. Since $|u| \ge h$ in A_h, we deduce that

$$(h-k)^2 |A_h|^{2/2^*} \le \left(\int_{A_k} (|u| - k)^{2^*} \, dx \right)^{2/2^*} \le C|A_k|^{1-2/p},$$

that is,

$$|A_h| \le C \frac{1}{(h-k)^{2^*}} |A_k|^{\frac{1/2-1/p}{1/2-1/N}}. \tag{B.50}$$

For $s \in \mathbb{N}$, $d > 0$, let $h_s = d - \frac{d}{2^s}$, $\theta_s = |A_{h_s}|$, $\alpha = 2^*$, $\beta = \frac{1/2-1/p}{1/2-1/N} > 1$. Then (B.50) applied with $k = h_s$, $h = h_{s+1}$ yields

$$\theta_{s+1} \le C \left(\frac{2^{s+1}}{d} \right)^\alpha \theta_s^\beta.$$

Fix at last $d^\alpha = C\theta_0^{\beta-1} 2^{\frac{\alpha\beta}{\beta-1}}$. We claim that

$$\theta_s \le \theta_0 2^{-\frac{\alpha}{\beta-1}s}, \qquad \text{for all } s \in \mathbb{N}. \tag{B.51}$$

Equation (B.51) is obviously satisfied for $s = 0$. Assume that it holds for some $s \in \mathbb{N}$. Then, by (B.50),

$$\theta_{s+1} \le C \left(\frac{2^{s+1}}{d} \right)^\alpha \theta_0^\beta 2^{-\frac{\alpha\beta}{\beta-1}s} = C \frac{2^\alpha \theta_0^\beta}{d^\alpha} 2^{-\frac{\alpha s}{\beta-1}}$$

$$= C \frac{2^\alpha \theta_0^\beta}{C\theta_0^{\beta-1} 2^{\frac{\alpha\beta}{\beta-1}}} 2^{-\frac{\alpha s}{\beta-1}} = \theta_0 2^{-\frac{\alpha}{\beta-1}(s+1)}. \tag{B.52}$$

We have just proved (B.51) by induction. Let at last $s \to +\infty$ in (B.51): $|A_d| = 0$ and (B.49) follows. $\qquad\square$

Theorem B.4.2 ([201], [212]) *Let $N \geq 1$, let Ω denote a domain of $\mathbb{R}^N$ and assume that $B(x_0, 2R) \subset\subset \Omega$. Assume that $u \in C^2(\Omega)$ satisfies*

$$\begin{cases} -\Delta u - V(x)u \leq 0 & \text{in } \Omega, \\ u > 0 & \text{in } \Omega, \end{cases} \tag{B.53}$$

where $V \in L_{loc}^{\frac{N}{2-\varepsilon}}(\Omega)$ for some $\varepsilon > 0$. Then, for every $t > 1$,

$$\|u\|_{L^\infty(B(x_0,R))} \leq C_{ST} R^{-N/t} \|u\|_{L^t(B(x_0,2R))}, \tag{B.54}$$

where C_{ST} is a constant depending only on N, t and $R^\varepsilon \|V\|_{L^{\frac{N}{2-\varepsilon}}(B(x_0,2R))}$.

Proof. Translating and scaling space if necessary, we may assume that $x_0 = 0$ and $R = 1$. Take $j \geq t/2 > 1/2$, $\varphi \in C_c^2(B_2)$ and multiply (B.53) with $u^{2j-1}\varphi^2$. Then,

$$\int_{\mathbb{R}^N} \nabla u \cdot \nabla \left(u^{2j-1}\varphi^2\right) dx \leq \int_{\mathbb{R}^N} V(x)u^{2j}\varphi^2 \, dx.$$

Expand the left-hand side:

$$(2j-1)\int_{\mathbb{R}^N} u^{2j-2}|\nabla u|^2 \varphi^2 \, dx \leq \int_{\mathbb{R}^N} V(x)u^{2j}\varphi^2 \, dx - \int_{\mathbb{R}^N} u^{2j-1}\nabla u \nabla \varphi^2 \, dx.$$

Integrate by parts the last term in the above:

$$\frac{2j-1}{j^2} \int_{\mathbb{R}^N} \left|\nabla u^j\right|^2 \varphi^2 \, dx \leq \int_{\mathbb{R}^N} V(x)u^{2j}\varphi^2 \, dx + \frac{1}{2j} \int_{\mathbb{R}^N} u^{2j}\Delta\varphi^2 \, dx.$$

Since $\left|\nabla(u^j\varphi)\right|^2 = \left|\nabla u^j\right|^2 \varphi^2 + \nabla\varphi \cdot \nabla(u^{2j}\varphi)$, it follows that

$$\frac{2j-1}{j^2} \int_{\mathbb{R}^N} \left|\nabla(u^j\varphi)\right|^2 \, dx \leq \int_{\mathbb{R}^N} V(x)u^{2j}\varphi^2 \, dx + \frac{1}{2j} \int_{\mathbb{R}^N} u^{2j}\Delta\varphi^2 \, dx$$
$$+ \frac{2j-1}{j^2} \int_{\mathbb{R}^N} \nabla\varphi \cdot \nabla(u^{2j}\varphi) \, dx.$$

Integrate by parts the last term in the above and multiply by $j^2/(2j-1)$. Then,

$$\int_{\mathbb{R}^N} \left|\nabla(u^j\varphi)\right|^2 \, dx \leq$$
$$C \left(j \int_{\mathbb{R}^N} |V(x)|u^{2j}\varphi^2 \, dx + \int_{\mathbb{R}^N} u^{2j}(|\Delta\varphi^2| + |\varphi\Delta\varphi|) \, dx\right),$$

for some constant C depending on t only.

Apply Sobolev's inequality to the left-hand side and Hölder's inequality to the first term on the right-hand side. Then,

$$\|u^j\varphi\|^2_{L^{\frac{2N}{N-2}}(\mathbb{R}^N)} \leq C\left(j\|V\|_{L^{\frac{N}{2-\varepsilon}}(B_2)}\|u^j\varphi\|^{2-\varepsilon}_{L^{\frac{2N}{N-2}}(\mathbb{R}^N)}\|u^j\varphi\|^{\varepsilon}_{L^2(\mathbb{R}^N)}+\right.$$

$$\left.\int_{\mathbb{R}^N}u^{2j}(|\Delta\varphi^2|+|\varphi\Delta\varphi|)\,dx\right)$$

$$\leq C'\left(j\|u^j\varphi\|^{2-\varepsilon}_{L^{\frac{2N}{N-2}}(\mathbb{R}^N)}\|u^j\varphi\|^{\varepsilon}_{L^2(\mathbb{R}^N)}+B\right),$$

where $C'=C'(t,N,\|V\|_{L^{\frac{N}{2-\varepsilon}}(B_2)})$ and where $B=\int_{\mathbb{R}^N}u^{2j}(|\Delta\varphi^2|+|\varphi\Delta\varphi|)\,dx$. If $B\geq j\|u^j\varphi\|^{2-\varepsilon}_{L^{\frac{2N}{N-2}}(\mathbb{R}^N)}\|u^j\varphi\|^{\varepsilon}_{L^2(\mathbb{R}^N)}$, then

$$\|u^j\varphi\|^2_{L^{\frac{2N}{N-2}}(\mathbb{R}^N)}\leq 2C'B, \tag{B.55}$$

while if $B<j\|u^j\varphi\|^{2-\varepsilon}_{L^{\frac{2N}{N-2}}(\mathbb{R}^N)}\|u^j\varphi\|^{\varepsilon}_{L^2(\mathbb{R}^N)}$, then

$$\|u^j\varphi\|^2_{L^{\frac{2N}{N-2}}(\mathbb{R}^N)}\leq 2C'j\|u^j\varphi\|^{2-\varepsilon}_{L^{\frac{2N}{N-2}}(\mathbb{R}^N)}\|u^j\varphi\|^{\varepsilon}_{L^2(\mathbb{R}^N)},$$

hence

$$\|u^j\varphi\|^2_{L^{\frac{2N}{N-2}}(\mathbb{R}^N)}\leq (Cj)^{\frac{2}{\varepsilon}}\|u^j\varphi\|^2_{L^2(\mathbb{R}^N)}. \tag{B.56}$$

From (B.55) and (B.56), we deduce that in all cases,

$$\|u^j\varphi\|^2_{L^{\frac{2N}{N-2}}(\mathbb{R}^N)}\leq (Cj)^{\frac{2}{\varepsilon}}\|u^j\varphi\|^2_{L^2(\mathbb{R}^N)}+C\int_{\mathbb{R}^N}u^{2j}(|\Delta\varphi^2|+|\varphi\Delta\varphi|)\,dx.$$

Given $1<\rho'<\rho<2$, we now choose our function φ such that $0\leq\varphi\leq 1$, $\varphi=1$ in $B_{\rho'}$, $\varphi=0$ in $\mathbb{R}^N\setminus B_\rho$, $|\nabla\varphi|\leq\frac{C}{\rho-\rho'}$, and $|\Delta\varphi|\leq\frac{C}{(\rho-\rho')^2}$. It follows that

$$\|u^j\|^2_{L^{\frac{2N}{N-2}}(B_{\rho'})}\leq (Cj)^{2/\varepsilon}\|u^j\|^2_{L^2(B_\rho)}+\frac{C}{(\rho-\rho')^2}\|u^j\|^2_{L^2(B_\rho)},$$

$$=\left((Cj)^{2/\varepsilon}+\frac{C}{(\rho-\rho')^2}\right)\|u^j\|^2_{L^2(B_\rho)}. \tag{B.57}$$

Setting $k = 2j$, (B.57) can be rephrased as

$$\|u\|_{L^{k\frac{N}{N-2}}(B_{\rho'})} \leq \left((Ck)^{2/\varepsilon} + \frac{C}{(\rho - \rho')^2}\right)^{1/k} \|u\|_{L^k(B_\rho)}. \qquad (B.58)$$

At last, choose sequences k_l and (ρ_l), which are defined by

$$k_l = t\left(\frac{N}{N-2}\right)^l, \qquad \text{for } l \in \mathbb{N} \qquad (B.59)$$

and

$$\rho_l = 1 + 2^{-l}, \qquad \text{for } l \in \mathbb{N}. \qquad (B.60)$$

Apply (B.57) with $k = k_l$, $\rho = \rho_l$, $\rho' = \rho_{l+1}$. It follows that

$$\|u\|_{L^{k_{l+1}}(B_{\rho_{l+1}})} \leq \left(C^{2l/\varepsilon} + C4^l\right)^{1/k_l} \|u\|_{L^{k_l}(B_{\rho_l})} \leq \tilde{C}^{\frac{l}{\varepsilon k_l}} \|u\|_{L^{k_l}(B_{\rho_l})}.$$

Iterating the above, we obtain

$$\|u\|_{L^{k_l}(B_{\rho_l})} \leq \tilde{C}^{\frac{1}{\varepsilon}\sum_{m=1}^{l-1}\frac{m}{k_m}} \|u\|_{L^t(B_2)}.$$

Passing to the limit as $l \to +\infty$ yields the desired inequality. $\qquad\square$

B.5   The inverse-square potential

In this section, we study elliptic regularity for an equation of the form

$$\begin{cases} -\Delta\phi - \dfrac{c}{|x - \xi|^2}\phi = g & \text{in } \Omega, \\ \phi = h & \text{on } \partial\Omega, \end{cases} \qquad (B.61)$$

where $c \in \mathbb{R}$, $\xi \in \Omega$ and f, g are given, say, smooth, functions. Note that the potential $V(x) = c/|x - \xi|^2 \notin L^p(\Omega)$ for any $p \geq N/2$, so standard elliptic regularity results, for example, Theorem B.4.2, cannot be applied to the operator $L = -\Delta - V(x)$. In fact, elliptic regularity *fails* for this operator: ϕ can be unbounded even if f, g are smooth. Still, regularity results can be recovered, provided the data f, g satisfy certain orthogonality relations. This is what we describe in this section.

B.5.1   The kernel of $L = -\Delta - \frac{c}{|x|^2}$

In Section C.4, we study the Laplace-Beltrami operator $-\Delta_{\mathscr{S}^{N-1}}$ on the sphere $\mathscr{S}^{N-1}$ and derive the following properties: the eigenvalues of $-\Delta_{\mathscr{S}^{N-1}}$ are given by

$$\lambda_k = k(N + k - 2), \quad k \geq 0,$$

and there exists an orthonormal basis of $L^2(\mathscr{S}^{N-1})$ formed with eigenvectors $\{\varphi_{k,l} : k \geq 0, l = 1, \ldots, m_k\}$, where m_k denotes the multiplicity of λ_k and $\varphi_{k,l}$, $l = 1, \ldots, m_k$ the eigenfunctions associated to λ_k. We choose the first functions to be

$$\varphi_{0,1} = \frac{1}{|\mathscr{S}^{N-1}|^{1/2}}, \quad \varphi_{1,l} = \frac{x_l}{(\int_{\mathscr{S}^{N-1}} x_l^2)^{1/2}} = \left(\frac{N}{|\mathscr{S}^{N-1}|}\right)^{1/2} x_l, \quad l = 1, \ldots, N.$$

Now, we seek solutions of

$$-\Delta w - \frac{c}{|x|^2} w = 0 \qquad \text{in } \mathbb{R}^N \setminus \{0\} \tag{B.62}$$

of the form $w(x) = f(r)\varphi_{k,l}(\omega)$, where $r = |x|$ and $\omega = x/r$ for $x \in \mathbb{R}^N \setminus \{0\}$. By Lemma C.4.2, this is equivalent to asking that f solves the following ordinary differential equation:

$$f'' + \frac{N-1}{r} f' + \frac{c - \lambda_k}{r^2} f = 0, \qquad \text{for } r > 0. \tag{B.63}$$

Equation (B.63) is of Euler type and it admits a basis of solutions of the form $f(r) = r^{-\alpha_k^{\pm}}$, where $-\alpha_k^{\pm}$ are the roots of the associated characteristic equation, that is,

$$\alpha_k^{\pm} = \frac{N-2}{2} \pm \sqrt{\left(\frac{N-2}{2}\right)^2 - c + \lambda_k}.$$

Note that $\alpha_k^{\pm}$ may have a nonzero imaginary part only for finitely many k's. If k_0 is the first integer k such that $\alpha_k^{\pm} \in \mathbb{R}$ then

$$\ldots < \alpha_{k_0+1}^- < \alpha_{k_0}^- \leq \frac{N-2}{2} \leq \alpha_{k_0}^+ < \alpha_{k_0+1}^+ < \ldots,$$

whereas, if $k < k_0$, we denote the imaginary part of α_k^+ by

$$b_k = \sqrt{c - \left(\frac{N-2}{2}\right)^2 - \lambda_k}.$$

258

For $k \geq 0$, $l = 1, \ldots, m_k$, we have just found a family of real-valued solutions of (B.62), denoted by $w^1 = w^1_{k,l}$, $w_2 = w^2_{k,l}$ and defined on $\mathbb{R}^N \setminus \{0\}$ by

$$
\begin{cases}
\text{if } (\tfrac{N-2}{2})^2 - c + \lambda_k > 0: & w^1(x) = |x|^{-\alpha^+_k} \varphi_{k,l}\left(\dfrac{x}{|x|}\right), \\[2ex]
& w^2(x) = |x|^{-\alpha^-_k} \varphi_{k,l}\left(\dfrac{x}{|x|}\right), \\[2ex]
\text{if } (\tfrac{N-2}{2})^2 - c + \lambda_k = 0: & w^1(x) = |x|^{-\frac{N-2}{2}} \log|x| \, \varphi_{k,l}\left(\dfrac{x}{|x|}\right), \\[2ex]
& w^2(x) = |x|^{-\frac{N-2}{2}} \varphi_{k,l}\left(\dfrac{x}{|x|}\right), \\[2ex]
\text{if } (\tfrac{N-2}{2})^2 - c + \lambda_k < 0: & w^1(x) = |x|^{-\frac{N-2}{2}} \sin(b_k \log|x|)\varphi_{k,l}\left(\dfrac{x}{|x|}\right), \\[2ex]
& w^2(x) = |x|^{-\frac{N-2}{2}} \cos(b_k \log|x|)\varphi_{k,l}\left(\dfrac{x}{|x|}\right).
\end{cases}
\tag{B.64}
$$

Each of the functions $W_{k,l}$ defined by

$$
\begin{cases}
\text{if } (\tfrac{N-2}{2})^2 - c + \lambda_k > 0: & W_{k,l}(x) = w^1(x) - w^2(x), \\[1ex]
\text{if } (\tfrac{N-2}{2})^2 - c + \lambda_k \leq 0: & W_{k,l}(x) = w^1(x),
\end{cases}
\tag{B.65}
$$

then solves

$$
\begin{cases}
-\Delta W_{k,l} - \dfrac{c}{|x|^2} W_{k,l} = 0 & \text{in } B \setminus \{0\}, \\[2ex]
W_{k,l} = 0 & \text{on } \partial B.
\end{cases}
$$

B.5.2 Functional setting

From the previous analysis, one can expect that solutions of an equation of the form $-\Delta u - \frac{c}{|x-\xi|^2} u = f$ behave near ξ like a (possibly negative) power of $|x - \xi|$. It is therefore convenient to work in the functional setting (see [12, 44, 186]) described below.

Given Ω a smooth bounded domain of $\mathbb{R}^N$, $\xi \in \Omega$, $k \in \mathbb{N}$, $\alpha \in (0,1)$, $r \in (0, \text{dist}(x, \partial\Omega)/2)$, and $u \in C^{k,\alpha}_{loc}(\overline{B} \setminus \{\xi\})$ we define:

$$
|u|_{k,\alpha,r,\xi} = \sup_{r \leq |x-\xi| \leq 2r} \sum_{j=0}^{k} r^j |\nabla^j u(x)| + r^{k+\alpha} \left[\sup_{r \leq |x-\xi|, |y-\xi| \leq 2r} \frac{|\nabla^k u(x) - \nabla^k u(y)|}{|x-y|^\alpha} \right].
$$

Let $d = \text{dist}(\xi, \partial\Omega)$ and for any $v \in \mathbb{R}$ let

$$\|u\|_{k,\alpha,v,\xi;\Omega} = \|u\|_{C^{k,\alpha}(\overline{\Omega}\setminus B_{d/2}(\xi))} + \sup_{0<r\leq\frac{d}{2}} r^{-v}|u|_{k,\alpha,r,\xi}.$$

Define the space

$$C^{k,\alpha}_{v,\xi}(\Omega) = \{u \in C^{k,\alpha}_{loc}(\overline{\Omega}\setminus\{\xi\}) : \|u\|_{k,\alpha,v,\xi;\Omega} < \infty\}.$$

One can easily check that $C^{k,\alpha}_{v,\xi}(\Omega)$ is a Banach space. It embeds continuously in the space of bounded functions whenever $v \geq 0$.

From here on, given $h \in C(\partial\Omega)$ and $g \in C(\Omega\setminus\{\xi\})$, we shall say that a function $\phi \in C^{k,\alpha}_{v,\xi}(\Omega)$ ($k \geq 2$) solves (B.61) whenever the boundary condition $\phi|_{\partial\Omega} = h$ holds and $-\Delta\phi(x) - \frac{c}{|x-\xi|^2}\phi(x) = g(x)$ for all $x \in \Omega\setminus\{\xi\}$.

B.5.3 The case $\xi = 0$

In this section, we investigate (B.61) in the case $\Omega = B$ and $\xi = 0$.

First observe that, by Hardy's inequality (Proposition C.1.1), in the case $c < H_N = \frac{(N-2)^2}{4}$, $N \geq 3$, the operator $L = -\Delta - \frac{c}{|x|^2}$ is coercive in $H^1_0(B)$. In particular, given $g \in H^{-1}(B)$, $h = 0$, (B.61) is uniquely solvable in $H^1_0(B)$. However, even if $g \in C^\infty(B)$, the solution ϕ need not be regular at the origin. Check that for $g = 1$ and $h = 0$, $\phi = \frac{1}{2N+c}\left(|x|^{-\alpha^-_0} - |x|^2\right)$. In other words, although the operator $L : H^1_0(B) \to H^{-1}(B)$ is coercive, thus invertible, L need not be invertible if acting on a space of bounded functions, for example, $L : C^{2,\alpha}_{0,0}(B) \to C^{0,\alpha}_{-2,0}(B)$. This situation is similar to Fredholm's alternative: in order to solve (B.61) in a given function space, one must request that the data is orthogonal to the kernel of the adjoint of the operator. Now, L is formally self-adjoint and it has a nontrivial kernel: the functions $W_{k,l}$ given in (B.65). In the next lemma, we present the orthogonality relations under which the equation is solvable in $C^{2,\alpha}_{v,0}(B)$ spaces.

Lemma B.5.1 *Let $N \geq 3$. Let $c, v \in \mathbb{R}$ and assume that*

$$\exists k_1 \geq k_0 \text{ such that } \quad -\alpha^-_{k_1} < v < -\alpha^-_{k_1+1}. \tag{B.66}$$

Let $g \in C^{0,\alpha}_{v-2,0}(B)$ and $h \in C^{2,\alpha}(\partial B)$ and consider

$$\begin{cases} -\Delta\phi - \dfrac{c}{|x|^2}\phi = g & \text{in } B, \\ \phi = h & \text{on } \partial B. \end{cases} \tag{B.67}$$

Then (B.67) *has a solution in* $C^{2,\alpha}_{v,0}(B)$ *if and only if*

$$\int_B g W_{k,l} \, dx = \int_{\partial B} h \frac{\partial W_{k,l}}{\partial n} \, d\sigma, \quad \forall k = 0, \dots, k_1, \, \forall l = 1, \dots, m_k. \quad \text{(B.68)}$$

Under this condition the solution $\phi \in C^{2,\alpha}_{v,0}(B)$ *to* (B.67) *is unique and it satisfies*

$$\|\phi\|_{2,\alpha,v,0;B} \leq C(\|g\|_{0,\alpha,v-2,0;B} + \|h\|_{C^{2,\alpha}(\partial B)}) \quad \text{(B.69)}$$

where C *is independent of* g *and* h.

Remark B.5.1 *Under the hypotheses of Lemma B.5.1 we have*

$$v > -\alpha^-_{k_1} \geq -\frac{N-2}{2}, \quad \text{(B.70)}$$

where the last inequality follows from the discussion in Section B.5.1. This implies that the integrals in the left-hand side of (B.68) *are finite.*

Remark B.5.2 *By taking* k_1 *sufficiently large, one can choose* $v \geq 0$ *in the previous lemma. In particular, the corresponding solution* ϕ *is bounded.*

Corollary B.5.1 *Assume that* (B.66), (B.67), *and* (B.68) *hold. In addition, assume that* $v \geq 0$.
  If $|x|^2 g$ *is continuous at the origin, then so is* ϕ.

Proof of Lemma B.5.1. Write ϕ as

$$\phi(x) = \sum_{k=0}^{\infty} \sum_{l=1}^{m_k} \phi_{k,l}(r)\varphi_{k,l}(\omega), \quad x = r\omega, \, 0 < r < 1, \, \omega \in \mathscr{S}^{N-1}.$$

Then ϕ solves $-\Delta\phi - \frac{c}{|x|^2}\phi = g$ in $B \setminus \{0\}$ if and only if $\phi_{k,l}$ satisfies the ordinary differential equation

$$\phi''_{k,l} + \frac{N-1}{r}\phi'_{k,l} + \frac{c - \lambda_k}{r^2}\phi_{k,l} = -g_{k,l} \quad 0 < r < 1, \quad \text{(B.71)}$$

for all $k \geq 0$ and $l = 1, \dots, m_k$, where

$$g_{k,l}(r) = \int_{\mathscr{S}^{N-1}} g(r\omega)\varphi_{k,l}(\omega) \, d\sigma, \quad 0 < r < 1.$$

Note that if $\phi \in C_{v,0}^{2,\alpha}(B)$ then there exists a constant $C > 0$ independent of r such that

$$|\phi_{k,l}(r)| \le Cr^v. \tag{B.72}$$

Furthermore, $\phi = h$ on ∂B if and only if $\phi_{k,l}(1) = h_{k,l}$ for all k, l, where

$$h_{k,l} = \int_{\mathscr{S}^{N-1}} h(\omega)\varphi_{k,l}(\omega)\,d\sigma.$$

Step 1. Clearly, $\sup_{0 \le t \le 1} t^{2-v}|g_{k,l}(t)| < \infty$ and observe that (B.68) still holds when g is replaced by $g_{k,l}\varphi_{k,l}$ and h by $h_{k,l}\varphi_{k,l}$. We claim that there is a unique $\phi_{k,l}$ that satisfies (B.71), (B.72), and

$$\phi_{k,l}(1) = h_{k,l}. \tag{B.73}$$

We also have

$$|\phi_{k,l}(r)| \le C_k r^v \left(\sup_{0 \le t \le 1} t^{2-v}|g_{k,l}(t)| \right), \quad 0 < r < 1. \tag{B.74}$$

Case $k = 0, \ldots, k_1$. A solution to (B.71) is given by:
- if $\alpha_{k,l}^{\pm} \notin \mathbb{R}$

$$\phi_{k,l}(r) = \frac{1}{b} \int_0^r s \left(\frac{s}{r} \right)^{\frac{N-2}{2}} \sin \left(b_k \log \frac{s}{r} \right) g_{k,l}(s)\,ds, \tag{B.75}$$

- if $\alpha_{k,l}^+ = \alpha_{k,l}^- = \frac{N-2}{2}$:

$$\phi_{k,l}(r) = \int_0^r s \left(\frac{s}{r} \right)^{\frac{N-2}{2}} \log \left(\frac{s}{r} \right) g_{k,l}(s)\,ds, \text{ and} \tag{B.76}$$

- if $\alpha_{k,l}^{\pm} \in \mathbb{R}$, $\alpha_{k,l}^{\pm} \ne \frac{N-2}{2}$:

$$\phi_{k,l}(r) = \frac{1}{\alpha_k^+ - \alpha_k^-} \int_0^r s \left(\left(\frac{s}{r} \right)^{\alpha_k^+} - \left(\frac{s}{r} \right)^{\alpha_k^-} \right) g_{k,l}(s)\,ds. \tag{B.77}$$

In each case, (B.74) holds and (B.73) follows from (B.68).

Concerning uniqueness, suppose that $\phi_{k,l}$ satisfies (B.71) with $g_{k,l} = 0$ and (B.73) with $h_{k,l} = 0$. Then $\phi_{k,l}$ is a linear combination of the functions w^1, w^2 defined in (B.64). By (B.66), (B.70), and (B.74), $\phi_{k,l}$ has to be zero.

Case $k \geq k_1 + 1$. Observe that (B.71) is equivalent to

$$-\Delta \tilde{\phi}_{k,l} + \frac{\lambda_k - c}{|x|^2} \tilde{\phi}_{k,l} = \tilde{g}_{k,l} \quad \text{in } B \setminus \{0\},$$

where $\tilde{\phi}_{k,l}(x) = \phi_{k,l}(|x|)$ and $\tilde{g}_{k,l}(x) = g_{k,l}(|x|)$. Since $\alpha_k^{\pm} \in \mathbb{R}$ we must have $\lambda_k - c \geq -(\frac{N-2}{2})^2$. By Hardy's inequality (Proposition C.1.1) and Lax-Milgram's lemma, the equation

$$\begin{cases} -\Delta \tilde{\phi}_{k,l} + \dfrac{\lambda_k - c}{|x|^2} \tilde{\phi}_{k,l} = \tilde{g}_{k,l} & \text{in } B \\ \tilde{\phi}_{k,l} = h_{k,l} & \text{on } \partial B, \end{cases} \tag{B.78}$$

has a unique solution $\tilde{\phi}_{k,l} \in H$, where H is the completion of $C_0^{\infty}(B)$ with the norm

$$\|\varphi\|_H^2 = \int_B |\nabla \varphi|^2 + \frac{\lambda_k - c}{|x|^2} \varphi^2,$$

see [215].

To show (B.74), observe that for some constant C depending only on N, λ_k, and ν,

$$A_{k,l}(r) = r^{\nu} C \left(\sup_{0 < t \leq 1} t^{2-\nu} |\tilde{g}_{k,l}(t)| + |h_{k,l}| \right)$$

is a supersolution to (B.78) and $-A_{k,l}$ is a subsolution. To see this, we emphasize that the condition $-\alpha_k^- > \nu > -(N-2)/2$ implies $\nu^2 + (N-2)\nu + c - \lambda_k < 0$. It follows that $|\tilde{\phi}_{k,l}(x)| \leq A_{k,l}(|x|)$ for $0 < |x| \leq 1$.

To show that $\tilde{\phi}_{k,l}$ is uniquely determined, we simply observe that any solution w of (B.78) such that $|w(x)| \leq C|x|^{\nu}$ must belong to H (where uniqueness holds). Indeed, by scaling, it can be checked that $|\nabla w(x)| \leq C|x|^{\nu-1}$ (see Claim B.5.1 and (B.82) later) and this together with (B.70) implies $w \in H^1(B)$, which is contained in H.

The computations above also yield the necessity of condition (B.68). Indeed, assuming a solution $\phi \in C_{\nu,0}^{2,\alpha}(B)$ exists, since $\phi_{k,l}$ satisfies the ODE (B.71) we see that for $k = 0, \ldots, k_1$ the difference between $\phi_{k,l}$ and one of the particular solutions (B.75), (B.76), or (B.77) can be written in the form $c_{k,l} r^{-\alpha_k^+} + d_{k,l} r^{-\alpha_k^-}$. Since $|\phi_{k,l}(r)| \leq Cr^{\nu}$ and $\nu > -\alpha_{k_1}^-$ we have $c_{k,l} = d_{k,l} = 0$ and this implies (B.68).

Step 2. Define for $m \geq 1$

$$\mathscr{G}_m = \left\{ g = \sum_{k=0}^{m} \sum_{l} g_{k,l}(r)\varphi_{k,l}(\omega) : |x|^{2-\nu}g(x) \in L^{\infty}(B) \right\}$$

and

$$\mathscr{H}_m = \left\{ h = \sum_{k=0}^{m} \sum_{l} h_{k,l}\varphi_{k,l}(\omega) : h_{k,l} \in \mathbb{R} \right\}.$$

Let $g_m \in \mathscr{G}_m$, $h_m \in \mathscr{H}_m$ be such that (B.68) hold. Write $g_m(x) = \sum_{k=0}^{m} \sum_{l} g_{k,l}(r)\varphi_{k,l}(\omega)$ and $h_m(\sigma) = \sum_{k=0}^{m} h_{k,l}\varphi_{k,l}(\omega)$. Let $\phi_{k,l}$ be the unique solution to (B.71), (B.72), and (B.73) associated to $g_{k,l}$, $h_{k,l}$, and define $\phi_m(x) = \sum_{k=0}^{m} \sum_{l} \phi_{k,l}(r)\varphi_{k,l}(\omega)$. We claim that there exists C independent of m such that

$$|\phi_m(x)| \leq C|x|^{\nu} \left(\sup_{B} |y|^{2-\nu}|g_m(y)| + \sup_{\partial B} |h_m| \right), \quad 0 < |x| < 1. \quad \text{(B.79)}$$

By the previous step, (B.79) holds for some constant C, which may depend on m. In particular, choosing $m = k_1$, we obtain a bound on the first components $\phi_{k,l}$, $k = 0 \ldots k_1$. Hence, it suffices to prove (B.79) in the case $g_{k,l} \equiv 0$ and $h_{k,l} = 0$, $k = 0, \ldots, k_1$. Working as in [186] (the argument already appeared in unpublished notes of Pacard), we argue by contradiction assuming that

$$\||\phi_m |x|^{-\nu}\|_{L^{\infty}(B)} \geq C_m(\||g_m |x|^{2-\nu}\|_{L^{\infty}(B)} + \|h_m\|_{L^{\infty}(\partial B)}),$$

where $C_m \to \infty$ as $m \to \infty$. Replacing ϕ_m by $\phi_m / \||\phi_m|x|^{-\nu}\|_{L^{\infty}(B)}$ if necessary, we may assume that

$$\||\phi_m |x|^{-\nu}\|_{L^{\infty}(B)} = 1,$$
$$\||g_m |x|^{2-\nu}\|_{L^{\infty}(B)} + \|h_m\|_{L^{\infty}(\partial B)} \to 0 \quad \text{as } m \to \infty. \quad \text{(B.80)}$$

Let $x_m \in B \setminus \{0\}$ be such that $|\phi_m(x_m)||x_m|^{-\nu} \in [\frac{1}{2}, 1]$. Let us show that $x_m \to 0$ as $m \to \infty$. Otherwise, up to a subsequence $x_m \to x_0 \neq 0$. By standard elliptic regularity, up to another subsequence, $\phi_m \to \phi$ uniformly on compact sets of $B \setminus \{0\}$ and hence

$$\begin{cases} -\Delta\phi - \dfrac{c}{|x|^2}\phi = 0 & \text{in } B \setminus \{0\}, \\[2mm] \phi = 0 & \text{on } \partial B. \end{cases}$$

264

Moreover, ϕ satisfies $|\phi(x_0)||x_0|^{-\nu} \in [\frac{1}{2}, 1]$ and $|\phi(x)| \leq |x|^\nu$ in B. Writing

$$\phi(x) = \sum_{k \geq k_1+1} \sum_l \phi_{k,l}(r)\varphi_{k,l}(\omega),$$

we see that $\phi_{k,l}$ solves (B.63). The growth restriction $|\phi_{k,l}(r)| \leq Cr^\nu$ and the explicit functions w^1, w^2 given by (B.64) rule out the cases $\alpha_k^\pm \notin \mathbb{R}$, $\alpha_k^- = \alpha_k^+$ and force $\phi_{k,l} = a_{k,l}r^{-\alpha_k^-}$. But $\phi_{k,l}(1) = 0$ so we deduce $\phi_{k,l} \equiv 0$ and hence $\phi \equiv 0$, contradicting $|\phi(x_0)||x_0|^{-\nu} \neq 0$.

The above argument shows that $x_m \to 0$. Define $r_m = |x_m|$ and

$$v_m(x) = r_m^{-\nu}\phi_m(r_m x), \quad x \in B_{1/r_m}.$$

Then $|v_m(x)| \leq |x|^\nu$ in B_{1/r_m}, $|v_m(\frac{x_m}{r_m})| \in [\frac{1}{2}, 1]$ and

$$-\Delta v_m(x) - \frac{c}{|x|^2}v_m(x) = r_m^{2-\nu}g(r_m x) \quad \text{in } B_{1/r_m}.$$

But

$$r_m^{2-\nu}\left|g(r_m x)\right| \leq \|g_m(y)|y|^{2-\nu}\|_{L^\infty(B)}|x|^{\nu-2} \to 0, \quad \text{as } m \to \infty$$

by (B.80). Passing to a subsequence, we have that $\frac{x_m}{r_m} \to x_0$ with $|x_0| = 1$, $v_m \to v$ uniformly on compact sets of $\mathbb{R}^N \setminus \{0\}$ and v satisfies

$$-\Delta v - \frac{c}{|x|^2}v = 0 \quad \text{in } \mathbb{R}^N \setminus \{0\}.$$

Furthermore, $|v(x)| \leq |x|^\nu$ in $\mathbb{R}^N \setminus \{0\}$ and $|v(x_0)| \neq 0$. Write

$$v(x) = \sum_{k=0}^\infty \sum_l v_{k,l}(r)\varphi_{k,l}(\sigma).$$

Then $|v_{k,l}(r)| \leq C_k r^\nu$ for $r > 0$. But $v_{k,l}$ has to be a linear combination of the functions w^1, w^2 given in (B.64), and none of these is bounded by Cr^ν for all $r > 0$. Thus $v \equiv 0$ yielding a contradiction.

Step 3. Fix an integer $d \geq 3(N-2)/2 + 1$. Suppose now that $g \in C^\infty(\overline{B} \setminus \{0\})$ and $|\nabla^i g(x)| \leq C|x|^{\nu-2-i}$ for $0 < |x| < 1$ and for $i = 0, \ldots, d$. Let $h \in C^\infty(\partial B)$ such that (B.68) holds. We will show that there exists $\phi \in C_{\nu,0}^{2,\alpha}(B)$ solving (B.67) and satisfying the estimate

$$\||\phi|x|^{-\nu}\|_{L^\infty(B)} \leq C\left(\||g|x|^{2-\nu}\|_{L^\infty(B)} + \|h\|_{L^\infty(\partial B)}\right). \tag{B.81}$$

To prove this, define for $m \in \mathbb{N}$

$$g_m(x) = \sum_{k=0}^{m}\sum_{l} g_{k,l}(r)\varphi_{k,l}(\sigma) \quad \text{and} \quad h_m(\sigma) = \sum_{k=0}^{m}\sum_{l} h_{k,l}\varphi_{k,l}(\sigma).$$

We have

$$\sum_{l}|g_{k,l}(r)| = \sum_{l}\left|\int_{\mathscr{S}^{N-1}} g(r\sigma)\varphi_{k,l}(\sigma)\,d\sigma\right| = \sum_{l}\frac{1}{\lambda_k}\left|\int_{\mathscr{S}^{N-1}} g(r\sigma)\Delta\varphi_{k,l}(\sigma)\,d\sigma\right|$$

$$\leq \frac{Cm_k r^{2d}}{\lambda_k^d}\sup_{|x|=r}|\nabla^{2d}g(x)|\,\|\varphi_{k,l}\|_{L^\infty(\mathscr{S}^{N-1})}$$

$$\leq Cr^{\nu-2}k^{-2d+2(N-2)},$$

where we used integration by parts d times to obtain the inequality and the facts: $\lambda_k \sim k^2$ as $k \to \infty$, $|\varphi_{k,l}| \leq Ck^{N-2}$ in $\mathscr{S}^{N-1}$ and $m_k \leq Ck^{N-2}$, where m_k is the multiplicity of λ_k, see Section C.4. It follows that $g_m(x)|x|^{2-\nu}$ converges uniformly in B to $g(x)|x|^{2-\nu}$ and hence $\|g_m|x|^{2-\nu}\|_{L^\infty(B)} \to \|g|x|^{2-\nu}\|_{L^\infty(B)}$ as $m \to \infty$. Similarly, h_m converges uniformly to h on ∂B and thus $\lim_{m\to\infty}\|h_m\|_{L^\infty(\partial B)} = \|h\|_{L^\infty(\partial B)}$.

Now $g_m \in \mathscr{G}_m$ and $h_m \in \mathscr{H}_m$ verify the orthogonality conditions (B.68). By the previous step, the associated solution ϕ_m satisfies

$$\|\phi_m|x|^{-\nu}\|_{L^\infty(B)} \leq C\left(\|g_m|x|^{2-\nu}\|_{L^\infty(B)} + \|h_m\|_{L^\infty(\partial B)}\right).$$

Using elliptic regularity, up to a subsequence, $\phi_m \to \phi$ uniformly in $B \setminus \{0\}$, for some ϕ satisfying the equations $-\Delta\phi - \frac{c}{|x|^2}\phi = g$ in $B \setminus \{0\}$, $\phi = h$ on ∂B and the estimate (B.81).

Claim B.5.1 *ϕ is a solution to the equation in the whole ball B.*

To see this, it suffices to prove that

$$|\nabla\phi(x)| \leq C|x|^{\nu-1} \quad \text{for } x \in B_{1/2}. \tag{B.82}$$

Recall that $\nu - 1 > -\frac{N}{2}$. This implies that $\phi \in H^1(B)$ and thus solves the equation in B.

Let $x_0 \in B_{1/2}$, $d = |x_0|$ and for $x \in B_{3/4}$, $v(x) = \phi(x_0 + dx)$. Then,

$$-\Delta v - \frac{cd^2}{|x_0 + dx|^2}v = d^2 g(x_0 + dx) \quad \text{in } B_{3/4}.$$

266

Observing that $0 \leq \frac{c\,d^2}{|x_0+dx|^2} \leq 16c$, it follows by elliptic regularity that for some constants C independent of d,

$$
\begin{aligned}
|\nabla v(0)| &\leq C \left(\|d^2 g(x_0+dx)\|_{L^\infty(B_{3/4})} + \|v\|_{L^\infty(B_{3/4})} \right) \\
&\leq C d^\nu \left(\|g|x|^{2-\nu}\|_{L^\infty(B)} + \|\phi|x|^{-\nu}\|_{L^\infty(B)} \right) \\
&\leq C|x_0|^\nu \left(\|g|x|^{2-\nu}\|_{L^\infty(B)} + \|h\|_{L^\infty(\partial B)} \right),
\end{aligned}
$$

where we used (B.81) in the last inequality. Hence, $|\nabla \phi(x_0)| \leq C'|x_0|^{\nu-1}$, which is the desired result.

Step 4. We assume now that $g \in C^{0,\alpha}_{\nu-2,0}(B)$ and $h \in C^{2,\alpha}(\partial B)$ satisfy (B.68). For $\varepsilon > 0$ let h_ε be the convolution product of h with a standard mollifier on the sphere ∂B. Let ρ_ε be a standard mollifier in $\mathbb{R}^N$ and define $g_\varepsilon(x) = |x|^{\nu-2}\rho_\varepsilon(x) * (g|x|^{2-\nu})$, where g is first extended by zero outside B. Since $g(x)|x|^{2-\nu} \in L^\infty(B)$, we have $g_\varepsilon \in C^\infty(\overline{B} \setminus \{0\})$ and

$$
|\nabla^i g_\varepsilon(x)| \leq C(i,\varepsilon)|x|^{\nu-2-i}.
$$

Moreover, $g_\varepsilon \to g$ a.e. in B, $h_\varepsilon \to h$ a.e. as $\varepsilon \to 0$ on ∂B and

$$
\|g_\varepsilon|x|^{2-\nu}\|_{L^\infty(B)} \leq \|g|x|^{2-\nu}\|_{L^\infty(B)} \quad \text{and} \quad \|h_\varepsilon\|_{L^\infty(\partial B)} \leq \|h\|_{L^\infty(\partial B)}.
$$

From this and (B.68), we deduce that for all $k = 0, \ldots, k_1$ and $l = 1, \ldots, m_k$,

$$
\int_B g_\varepsilon W_{k,l}\, dx - \int_{\partial B} h_\varepsilon \frac{\partial W_{k,l}}{\partial n}\, d\sigma \to 0 \quad \text{as } \varepsilon \to 0.
$$

Let

$$
a^{(\varepsilon)}_{k,l} = \frac{1}{\int_{\partial B} W_{k,l} \frac{\partial W_{k,l}}{\partial n}\, d\sigma} \left(\int_B g_\varepsilon W_{k,l}\, dx - \int_{\partial B} h_\varepsilon \frac{\partial W_{k,l}}{\partial n}\, d\sigma \right)
$$

and

$$
\tilde{h}_\varepsilon = h_\varepsilon + \sum_{k=0}^{k_1} \sum_{l=1}^{m_k} a^{(\varepsilon)}_{k,l} W_{k,l}.
$$

Then g_ε, $\tilde{h}_\varepsilon$ satisfy the orthogonality conditions (B.68). Let $\phi_\varepsilon \in C^{2,\alpha}_{\nu,0}(B)$ denote the solution to (B.67) with data g_ε, $\tilde{h}_\varepsilon$. We have

$$
\begin{aligned}
\|\phi_\varepsilon|x|^{-\nu}\|_{L^\infty(B)} &\leq C \left(\|g_\varepsilon|x|^{2-\nu}\|_{L^\infty(B)} + \|\tilde{h}_\varepsilon\|_{L^\infty(\partial B)} \right) \\
&\leq C \left(\|g|x|^{2-\nu}\|_{L^\infty(B)} + \|h\|_{L^\infty(\partial B)} \right).
\end{aligned}
$$

As in the previous step, from here we deduce that $\phi = \lim_{\varepsilon \to 0} \phi_\varepsilon$ is a solution to (B.67) with data g, h. In addition, (B.81) holds.

Finally, the estimate (B.69) is obtained by scaling, working as in Claim B.5.1. $\qquad \square$

Proof of Corollary B.5.1. Let (α_n) denote an arbitrary sequence of real numbers converging to zero, $\tilde{g}(x) = |x|^2 g(x)$ and $\phi_n(x) = \phi(\alpha_n x)$ for $x \in B_{1/\alpha_n}(0)$. Then ϕ_n solves

$$-\Delta \phi_n - \frac{c}{|x|^2} \phi_n = \frac{\tilde{g}(\alpha_n x)}{|x|^2} \qquad \text{in } B_{1/\alpha_n}(0).$$

Also, (ϕ_n) is uniformly bounded so that up to a subsequence, it converges in the topology of $C^{1,\alpha}(\mathbb{R}^N \setminus \{0\})$ to a bounded solution Φ of

$$-\Delta \Phi - \frac{c}{|x|^2} \Phi = \frac{\tilde{g}(0)}{|x|^2} \qquad \text{in } \mathbb{R}^N \setminus \{0\}.$$

Now, $\Phi + \tilde{g}(0)/c$ is bounded and solves (B.62), so it must be identically zero. It follows that the whole sequence (ϕ_n) converges to $-\tilde{g}(0)/c$. Now let (x_n) denote an arbitrary sequence of points in $\mathbb{R}^N$ converging to 0 and $\alpha_n = |x_n|$. Then, $\phi(x_n) = \phi_n(\frac{x_n}{|x_n|})$ and up to a subsequence, $\phi(x_n) \to -\tilde{g}(0)/c$. Again, since the limit of such a subsequence is unique, the whole sequence converges. $\qquad \square$

B.5.4   The case $\xi \neq 0$

The purpose of this section is to extend Lemma B.5.1 to a general bounded, smooth domain Ω of $\mathbb{R}^N$, $N \geq 3$ and general $\xi \in \Omega$, by redefining the functions $W_{k,l}$, which appear in (B.68). For this we restrict ourselves to the values of c in the range

$$0 < c < \frac{(N-2)^2}{4},$$

which guarantees $\alpha_0^- < \frac{N-2}{2} < \alpha_0^+$.

Take $g \in C^{0,\alpha}_{\nu-2,0}(\Omega) \cap H^{-1}(\Omega)$ and $h \in C^{2,\alpha}(\partial \Omega)$. Hardy's inequality and the condition $c < \frac{(N-2)^2}{4}$ ensure that equation

$$\begin{cases} -\Delta \phi - \dfrac{c}{|x-\xi|^2} \phi = g & \text{in } \Omega, \\[2ex] \phi = h & \text{on } \partial \Omega, \end{cases} \tag{B.83}$$

268

has a unique solution $\phi \in H^1(\Omega)$. We define $W_{k,l,\xi}$, which will play the same role as in (B.68), to be smooth functions in $\Omega \setminus \{\xi\}$ satisfying

$$
\begin{cases}
-\Delta W_{k,l,\xi} - \dfrac{c}{|x-\xi|^2} W_{k,l,\xi} = 0 & \text{in } \Omega \setminus \{\xi\}, \\[2mm]
\hspace{2.5cm} W_{k,l,\xi} = 0 & \text{on } \partial\Omega, \\[2mm]
\hspace{1cm} W_{k,l,\xi}(x) \sim |x-\xi|^{-\alpha_k^+} \varphi_{k,l}\left(\dfrac{x-\xi}{|x-\xi|}\right) & x \sim \xi.
\end{cases}
\tag{B.84}
$$

Indeed, let

$$
W_{k,l,\xi}(x) = |x-\xi|^{-\alpha_k^+} \varphi_{k,l}\left(\frac{x-\xi}{|x-\xi|}\right) - \psi_{k,l,\xi}(x),
\tag{B.85}
$$

where $\psi_{k,l,\xi} \in H^1(\Omega)$ is the unique solution to

$$
\begin{cases}
-\Delta \psi_{k,l,\xi} - \dfrac{c}{|x-\xi|^2} \psi_{k,l,\xi} = 0 & \text{in } \Omega, \\[2mm]
\hspace{2cm} \psi_{k,l,\xi} = |x-\xi|^{-\alpha_k^+} \varphi_{k,l}\left(\dfrac{x-\xi}{|x-\xi|}\right) & \text{on } \partial\Omega.
\end{cases}
$$

Observe that for $C > 0$ large enough, $C|x-\xi|^{-\alpha_0^-}$ and $-C|x-\xi|^{-\alpha_0^-}$ are respectively a super- and a subsolution of the above equation, hence by the maximum principle (which is valid in virtue of Hardy's inequality and the restriction $c < \frac{(N-2)^2}{4}$), $|\psi| \le C|x-\xi|^{-\alpha_0^-}$ and $W_{k,l,\xi}$ satisfies (B.84).

Remark B.5.3 *If $\Omega = B_1(0)$ and $\xi = 0$, our definition is consistent with (B.65), since*

$$
\begin{aligned}
\psi_{k,l,\xi} &= |x|^{-\alpha_k^-} \varphi_{k,l}(x/|x|) \quad \text{and} \\
W_{k,l,\xi} &= (|x|^{-\alpha_k^+} - |x|^{-\alpha_k^-}) \varphi_{k,l}(x/|x|).
\end{aligned}
\tag{B.86}
$$

Theorem B.5.1 *Let $N \ge 3$ and $0 < c < (N-2)^2/4$. Assume that*

$$
\exists k_1 \ge k_0 \quad \text{such that} \quad -\alpha_{k_1+1}^- > \nu > -\alpha_{k_1}^-.
\tag{B.87}
$$

Let Ω a smooth bounded domain of $\mathbb{R}^N$, $\xi \in \Omega$, $g \in C^{0,\alpha}_{\nu-2,\xi}(\Omega) \cap H^{-1}(\Omega)$ and $h \in C^{2,\alpha}(\partial\Omega)$. If

$$
\int_\Omega g W_{k,l,\xi}\, dx = \int_{\partial\Omega} h \frac{\partial W_{k,l,\xi}}{\partial n}\, d\sigma, \quad \forall k = 0, \dots, k_1, \; \forall l = 1, \dots, m_k
\tag{B.88}
$$

then there exists a unique $\phi \in C^{2,\alpha}_{\nu,\xi}(\Omega) \cap H^1(\Omega)$ solution to

$$
\begin{cases}
-\Delta\phi - \dfrac{c}{|x-\xi|^2}\phi = g & \text{in } \Omega \\
\phi = h & \text{on } \partial\Omega,
\end{cases}
\tag{B.89}
$$

and it satisfies

$$
\|\phi\|_{2,\alpha,\nu,0} \le C(\|g\|_{0,\alpha,\nu-2,\xi;\Omega} + \|h\|_{C^{2,\alpha}(\partial\Omega)})
\tag{B.90}
$$

where C is independent of g and h.

Proof of Theorem B.5.1.

By translating the domain, we consider from here on $\xi = 0$. By Lemma B.5.1, a straightforward scaling argument implies that Theorem B.5.1 holds when $\Omega = B_R(0)$ and $\xi = 0$. In this case, $W_{k,l,0}$ takes the form

$$
\widetilde{W}_{k,l}(x) = \left(\left(\frac{|x|}{R}\right)^{-\alpha_k^+} - \left(\frac{|x|}{R}\right)^{-\alpha_k^-} \right) \varphi_{k,l}\left(\frac{x}{|x|}\right).
\tag{B.91}
$$

This is obtained by scaling the functions in (B.86) and is the same as in definition (B.85) except for a multiplicative constant. Take $R > 0$ small, so that $B_R(0) \subset \Omega$. Then the unique solution $\phi \in H^1(\Omega)$ of (B.89) satisfies (B.90) if

$$
\int_{B_R} g\widetilde{W}_{k,l}\, dx = \int_{\partial B_R} \phi \frac{\partial \widetilde{W}_{k,l}}{\partial n}\, d\sigma, \quad \forall k = 0,\ldots,k_1, \forall l = 1,\ldots,m_k,
\tag{B.92}
$$

where $\widetilde{W}_{k,l}$ is defined in (B.91). Since $\widetilde{W}_{k,l}$ satisfies

$$
-\Delta\widetilde{W}_{k,l} - \frac{c}{|x|^2}\widetilde{W}_{k,l} = 0 \quad \text{in } \mathbb{R}^N \setminus \{0\},
$$

multiplying this equation by ϕ and integrating in $\Omega \setminus B_R$, we obtain

$$
\int_{\partial\Omega} \left(\frac{\partial \widetilde{W}_{k,l}}{\partial n}\phi - \widetilde{W}_{k,l}\frac{\partial\phi}{\partial n} \right) d\sigma - \int_{\partial B_R} \frac{\partial \widetilde{W}_{k,l}}{\partial n}\phi\, d\sigma = \int_{\Omega\setminus B_R} g\widetilde{W}_{k,l}\, dx.
\tag{B.93}
$$

Adding (B.92) and (B.93) we see that (B.92) is equivalent to

$$
\int_{\partial\Omega} \left(\frac{\partial \widetilde{W}_{k,l}}{\partial n}\phi - \widetilde{W}_{k,l}\frac{\partial\phi}{\partial n} \right) d\sigma = \int_{\Omega} g\widetilde{W}_{k,l}\, dx.
\tag{B.94}
$$

Let $\tilde{\psi}_{k,l} \in H^1(\Omega)$ be the solution to

$$\begin{cases} -\Delta\tilde{\psi}_{k,l} - \dfrac{c}{|x|^2}\tilde{\psi}_{k,l} = 0 & \text{in } \Omega, \\[2mm] \qquad\qquad\quad \tilde{\psi}_{k,l} = \widetilde{W}_{k,l} & \text{on } \partial\Omega. \end{cases}$$

Multiplying this equation by ϕ and integrating by parts yields

$$\int_{\partial\Omega}\left(\frac{\partial\tilde{\psi}_{k,l}}{\partial n}\phi - \frac{\partial\phi}{\partial n}\tilde{\psi}_{k,l}\right) d\sigma = \int_{\Omega} g\tilde{\psi}_{k,l}\, dx.$$

Subtracting this equation from (B.94) we obtain that (B.92) is equivalent to

$$\int_{\partial\Omega}\frac{\partial(\widetilde{W}_{k,l} - \tilde{\psi}_{k,l})}{\partial n}\phi\, d\sigma = \int_{\Omega} g(\widetilde{W}_{k,l} - \tilde{\psi}_{k,l})\, dx.$$

Up to multiplicative constant $\widetilde{W}_{k,l} - \tilde{\psi}_{k,l}$ is the same as $W_{k,l,0}$ as defined in (B.85). $\qquad\qquad\qquad\qquad\qquad\qquad\qquad\qquad\qquad\qquad\qquad\qquad\square$

Appendix C

Geometric tools

In this chapter, we gather the main geometric results used in the book.

C.1 Functional inequalities

We begin our discussion with three fundamental inequalities that are deeply related to the geometry of $\mathbb{R}^N$.

C.1.1 The isoperimetric inequality

Theorem C.1.1 *Let $N \geq 2$. Among all sets $\Omega \subset \mathbb{R}^N$ having a smooth boundary of given finite $(N-1)$ Hausdorff measure, the ball has the largest N-dimensional volume, that is,*

$$|\Omega|^{\frac{N-1}{N}} \leq C |\partial \Omega|, \tag{C.1}$$

with $C = |B_1|^{\frac{N-1}{N}} / |\partial B_1|$. Equality holds if and only if Ω is the ball of radius $r = \left(\frac{|\partial \Omega|}{|\partial B_1|} \right)^{\frac{1}{N-1}}$.

Proof. We follow [189]. For $x \in \Omega$, consider the sets

$$A_i(x) = \{z \ : \ z_j = x_j \text{ for } j < i \text{ and } z_i \leq x_i\}$$

and

$$B_i(x) = \{z \ : \ z_j = x_j \text{ for } j < i\},$$

273

with the convention that $B_1 = \Omega$. Consider the map $y_\Omega : x \to y$ defined by

$$y_i = \frac{\int_{A_i(x)\cap\Omega} dz_i \ldots dz_N}{\int_{B_i(x)\cap\Omega} dz_i \ldots dz_N}.$$

Since $A_i \subset B_i$, we have $0 \leq y_i \leq 1$, that is, y_Ω maps Ω into the cube $[0,1]^N$. Furthermore, y_Ω is a triangular map, that is, for each i, y_i is a function of $x_1, \ldots, x_i$ only. Also, each partial derivative $\partial y_i / \partial x_i$ is nonnegative and equal to

$$\frac{\partial y_i}{\partial x_i}(x) = \begin{cases} \dfrac{\int_{B_{i+1}(x)\cap\Omega} dz_{i+1} \ldots dz_N}{\int_{B_i(x)\cap\Omega} dz_i \ldots dz_N} & \text{if } 1 \leq i < N, \\[4mm] \dfrac{1}{\int_{B_N(x)\cap\Omega} dz_N} & \text{if } i = N, \end{cases}$$

for every $x \in \Omega$. Thus, the Jacobian of y_Ω is equal to

$$J(x) = \prod_{i=1}^{N} \frac{\partial y_i}{\partial x_i} = \frac{1}{|\Omega|}.$$

In the particular case where $\Omega = B$ is the unit ball, the map y_B is invertible. Its inverse $z : (0,1)^N \to B$ is still triangular, with Jacobian equal to $|B|$. Now, consider the map

$$F = z \circ y_\Omega : \Omega \to B.$$

By construction, F has nonnegative partial derivatives $\partial F_i / \partial x_i$ and its Jacobian is given by

$$J_F(x) = \frac{|B|}{|\Omega|},$$

for all $x \in \Omega$. Thus, the divergence $\nabla \cdot F$ satisfies

$$\frac{1}{N} \nabla \cdot F \geq J_F^{1/N} = \left(\frac{|B|}{|\Omega|} \right)^{1/N}.$$

Furthermore, since $|F| \leq 1$,

$$N \left(\frac{|B|}{|\Omega|} \right)^{1/N} |\Omega| \leq \int_\Omega \nabla \cdot F \, dx = \int_{\partial\Omega} F \cdot n \, d\sigma \leq |\partial\Omega|,$$

and equality holds only if Ω is a ball. $\qquad\square$

C.1.2   The Sobolev inequality

The celebrated Sobolev inequality reads as follows.

Theorem C.1.2 *Let $N \geq 2$ and $p \in [1, N)$. Then, there exists a constant $C = C(N, p) > 0$ such that*

$$\|\varphi\|_{L^{p^*}(\mathbb{R}^N)} \leq C \|\nabla \varphi\|_{L^p(\mathbb{R}^N)}, \qquad \text{for all } \varphi \in C_c^1(\mathbb{R}^N), \tag{C.2}$$

where

$$\frac{1}{p^*} = \frac{1}{p} - \frac{1}{N}.$$

Proof.
Step 1. We begin by proving the inequality when $p = 1$. Let $\varphi \in C_c^\infty(\mathbb{R}^N)$. We may always assume that $\varphi \geq 0$. Letting χ_A be the characteristic function of the set A, we have

$$\varphi(x) = \int_0^{+\infty} \chi_{[\varphi(x) > t]} \, ds.$$

So,

$$\|\varphi\|_{L^{\frac{N}{N-1}}(\mathbb{R}^N)} \leq \int_0^{+\infty} \|\chi_{[\varphi(\cdot) > t]}\|_{L^{\frac{N}{N-1}}(\mathbb{R}^N)} \, ds = \int_0^{+\infty} \left| [x \,:\, \varphi(x) > t] \right|^{\frac{N-1}{N}} \, ds.$$

By Sard's theorem (see Theorem C.3.1), for almost every t, the level set $\Omega_t = [x \,:\, \varphi(x) > t]$ has a smooth boundary of finite perimeter. The isoperimetric inequality then implies that

$$\left| [x \,:\, \varphi(x) > t] \right|^{\frac{N-1}{N}} \leq C \left| [x \,:\, \varphi(x) = t] \right|,$$

hence

$$\|\varphi\|_{L^{\frac{N}{N-1}}(\mathbb{R}^N)} \leq C \int_0^{+\infty} \left| [x \,:\, \varphi(x) = t] \right| \, dt = C \int_{\mathbb{R}^N} |\nabla \varphi| \, dx,$$

where we used the coarea formula, see Equation (C.23).
Step 2. It remains to prove the case $1 < p < N$. We show that the inequality can be derived from the case $p = 1$. Take $\varphi \in C_c^1(\mathbb{R}^N)$ and let $\psi = |\varphi|^{t-1} \varphi$, where $t = p^*/1^*$, for some $p \in (1, N)$. In particular, $t > 1$ so that $\psi \in C_c^1(\mathbb{R}^N)$ and we may apply the Sobolev inequality (C.2) with $p = 1$ to get

$$\left(\int_{\mathbb{R}^N} |\varphi|^{p^*} \, dx \right)^{1/1^*} = \left(\int_{\mathbb{R}^N} |\psi|^{1^*} \, dx \right)^{1/1^*} \leq C \int_{\mathbb{R}^N} |\nabla \psi| \, dx. \tag{C.3}$$

Now, $\left|\nabla\psi\right| = t\left|\varphi\right|^{t-1}\left|\nabla\varphi\right|$. By Hölder's inequality, it follows that

$$\int_{\mathbb{R}^N}\left|\nabla\psi\right| dx \leq t\left(\int_{\mathbb{R}^N}\left|\nabla\varphi\right|^p dx\right)^{1/p}\left(\int_{\mathbb{R}^N}\left|\varphi\right|^{(t-1)p'} dx\right)^{1/p'}.$$

Observe that by the definition of p^*,

$$\frac{1}{1^*} - \frac{1}{p^*} = \frac{1}{1} - \frac{1}{p} = \frac{1}{p'}. \tag{C.4}$$

In particular,

$$(t-1)p' = \left(\frac{p^*}{1^*} - 1\right)p' = p^*\left(\frac{1}{1^*} - \frac{1}{p^*}\right)p' = p^*.$$

Hence,

$$\int_{\mathbb{R}^N}\left|\nabla\psi\right| dx \leq t\left(\int_{\mathbb{R}^N}\left|\nabla\varphi\right|^p dx\right)^{1/p}\left(\int_{\mathbb{R}^N}\left|\varphi\right|^{p^*} dx\right)^{1/p'}.$$

Plugging this into (C.3), we obtain

$$\left(\int_{\mathbb{R}^N}\left|\varphi\right|^{p^*} dx\right)^{1/1^*-1/p'} \leq C\left(\int_{\mathbb{R}^N}\left|\nabla\varphi\right|^p dx\right)^{1/p}.$$

By (C.4), $1/1^* - 1/p' = 1/p^*$. The desired inequality follows. $\qquad\square$

C.1.3   The Hardy inequality

Proposition C.1.1 *For $N \geq 3$, let $H_N = \frac{(N-2)^2}{4}$. Then,*

$$H_N\int_{\mathbb{R}^N}\frac{\varphi^2}{\left|x\right|^2} dx \leq \int_{\mathbb{R}^N}\left|\nabla\varphi\right|^2 dx, \qquad \text{for all } \varphi \in C_c^1(\mathbb{R}^N).$$

Furthermore, the constant H_N is sharp.

Proof. Note that the inequality is invariant under rotation: replacing φ by $\varphi(Tx)$ where T is an orthogonal transformation leaves the inequality unchanged. Similarly, the inequality is preserved when replacing φ by

$R^{\frac{N-2}{2}}\varphi(Rx)$. It is therefore natural to look for a minimizer that shares both the rotation invariance and the scaling property, that is,

$$\varphi_0 = |x|^{-\frac{N-2}{2}}.$$

Unfortunately, as can be seen by direct inspection, φ_0 makes both integrals in the inequality infinite. So we use the change of unknown $\varphi = \varphi_0\psi$, where we assume for the time being that $\psi \in C_c^1(\mathbb{R}^N \setminus \{0\})$. Then, $\varphi \in C_c^1(\mathbb{R}^N \setminus \{0\})$ and

$$
\begin{aligned}
\int_{\mathbb{R}^N} |\nabla\varphi|^2 \, dx &= \int_{\mathbb{R}^N} |\nabla(\varphi_0\psi)|^2 \, dx = \int_{\mathbb{R}^N} |\psi\nabla\varphi_0 + \varphi_0\nabla\psi|^2 \, dx \\
&= \int_{\mathbb{R}^N} \psi^2 |\nabla\varphi_0|^2 \, dx + 2\int_{\mathbb{R}^N} \varphi_0\psi\nabla\varphi_0\nabla\psi \, dx + \int_{\mathbb{R}^N} \varphi_0^2 |\nabla\psi|^2 \, dx \\
&= \frac{(N-2)^2}{4}\int_{\mathbb{R}^N} \psi^2 \frac{\varphi_0^2}{|x|^2} \, dx + \frac{1}{2}\int_{\mathbb{R}^N} \nabla(\psi^2)\cdot\nabla(\varphi_0^2) \, dx + \\
&\quad + \int_{\mathbb{R}^N} \varphi_0^2 |\nabla\psi|^2 \, dx \\
&\geq \frac{(N-2)^2}{4}\int_{\mathbb{R}^N} \frac{\varphi^2}{|x|^2} \, dx - \frac{1}{2}\int_{\mathbb{R}^N} \psi^2\Delta(\varphi_0^2) \, dx \\
&\geq \frac{(N-2)^2}{4}\int_{\mathbb{R}^N} \frac{\varphi^2}{|x|^2} \, dx.
\end{aligned}
$$

In the last inequality, we used the fact that $\varphi_0^2 = |x|^{2-N}$ is a constant multiple of the fundamental solution of the Laplace equation, hence it is harmonic on the support of ψ.

We have just established Hardy's inequality for test functions $\varphi \in C_c^1(\mathbb{R}^N \setminus \{0\})$. Now take a cutoff function $\eta \in C^1(\mathbb{R})$, such that $\eta(r) \equiv 0$ for $r \leq 1$ and $\eta \equiv 1$ for $r \geq 2$. Given, $\varphi \in C_c^1(\mathbb{R}^N)$, let $\eta_n(x) = \eta(n|x|)$ and $\varphi_n = \eta_n\varphi$. Then, Hardy's inequality holds for φ_n, that is,

$$H_N \int_{\mathbb{R}^N} \frac{\varphi_n^2}{|x|^2} \, dx \leq \int_{\mathbb{R}^N} |\nabla\varphi_n|^2 \, dx \tag{C.5}$$

and

$$\int_{\mathbb{R}^N} |\nabla \varphi_n(x)|^2 \, dx = \int_{\mathbb{R}^N} |\eta_n \nabla \varphi + n\varphi \nabla \eta(n|x|)|^2 \, dx$$

$$= \int_{\mathbb{R}^N} \eta_n^2 |\nabla \varphi|^2 \, dx + \int_{\mathbb{R}^N} \left(2n\varphi \eta_n \nabla \varphi \nabla \eta(n|x|) + n^2 \varphi^2 |\nabla \eta|^2 (n|x|) \right) \, dx$$

$$=: \int_{\mathbb{R}^N} \eta_n^2 |\nabla \varphi|^2 \, dx + E_n. \quad \text{(C.6)}$$

We control the error term by

$$E_n \leq \left(2n\|\varphi\|_\infty \|\eta\|_\infty \|\nabla \eta\|_\infty + n^2 \|\varphi\|_\infty^2 \|\nabla \eta\|_\infty^2 \right) \int_{[x\,:\,n|x|\leq 2]} dx \leq Cn^{2-N}.$$

Using this together with (C.6) in (C.5), we obtain

$$H_N \int_{\mathbb{R}^N} \eta_n^2 \frac{\varphi^2}{|x|^2} \, dx \leq \int_{\mathbb{R}^N} \eta_n^2 |\nabla \varphi|^2 \, dx + Cn^{2-N}.$$

We may then easily pass to the limit by monotone convergence. It remains to prove that the constant H_N is optimal. Take a larger constant $H > H_N$ and test the equation with $\varphi_n = \varphi_0 \psi_n = |x|^{-(N-2)/2} \psi_n$, where this time $\psi_n \in C_c^1(\mathbb{R}^N \setminus \{0\})$ is given by $\psi_n = \eta_n(x)(1 - \eta(|x|))$. Working as previously, we obtain

$$\int_{\mathbb{R}^N} |\nabla \varphi_n|^2 \, dx - H \int_{\mathbb{R}^N} \frac{\varphi_n^2}{|x|^2} \, dx = (H_N - H) \int_{\mathbb{R}^N} \frac{\varphi_n^2}{|x|^2} \, dx + \int_{\mathbb{R}^N} \varphi_0^2 |\nabla \psi_n|^2 \, dx$$

$$\leq (H_N - H) \int_{\mathbb{R}^N} \frac{\psi_n^2}{|x|^N} \, dx + n^2 \int_{\mathbb{R}^N} |x|^{2-N} |\nabla \eta|^2 (n|x|) \, dx + C$$

$$= (H_N - H) \int_{\mathbb{R}^N} \frac{\psi_n^2}{|x|^N} \, dx + \int_{\mathbb{R}^N} |y|^{2-N} |\nabla \eta|^2 \, dy + C$$

$$\leq (H_N - H) \int_{[x\,:\,1/n\leq|x|\leq 1]} \frac{1}{|x|^N} \, dx + C.$$

The right-hand side converges to $-\infty$ as $n \to +\infty$, which implies the optimality of H_N. $\qquad \square$

C.2 Submanifolds of $\mathbb{R}^N$

Here is a quick review of some basic facts in differential geometry. See, for example, [170, 171] for a thorough introduction to the subject.

C.2.1 Metric tensor, tangential gradient

Definition C.2.1 *Let $N \geq m \geq 1$. A set $M \subset \mathbb{R}^N$ is an m-dimensional C^2 submanifold of $\mathbb{R}^N$, if given any point $x_0 \in M$, there exists open sets $D \subset \mathbb{R}^m$, $\Omega \subset \mathbb{R}^N$ and a C^2 map*

$$x : \begin{cases} D \to \Omega \\ t = (t_1, \ldots, t_m) \mapsto x(t) = (x_1(t), \ldots, x_N(t)) \end{cases}$$

such that

$$x_0 \in \Omega \cap M = x(D),$$

and such that the vectors $\frac{\partial x}{\partial t_i}(t)$, $i = 1, \ldots, m$ are linearly independent for each $t \in D$. The mapping x is called a representation of M at x_0 and the vector space spanned by $\frac{\partial x}{\partial t_i}(t_0)$, $i = 1, \ldots, m$, where $x(t_0) = x_0$, is called the tangent space $T_{x_0} M$ to M at x_0.

Remark C.2.1 *As follows from the implicit function theorem, if $a \in \mathbb{R}$ is a regular value of a function $u \in C^2(\mathbb{R}^N)$, then the level set $L_a = \{ y \in \mathbb{R}^N : u(y) = a \}$ is an $N - 1$ submanifold of $\mathbb{R}^N$.*

Definition C.2.2 *Let $N \geq m \geq 1$. Given an m-dimensional C^2 submanifold $M \subset \mathbb{R}^N$, the first fundamental form or metric tensor of M is the matrix-valued function $(g_{ij})_{i,j=1,\ldots,m}$ defined for $t \in D$ by*

$$g_{ij}(t) = \frac{\partial x(t)}{\partial t_i} \cdot \frac{\partial x(t)}{\partial t_j} = \sum_{k=1}^{N} \frac{\partial x_k(t)}{\partial t_i} \frac{\partial x_k(t)}{\partial t_j}.$$

Remark C.2.2 *It can be easily checked that given $x_0 = x(t_0) \in M$, the relation*

$$g(v, w) = \sum_{i,j=1,\ldots m} g_{ij}(t_0) v_i w_j,$$

where $v = \sum_{i=1}^{m} v_i \frac{\partial x}{\partial t_i}(t_0)$, $w = \sum_{i=1}^{m} w_i \frac{\partial x}{\partial t_i}(t_0)$ are two arbitrary vectors belonging to $T_{x_0} M$ defines an inner product on $T_{x_0} M$. Furthermore, g is independent of the choice of representation x. Using a standard abuse of notations, we shall also write $g = \det(g_{ij})$, while $(g^{ij}) = (g_{ij})^{-1}$ denotes the inverse matrix of (g_{ij}).

Thanks to the metric g, one can easily compute the orthogonal projection (with respect to the standard inner product) of a vector in $\mathbb{R}^N$ on the tangent space $T_{x_0} M$.

Proposition C.2.1 *Let* $N \geq m \geq 1$. *Given an m-dimensional C^2 submanifold* $M \subset \mathbb{R}^N$ *and a point* $x_0 = x(t_0) \in M$, *the matrix* $\tilde{G}(x_0) = (\tilde{g}^{ij}(x_0))_{i,j=1,...,N}$ *defined by*

$$\tilde{g}^{ij}(x_0) = \sum_{r,s=1}^{m} g^{rs}(t_0) \frac{\partial x_i}{\partial t_r}(t_0) \frac{\partial x_j}{\partial t_s}(t_0)$$

represents the orthogonal projection on $T_{x_0}M$ in the canonical basis of $\mathbb{R}^N$, that is,

$$\tilde{G}(x_0).v = \begin{cases} v & \text{for all } v \in T_{x_0}M, \\ 0 & \text{for all } v \in (T_{x_0}M)^{\perp}. \end{cases}$$

In particular, for all $x \in M$,

$$\sum_{i=1,...,N} \tilde{g}^{ii}(x) = m,$$

$$0 \leq \sum_{i,j=1}^{N} \tilde{g}^{ij}(x)v_i v_j \leq |v|^2, \quad \text{for all } v = (v_1,...,v_N) \in \mathbb{R}^N,$$

$$\tilde{g}^{ij}(x) = \tilde{g}^{ji}(x), \quad \text{for all } i,j = 1,...,N.$$

Proof. For any $v = (v_1,...,v_N) \in \mathbb{R}^N$,

$$\sum_{j=1}^{N} \tilde{g}^{ij}v_j = \sum_{r,s=1}^{m} \sum_{j=1}^{N} g^{rs} \frac{\partial x_i}{\partial t_r} \frac{\partial x_j}{\partial t_s} v_j = \sum_{r,s=1}^{m} g^{rs} \frac{\partial x_i}{\partial t_r} \frac{\partial x}{\partial t_s} \cdot v.$$

So, $(\tilde{g}^{ij})$ leaves invariant all the vectors $\frac{\partial x}{\partial t_1},...,\frac{\partial x}{\partial t_m}$, while it vanishes on the orthogonal complement of those vectors. $\square$

Using the projection matrix, the tangential gradient is simply defined as the projection of the gradient on the tangent space.

Proposition C.2.2 *Let* $N \geq m \geq 1$. *Given an m-dimensional C^2 submanifold* $M \subset \mathbb{R}^N$, *a point* $x_0 = x(t_0) \in M$, *an open set* $\Omega \subset \mathbb{R}^N$ *such that* $x_0 \in \Omega$, *and a function* $\varphi \in C^1(\Omega,\mathbb{R})$, *the tangential gradient of φ at x_0 is defined as the orthogonal projection of $\nabla\varphi$ on $T_{x_0}M$. We have*

$$\nabla_T\varphi(x_0) := \tilde{G}(x_0).\nabla\varphi(x_0) = \sum_{r=1}^{m} \left(\sum_{s=1}^{m} g^{rs}(t_0) \frac{\partial \varphi}{\partial t_s}(t_0) \right) \frac{\partial x}{\partial t_r}(t_0).$$

Proof. Let e_i, $i = 1, \ldots, N$, denote the canonical basis of $\mathbb{R}^N$. We compute

$$
\tilde{G}(x_0).\nabla\varphi(x_0) = \sum_{i,j=1}^{N} \tilde{g}^{ij} \frac{\partial \varphi}{\partial x_j} e_i
$$

$$
= \sum_{i,j=1}^{N} \sum_{r,s=1}^{m} g^{rs} \frac{\partial x_i}{\partial t_r} \frac{\partial x_j}{\partial t_s} \frac{\partial \varphi}{\partial x_j} e_i
$$

$$
= \sum_{i=1}^{N} \sum_{r,s=1}^{m} g^{rs} \frac{\partial x_i}{\partial t_r} \frac{\partial \varphi}{\partial t_s} e_i
$$

$$
= \sum_{r,s=1}^{m} g^{rs} \frac{\partial \varphi}{\partial t_s} \frac{\partial x}{\partial t_r}.
$$

$\square$

C.2.2   Surface area of a submanifold

Let $x : D \subset \mathbb{R}^m \to \Omega$ denote a representation of an m-dimensional C^2 submanifold $M \subset \mathbb{R}^N$. Given a smoothly bounded subdomain $\omega \subset\subset D$, the image of ω on M has its m-dimensional area equal to

$$
\int_{x(\omega)} d\sigma = \int_{\omega} \sqrt{g} \, dt_1 \ldots dt_m.
$$

Using a standard partition of unity, one can use the above formula to compute the integral over M of any function, which is, say, continuous in a neighborhood of M.

The following elementary property of the area element will be useful in the sequel.

Lemma C.2.1 *Let $1 \le m \le N$. Let M denote an m-dimensional C^2 submanifold of the Euclidean space $\mathbb{R}^N$ and let $d\sigma = \sqrt{g} \, dt_1 \ldots dt_m$ denote its volume element. Also let ω_m denote the Lebesgue measure of the unit ball in $\mathbb{R}^m$. Then, for all $x_0 \in M$,*

$$
\lim_{\rho \to 0^+} \frac{\sigma(S_\rho(x_0))}{\rho^m} = \omega_m,
$$

where

$$
S_\rho(x_0) = \{x \in M \ : \ |x - x_0| \le \rho\}.
$$

Proof. Let x be a representation of M at $x_0 = x(t_0)$. By Taylor's formula,

$$x(t) = x_0 + \sum_{i=1}^{m}(t-t_0)_i \frac{\partial x}{\partial t_i}(t_0) + o(|t-t_0|). \tag{C.7}$$

Without loss of generality, we may assume that the canonical basis of $\mathbb{R}^m$ is orthonormal for the innner product $g(t_0)$, that is, there exists $\lambda_1,\ldots,\lambda_m > 0$ such that $g_{ij}(t_0) = \lambda_i \delta_{ij}$. Using this and (C.7), we obtain

$$|x(t) - x_0|^2 = \sum_{i,j=1}^{m}(t-t_0)_i \frac{\partial x}{\partial t_i}(t_0)(t-t_0)_j \frac{\partial x}{\partial t_j}(t_0) + o(|t-t_0|^2) \tag{C.8}$$

$$= \sum_{i,j=1}^{m} g_{ij}(t_0)(t-t_0)_i(t-t_0)_j + o(|t-t_0|^2) \tag{C.9}$$

$$= \sum_{i=1}^{m}\lambda_i(t-t_0)_i^2 + o(|t-t_0|^2). \tag{C.10}$$

Given $\varepsilon > 0$, we deduce that for ρ sufficiently small

$$\left\{ t \in \mathbb{R}^m : \sum_{i=1}^{m}\lambda_i(t-t_0)_i^2 \leq (1-\varepsilon)\rho^2 \right\} \subset x^{-1}(S_\rho(x_0))$$

$$\subset \left\{ t \in \mathbb{R}^m : \sum_{i=1}^{m}\lambda_i(t-t_0)_i^2 \leq (1+\varepsilon)\rho^2 \right\}.$$

It follows that

$$\lim_{\rho \to 0^+} \frac{\sigma(S_\rho(x_0))}{\rho^m} = \lim_{\rho \to 0^+} \frac{\int_{\{t:\sum_{i=1}^{m}\lambda_i(t-t_0)_i^2 \leq \rho^2\}} \sqrt{g}\, dt_1 \ldots dt_m}{\rho^m}$$

$$= \lim_{\rho \to 0^+} \frac{\int_{\{s:\sum_{i=1}^{m}s_i^2 \leq \rho^2\}} ds_1 \ldots ds_m}{\rho^m}$$

$$= \omega_m,$$

where we used the change of variable $s_i = \sqrt{\lambda_i}(t-t_0)_i$, $i = 1,\ldots,m$. $\qquad\square$

C.2.3   Curvature, Laplace-Beltrami operator

Tangent vector and curvature vector of a curve

A regular curve on a submanifold M is a C^1 mapping $x : I = (\alpha,\beta) \to M$, such that $|x'(\tau)| > 0$, for all $\tau \in I$. Let $t_1(\tau),\ldots,t_m(\tau)$ denote the coordinates of

$x(\tau)$ in a given representation of M. Then,

$$|x'(\tau)|^2 = \sum_{i,j=1}^{m} g_{ij} t_i'(\tau) t_j'(\tau)$$

and the length of the curve $x(\tau)$ is given by

$$L = \int_{\alpha}^{\beta} |x'(\tau)| \, d\tau.$$

Now set $s(\tau) = \int_{\alpha}^{\tau} |x'(\tau)| \, d\tau$. Then, since the curve is regular, the mapping $s : (\alpha, \beta) \to (0, L)$ is invertible. s is called the arc-length parameter and the mapping

$$\begin{cases} (0, L) \to \mathbb{R}^N \\ \quad s \mapsto x(\tau(s)) \end{cases}$$

where $\tau(s)$ is the inverse map of $s(\tau)$, the parametrization of the curve by arc length. The unit tangent vector to the curve is given by

$$T = \frac{dx}{ds} = \frac{x'(\tau)}{s'(\tau)}$$

and the curvature vector of the curve is given by

$$N = \frac{dT}{ds} = \frac{d^2 x}{ds^2}.$$

Note that since $|T|^2 = 1$, $T \cdot \frac{dT}{ds} = 0$, that is, the unit tangent vector and the curvature vector are orthogonal.

Normal curvature, second fundamental form

Given a submanifold M, a point $x_0 \in M$, the orthogonal complement of the tangent space (with respect to the standard inner product in $\mathbb{R}^N$)

$$N_{x_0}(M) = (T_{x_0} M)^{\perp}$$

is called the normal space of M at x_0. Given a regular curve $x(s)$ parametrized by arc-length, its unit tangent vector and curvature vector can be computed in the coordinates of a representation as follows:

$$\frac{dx}{ds} = \sum_{i=1}^{m} \frac{dt_i}{ds} \frac{\partial x}{\partial t_i}$$

and

$$\frac{d^2x}{ds^2} = \sum_{i=1}^{m} \frac{d^2t_i}{ds^2}\frac{\partial x}{\partial t_i} + \sum_{i,j=1}^{m} \frac{dt_i}{ds}\frac{dt_j}{ds}\frac{\partial^2 x}{\partial t_i \partial t_j}.$$

Hence, given a normal vector $N \in N_{x_0}M$,

$$\frac{d^2x}{ds^2}\cdot N = \sum_{i,j=1}^{m}\left(\frac{\partial^2 x}{\partial t_i \partial t_j}\cdot N\right)\frac{dt_i}{ds}\frac{dt_j}{ds}.$$

The right-hand side of the above expression can be seen as a quadratic form acting on the vector $T = \frac{dx}{ds}$. This quadratic form is called the second fundamental form of M with respect to the normal N and it is represented in the basis $(\frac{\partial x}{\partial t_i})$ of $T_{x_0}M$ by the matrix

$$B_{ij} = B_{ij}(N) = \frac{\partial^2 x}{\partial t_i \partial t_j}\cdot N. \tag{C.11}$$

The matrix B_{ij} can also be expressed as

$$B_{ij} = -\frac{\partial N}{\partial \tau_i}\cdot \tau_j, \tag{C.12}$$

where $\tau_i = \frac{\partial x}{\partial t_i}$. Indeed,

$$\frac{\partial N}{\partial \tau_i}\cdot \tau_j = \sum_{k=1}^{N}\frac{\partial N_k}{\partial \tau_i}\frac{\partial x_k}{\partial t_j} = \sum_{k=1}^{N}(\nabla N_k\cdot\tau_i)\frac{\partial x_k}{\partial t_j} = \sum_{k,l=1}^{N}\frac{\partial N_k}{\partial x_l}\frac{\partial x_l}{\partial t_i}\frac{\partial x_k}{\partial t_j} = \frac{\partial N}{\partial t_i}\cdot\frac{\partial x}{\partial t_j},$$

while, since $N\cdot\tau_j = 0$,

$$0 = \frac{\partial}{\partial t_i}(N\cdot\tau_j) = \frac{\partial N}{\partial t_i}\cdot\frac{\partial x}{\partial t_j} + N\cdot\frac{\partial^2 x}{\partial t_j \partial t_i} = \frac{\partial N}{\partial t_i}\cdot\frac{\partial x}{\partial t_j} + B_{ij}.$$

Equation (C.12) follows.

Letting $T = dx/ds$, the second fundamental form calculated at T, that is, the quantity

$$k(N,T) = \frac{d^2x}{ds^2}\cdot N$$

is called the normal curvature to M in the direction of T, with respect to N. Take an orthonormal basis of $T_{x_0}M$ and let (B_{ij}) be the matrix representing the second fundamental form in this basis. Then the eigenvalues $k_i = k_i(N)$, $i = 1, \ldots m$ of (B_{ij}) are called the principal curvatures of M with respect to the normal N. Their arithmetic mean

$$H(N) = \frac{k_1(N) + \cdots + k_m(N)}{m}$$

is the mean curvature of M with respect to N. Since the quantity $H(N)$ is linear in N, there exists a unique vector $H \in N_{x_0}M$ such that

$$H(N) = H \cdot N.$$

H is called the mean curvature vector of M.

Laplace-Beltrami operator and mean curvature vector

Proposition C.2.3 *Let $x : D \subset \mathbb{R}^m \to \Omega$ denote a representation of an m-dimensional C^2 submanifold $M \subset \mathbb{R}^N$. Let $\varphi \in C^2(\Omega)$. The Laplace-Beltrami operator acting on φ is defined by*

$$\int_M (-\Delta_M \varphi)\psi \, d\sigma := \int_M \nabla_T \varphi \cdot \nabla_T \psi \, d\sigma, \qquad \text{for all } \psi \in C^1_c(\Omega).$$

In coordinates, given a point $x_0 = x(t_0) \in \Omega \cap M$,

$$\Delta_M \varphi(x_0) = \frac{1}{\sqrt{g(t_0)}} \sum_{i=1}^{m} \frac{\partial}{\partial t_i} \left(\sqrt{g(t)} \sum_{j=1}^{m} g^{ij}(t) \frac{\partial \varphi}{\partial t_j}(t) \right) \Bigg|_{t=t_0}.$$

Proof. By definition of the tangential gradient,

$$\nabla_T \varphi \cdot \nabla_T \psi = \left(\sum_{r,s=1}^{m} g^{rs} \frac{\partial \varphi}{\partial t_s} \frac{\partial x}{\partial t_r} \right) \cdot \left(\sum_{\mu,\nu=1}^{m} g^{\mu\nu} \frac{\partial \psi}{\partial t_\mu} \frac{\partial x}{\partial t_\nu} \right)$$

$$= \sum_{r,s,\mu,\nu=1}^{m} g^{rs} g^{\mu\nu} g_{r\nu} \frac{\partial \varphi}{\partial t_s} \frac{\partial \psi}{\partial t_\mu} = \sum_{r,s=1}^{m} g^{rs} \frac{\partial \varphi}{\partial t_s} \frac{\partial \psi}{\partial t_r}.$$

Hence,

$$\int_M \nabla_T \varphi \cdot \nabla_T \psi \, d\sigma = \sum_{r,s=1}^m \int_D g^{rs} \frac{\partial \varphi}{\partial t_s} \frac{\partial \psi}{\partial t_r} \sqrt{g} \, dt_1 \ldots dt_m$$

$$= -\int_D \psi \sum_{r=1}^m \frac{\partial}{\partial t_r} \left(\sqrt{g} \sum_{s=1}^m g^{rs} \frac{\partial \varphi}{\partial t_s} \right) dt_1 \ldots dt_m$$

$$= -\int_M \psi \frac{1}{\sqrt{g}} \sum_{r=1}^m \frac{\partial}{\partial t_r} \left(\sqrt{g} \sum_{s=1}^m g^{rs} \frac{\partial \varphi}{\partial t_s} \right) d\sigma,$$

as claimed. $\qquad\qquad\square$

Proposition C.2.4 *Let* $x : D \subset \mathbb{R}^m \to \Omega$ *denote a representation of an m-dimensional* C^2 *submanifold* $M \subset \mathbb{R}^N$. *Then, the mean curvature vector of M satisfies*

$$H = \Delta_M x.$$

Proof. We begin by proving that $\Delta_M x$ belongs to the normal space $N_{x_0} M$, at any point $x = x_0 \in M$. It suffices to show that $\Delta_M x \cdot \frac{\partial x}{\partial t_k} = 0$, for $k = 1, \ldots, m$. Now,

$$\sqrt{g} \Delta_M x \cdot \frac{\partial x}{\partial t_k} = \sum_i \frac{\partial}{\partial t_i} \left(\sqrt{g} \sum_j g^{ij} \frac{\partial x}{\partial t_j} \right) \cdot \frac{\partial x}{\partial t_k}$$

$$= \sum_i \frac{\partial}{\partial t_i} \left(\sqrt{g} \sum_j g^{ij} \frac{\partial x}{\partial t_j} \cdot \frac{\partial x}{\partial t_k} \right) - \sum_i \sqrt{g} \sum_j g^{ij} \frac{\partial x}{\partial t_j} \frac{\partial^2 x}{\partial t_i \partial t_k}$$

$$= \sum_i \frac{\partial}{\partial t_i} \left(\sqrt{g} \sum_j g^{ij} g_{jk} \right) - \frac{1}{2} \sum_i \sqrt{g} \sum_j g^{ij} \frac{\partial g_{ij}}{\partial t_k}$$

$$= \sum_i \frac{\partial \sqrt{g}}{\partial t_i} \delta_{ik} - \frac{1}{2} \sqrt{g} \sum_{ij} g^{ij} \frac{\partial g_{ij}}{\partial t_k}.$$

Now, using the letter G to denote the matrix (g_{ij}),

$$\frac{\partial \sqrt{g}}{\partial t_i} = \frac{1}{2\sqrt{g}} \frac{dg}{dt_i} = \frac{1}{2\sqrt{g}} g \operatorname{tr} \left(G^{-1} \frac{dG}{dt} \right) = \frac{\sqrt{g}}{2} \sum_{ij} g^{ij} \frac{dg_{ij}}{dt},$$

and so $\Delta x \cdot \frac{\partial x}{\partial t_k} = 0$.

It remains to prove that $\Delta_M x = H$. To see this, simply note that

$$\Delta_M x = \frac{1}{\sqrt{g}} \sum_{ij} \frac{\partial}{\partial t_i} \left(\sqrt{g} g^{ij} \right) \frac{\partial x}{\partial t_j} + \sum_{ij} g^{ij} \frac{\partial^2 x}{\partial t_i \partial t_j}$$

and so, for any normal vector $N \in T_{x_0} M$,

$$\Delta_M x \cdot N = \sum_{ij} g^{ij} \frac{\partial^2 x}{\partial t_i \partial t_j} \cdot N = H \cdot N.$$

□

As an immediate consequence of Proposition C.2.4, we obtain the following useful formula.

Lemma C.2.2 *Let* $1 \leq m \leq N$. *Let* $x : D \subset \mathbb{R}^m \to \Omega$ *denote a representation of an m-dimensional C^2 submanifold $M \subset \mathbb{R}^N$. For all $\varphi \in C_c^1(\Omega)$, there holds*

$$\int_M \left(\nabla_T \varphi + H \varphi \right) \, d\sigma = 0.$$

C.2.4 The Sobolev inequality on submanifolds

This section is dedicated to the proof of the following theorem.

Theorem C.2.1 ([5, 153]) *Let $1 \leq m \leq N$. Let M denote an m-dimensional C^2 submanifold of the Euclidean space $\mathbb{R}^N$ and U an open set containing M. For all $p \in [1, m)$, there exists a constant $C = C(m, p) > 0$ such that for all $\varphi \in C_c^1(U)$,*

$$\left(\int_M |\varphi|^{p^*} \, d\sigma \right)^{1/p^*} \leq C \left(\int_M \left(|\nabla_T \varphi|^p + |H\varphi|^p \right) \, d\sigma \right)^{1/p}, \qquad \text{(C.13)}$$

where $1/p^* = 1/p - 1/m$, where H is the mean curvature vector of M and where $\nabla_T \varphi$ is the projection of the gradient of φ on the tangent space of M.

Remark C.2.3

- *Note that the constant C appearing in (C.13) is independent of the given manifold M. The geometry of M enters only through its mean curvature vector H.*

- *The inequality is local: using a standard partition of unity, it suffices to prove it on a neighborhood of a point $x_0 \in M$. At the expense of making global assumptions on the manifold M (for example, if M is compact or if the following three assumptions hold: M is complete, M has a nonnegative Ricci curvature, and M has maximal volume growth), one can derive the standard (global) Sobolev inequality, that is,*

$$\left(\int_M |\varphi|^{p^*} \, d\sigma \right)^{1/p^*} \leq C \left(\int_M |\nabla_T \varphi|^p \, d\sigma \right)^{1/p}$$

holds for all $\varphi \in C^1_c(M)$, with $C = C(m,p,M)$. See [189].

The proof of Theorem C.2.1 uses the following two lemmata.

Lemma C.2.3 *Suppose $\lambda \in C^1(\mathbb{R})$ is a nondecreasing function such that $\lambda(t) \leq 0$ for $t \leq 0$. Let $\varphi \in C^1_c(U)$, $\varphi \geq 0$. For $x_0 \in M$, define $\varphi_{x_0}, \psi_{x_0} \in C^1(0, +\infty)$ by*

$$\varphi_{x_0}(\rho) = \int_M \varphi(x) \lambda(\rho - r) \, d\sigma(x)$$

and

$$\psi_{x_0}(\rho) = \int_M \left[|\nabla_T \varphi(x) + |H(x)| \varphi(x) \right] \lambda(\rho - r) \, d\sigma(x),$$

where $r = |x - x_0|$. Then,

$$-\frac{d}{d\rho} \left(\frac{\varphi_{x_0}(\rho)}{\rho^m} \right) \leq \frac{\psi_{x_0}(\rho)}{\rho^m}, \qquad \text{for all } \rho > 0.$$

Proof. Using Lemma C.2.2 with $(x - x_0)_i \lambda(\rho - r)\varphi$ in place of φ and summing over $i = 1, \ldots, N$, we have

$$\int_M \sum_{i=1}^N \delta_i [(x - x_0)_i \lambda(\rho - r)\varphi] \, d\sigma = - \int_M \varphi \lambda(\rho - r) \left[\sum_{i=1}^N (x - x_0)_i H_i \right] d\sigma,$$

where δ_i, H_i are the components of ∇_T, H in the canonical base of $\mathbb{R}^N$. Now,

$$\delta_i [(x-x_0)_i \lambda(\rho-r)\varphi] = \tilde{g}^{ii} \lambda(\rho-r)\varphi - r\lambda'(\rho-r)\varphi \sum_{j=1}^N \frac{(x - x_0)_i}{r} \frac{(x - x_0)_j}{r} \tilde{g}^{ij}$$

$$+ \lambda(\rho - r)(x - x_0)_i \delta_i \varphi, \qquad \text{for all } i = 1, \ldots, N,$$

288

and hence, using Proposition C.2.1 together with

$$\sum_{i=1}^{N} \left(\frac{(x - x_0)_i}{r} \right)^2 ,$$

we obtain

$$m\varphi_{x_0}(\rho) - \int_M r\lambda'(\rho - r)\varphi \, d\sigma \leq \int_M r\lambda(\rho - r)[|\nabla_T \varphi| + |H|\varphi] \, d\sigma.$$

Since $\lambda(\rho - r) = 0$ when $r \geq \rho$, so that

$$r\lambda(\rho - r) \leq \rho\lambda(\rho - r) \quad \text{and} \quad r\lambda'(\rho - r) \leq \rho\lambda'(\rho - r),$$

we obtain

$$m\varphi_{x_0}(\rho) - \rho\varphi'_{x_0}(\rho) \leq \rho\psi_{x_0}(\rho).$$

This last inequality can be rewritten in the form

$$-\frac{d}{d\rho} \left(\frac{\varphi_{x_0}(\rho)}{\rho^m} \right) \leq \frac{\psi_{x_0}(\rho)}{\rho^m},$$

as requested. $\qquad\qquad\qquad\qquad\qquad\qquad\qquad\qquad\qquad\qquad\qquad\qquad\qquad\qquad$ $\square$

Lemma C.2.4 *Let φ be as in Lemma C.2.3 and let $x_0 \in M$ such that $\varphi(x_0) \geq 1$. Define $\overline{\varphi}_{x_0}, \overline{\psi}_{x_0}$ on $(0, +\infty)$ by*

$$\overline{\varphi}_{x_0}(\rho) = \int_{S_\rho(x_0)} \varphi \, d\sigma,$$

$$\overline{\psi}_{x_0}(\rho) = \int_{S_\rho(x_0)} [|\nabla_T \varphi| + |H|\varphi] \, d\sigma,$$

where $S_\rho(x_0) = \{x \in M : |x - x_0| \leq \rho\}$.
Then, there exists ρ such that $0 < \rho < 2\left[\omega_m^{-1} \int_M \varphi \, d\sigma \right]^{1/m}$ and

$$\overline{\varphi}_{x_0}(4\rho) \leq 4^m \left[\omega_m^{-1} \int_M \varphi \, d\sigma \right]^{1/m} \overline{\psi}_{x_0}(\rho),$$

where ω_m denotes the Lebesgue measure of the unit ball in $\mathbb{R}^m$.

Proof. Let $\varphi_{x_0}, \psi_{x_0}$ be as in Lemma C.2.3, so that

$$-\frac{d}{d\rho}\left(\frac{\varphi_{x_0}(\rho)}{\rho^m}\right) \leq \frac{\psi_{x_0}(\rho)}{\rho^m}. \tag{C.14}$$

Let

$$\rho_0 = 2\left[\omega_m^{-1}\int_M \varphi\, d\sigma\right]^{1/m} > 0.$$

Assume that $t \in (0, \rho_0)$ and integrate (C.14) over the interval (t, ρ_0). This yields

$$t^{-m}\varphi_{x_0}(t) \leq \rho_0^{-m}\varphi_{x_0}(\rho_0) + \int_t^{\rho_0} \rho^{-m}\psi_{x_0}(\rho)\, d\rho$$

$$\leq \rho_0^{-m}\varphi_{x_0}(\rho_0) + \int_0^{\rho_0} \rho^{-m}\psi_{x_0}(\rho)\, d\rho. \tag{C.15}$$

Now let $\varepsilon \in (0, t)$ and suppose that the function λ appearing in the definition of $\varphi_{x_0}, \psi_{x_0}$ is chosen so that $\lambda(t) = 1$ for all $t \geq \varepsilon$. It follows from (C.15) that

$$t^{-m}\overline{\varphi}_{x_0}(t - \varepsilon) \leq \rho_0^{-m}\overline{\varphi}_{x_0}(\rho_0) + \int_0^{\rho_0} \rho^{-m}\overline{\psi}_{x_0}(\rho)\, d\rho.$$

Since $t < \rho_0$ and $\varepsilon \in (0, t)$ are arbitrary, this gives

$$\sup_{t\in(0,\rho_0)} t^{-m}\overline{\varphi}_{x_0}(t) \leq \rho_0^{-m}\overline{\varphi}_{x_0}(\rho_0) + \int_0^{\rho_0} \rho^{-m}\overline{\psi}_{x_0}(\rho)\, d\rho. \tag{C.16}$$

Assume that, contrary to the statement of the lemma, $\overline{\psi}_{x_0}(\rho) < 2.4^{-m}\rho_0^{-1}\overline{\varphi}_{x_0}(4\rho)$, for all $\rho \in (0, \rho_0)$. Then,

$$\int_0^{\rho_0} \rho^{-m}\overline{\psi}_{x_0}(\rho)\, d\rho \leq 2.4^{-m}\rho_0^{-1}\int_0^{\rho_0} \rho^{-m}\overline{\varphi}_{x_0}(4\rho)\, d\rho$$

$$= \frac{1}{2}\rho_0^{-1}\int_0^{4\rho_0} t^{-m}\overline{\varphi}_{x_0}(t)\, dt$$

$$\leq \frac{1}{2}\rho_0^{-1}\left[\int_0^{\rho_0} t^{-m}\overline{\varphi}_{x_0}(t)\, dt + \int_{\rho_0}^{+\infty} t^{-m}\overline{\varphi}_{x_0}(t)\, dt\right]$$

$$\leq \frac{1}{2}\rho_0^{-1}\left[\rho_0 \sup_{t\in(0,\rho_0)} t^{-m}\overline{\varphi}_{x_0}(t) + \frac{1}{m-1}\rho_0^{1-m}\int_M \varphi\, d\sigma\right],$$

where we used the fact that $\overline{\varphi}_{x_0}(t) \leq \int_M \varphi \, d\sigma$ for all $t > 0$. Hence, it follows from (C.16) that

$$\frac{1}{2} \sup_{t \in (0, \rho_0)} t^{-m} \overline{\varphi}_{x_0}(t) \leq \rho_0^{-m} \left(1 + \frac{1}{2(m-1)} \right) \int_M \varphi \, d\sigma,$$

so that, using the definition of ρ,

$$\sup_{t \in (0, \rho_0)} t^{-m} \overline{\varphi}_{x_0}(t) \leq 2^{1-m} \omega_m \left(1 + \frac{1}{2(m-1)} \right) < \omega_m.$$

Using Lemma C.2.1 and the assumption $\varphi(x_0) \geq 1$, we obtain a contradiction. $\square$

Proof of Theorem C.2.1.
Step 1. We begin by proving that the case $p \in (1, m)$ can be derived from the case $p = 1$. Assume that (C.13) holds in the latter case, take $\varphi \in C_c^1(U)$ and let $\psi = |\varphi|^{t-1} \varphi$, where $t = p^*/1^*$, for some $p \in (1, m)$. In particular, $t > 1$ so that $\psi \in C_c^1(U)$ and we may apply (C.13) to get

$$\left(\int_M |\varphi|^{p^*} d\sigma \right)^{1/1^*} = \left(\int_M |\psi|^{1^*} d\sigma \right)^{1/1^*} \leq C \int_M \left(|\nabla_T \psi| + |H \psi| \right) d\sigma. \tag{C.17}$$

Now,

$$|\nabla_T \psi| + |H \psi| = |\varphi|^{t-1} \left(t |\nabla_T \varphi| + |H \varphi| \right) \leq t |\varphi|^{t-1} \left(|\nabla_T \varphi| + |H \varphi| \right).$$

By Hölder's inequality, it follows that

$$\int_M \left(|\nabla_T \psi| + |H \psi| \right) d\sigma \leq$$
$$C(p) \left(\int_M \left(|\nabla_T \varphi|^p + |H \varphi|^p \right) d\sigma \right)^{1/p} \left(\int_M |\varphi|^{(t-1)p'} d\sigma \right)^{1/p'}.$$

Observe that by the definition of p^*,

$$\frac{1}{1^*} - \frac{1}{p^*} = \frac{1}{1} - \frac{1}{p} = \frac{1}{p'}. \tag{C.18}$$

In particular,

$$(t-1)p' = \left(\frac{p^*}{1^*} - 1 \right) p = p^* \left(\frac{1}{1^*} - \frac{1}{p^*} \right) p' = p^*.$$

Hence,

$$\int_M \left(|\nabla_T \psi| + |H\psi| \right) d\sigma \le$$

$$C(p) \left(\int_M \left(|\nabla_T \varphi|^p + |H\varphi|^p \right) d\sigma \right)^{1/p} \left(\int_M |\varphi|^{p^*} d\sigma \right)^{1/p'}.$$

Plugging this in (C.17), we obtain

$$\left(\int_M |\varphi|^{p^*} d\sigma \right)^{1/1^* - 1/p'} \le C(m,p) \left(\int_M \left(|\nabla_T \varphi|^p + |H\varphi|^p \right) d\sigma \right)^{1/p}.$$

By (C.18), $1/1^* - 1/p' = 1/p^*$ and the desired inequality follows. It remains to prove (C.13) in the case $p = 1$.

Step 2. We now assume that $p = 1$ and, without loss of generality $\varphi \ge 0$. Using a covering argument and Lemma C.2.4, we prove that

$$\sigma(\{x \in M \, : \, \varphi(x) \ge 1\}) \le 4^m \left(\omega_m^{-1} \int_M \varphi \, d\sigma \right)^{1/m} \int_M [|\nabla_T \varphi| + |H|\varphi] \, d\sigma.$$

Let

$$A = \{x \in M \, : \, \varphi(x) \ge 1\}$$

and assume for the time being that A is nonempty. For $x \in A$, let $\overline{\varphi}_x, \overline{\psi}_x$ be as in Lemma C.2.4 and let $J = 4^m \left(\omega_m^{-1} \int_M \varphi \, d\sigma \right)^{1/m}$. For $i = 1, 2, \ldots$, let $\rho_i = 4.2^{-i} J$ and

$$A_i = \left\{ x \in A \, : \, \overline{\varphi}_x(4\rho) \le J \overline{\psi}_x(\rho) \text{ for some } \rho \in \left(\frac{1}{2} \rho_i, \rho_i \right] \right\}.$$

It follows from Lemma C.2.4 that $A = \cup_{i=1}^{+\infty} A_i$.

Next, define inductively a sequence $\mathscr{F}_0, \mathscr{F}_1, \ldots$ of subsets of A as follows:

(i) $\mathscr{F}_0 = \emptyset$.

(ii) Let $k \ge 1$ and assume that $\mathscr{F}_0, \ldots, \mathscr{F}_{k-1}$ have been defined. Let $B_k = A_k \setminus \cup_{i=0}^{k-1} \cup_{x \in \mathscr{F}_i} S_{2\rho_i}(x)$. If $B_k = \emptyset$, then set $\mathscr{F}_k = \emptyset$. If $B_k \ne \emptyset$, then choose $\mathscr{F}_k$ to be a finite subset of B_k such that $B_k \subset \cup_{x \in \mathscr{F}_k} S_{2\rho_k}(x)$ and such that the sets $S_{\rho_k}(x)$, $x \in \mathscr{F}_k$, are pairwise disjoint.

Then, it is not difficult to check that the following properties hold:

(a) $\mathscr{F}_i \subset A_i$, for $i = 1, 2, \ldots,$

(b) $A \subset \cup_{i=1}^{+\infty} \cup_{x \in \mathscr{F}_i} S_{2\rho_i}(x)$, and

(c) the sets of the countable collections $S_{\rho_i}(x)$, $x \in \mathscr{F}_i$, $i = 1, 2, \ldots$, are pairwise disjoint.

By property (a) we have, for each $x \in \mathscr{F}_i$,

$$\overline{\varphi}_x(4\rho) \le J\overline{\psi}_x(\rho)$$

for some $\rho \in \left(\frac{1}{2}\rho_i, \rho_i \right]$. Thus,

$$\overline{\varphi}_x(2\rho_i) \le J\overline{\psi}_x(\rho_i),$$

for each $x \in \mathscr{F}_i$. Summing over all $x \in \mathscr{F}_i$, $i = 1, 2, \ldots$, and using properties (b) and (c), we then have

$$\sigma(M_1) \le J \int_M [|\nabla_T \varphi| + |H|\varphi] \, d\sigma, \tag{C.19}$$

where $M_1 = \{x \in M \ : \ \varphi(x) \ge 1\}$, provided that $A \ne \emptyset$. When $A = \emptyset$, (C.19) is trivially true.

Step 3. Completion of the proof.

Let $\alpha, \varepsilon > 0$ be arbitrary constants, let $\lambda \in C^1(\mathbb{R})$ be a nondecreasing function such that $\lambda(t) = 0$ for $t \le -\varepsilon$, $\lambda(t) = 1$ for $t \ge 0$, and use (C.19) with $\lambda(\varphi - \alpha)$ in place of φ. Then, (C.19) gives

$$\sigma(M_\alpha) \le 4^m \omega_m^{-1/m} \left[\int_M \lambda(\varphi - \alpha) \, d\sigma \right]^{1/m}$$
$$\times \int_M [\lambda'(\varphi - \alpha)|\nabla_T \varphi| + \lambda(\varphi - \alpha)|H|] \, d\sigma, \tag{C.20}$$

where $M_\alpha = \{x \in M \ : \ \varphi(x) \ge \alpha\}$. Multiplying each side of (C.20) by $\alpha^{1/(m-1)}$ and using the fact that $\lambda(\varphi - \alpha) = 0$ for $\alpha \ge \varphi + \varepsilon$, we then get

$$\alpha^{1/(m-1)} \sigma(M_\alpha) \le 4^m \omega_m^{-1/m} \left[\int_M (\varphi + \varepsilon)^{m/(m-1)} \, d\sigma \right]^{1/m}$$
$$\times \int_M [\lambda'(\varphi - \alpha)|\nabla_T \varphi| + \lambda(\varphi - \alpha)|H|] \, d\sigma. \tag{C.21}$$

Finally, we obtain the desired inequality by first integrating over $(0, +\infty)$ with respect to α, using the fact that

$$\int_0^{+\infty} \lambda(\varphi - \alpha) \, d\alpha = \int_0^{+\infty} \alpha \lambda'(\varphi - \alpha) \, d\alpha$$

$$\leq (\varphi + \varepsilon) \int_0^{+\infty} \lambda'(\varphi - \alpha) \, d\alpha \leq \varphi + \varepsilon,$$

and

$$\int_0^{+\infty} \alpha^{1/(m-1)} \sigma(M_\alpha) \, d\alpha = \frac{m-1}{m} \int_M \varphi^{m/(m-1)} \, d\sigma,$$

and then letting $\varepsilon \to 0$. $\qquad\qquad\qquad\qquad\qquad\qquad\qquad\qquad\square$

C.3 Geometry of level sets

Consider an open set $\Omega \subset \mathbb{R}^N$ and a value $a \in \mathbb{R}$. The level set of the real valued map $u \in C^1(\Omega)$ is given by

$$L_a = \{y \in \Omega : u(y) = a\}.$$

For example, if Ω is the unit ball and if u is a radial function, the level set L_a is the union of all the hyperspheres of radius R such that $u(R) = a$.

The classical Morse-Sard theorem asserts that for any $u \in C^N(\Omega)$ and almost all $a \in \mathbb{R}$, the level set L_a is a regular hypersurface.

Theorem C.3.1 ([164], [192]) *Let $N \geq 1$ and let Ω denote an open set in $\mathbb{R}^N$. Assume that $u \in C^N(\Omega)$. Let X denote the set of critical points of u, that is,*

$$X = \{x \in \Omega : \nabla u(x) = 0\}.$$

Then $u(X)$ has Lebesgue measure 0 in $\mathbb{R}$.

For a proof, see, for example, [162].

Definition C.3.1 *A value $a \in \mathbb{R}$ such that $\nabla u(x) \neq 0$ for all $x \in L_a$ is called a regular value of u.*

Using the implicit function theorem, we obtain the following corollary.

Corollary C.3.1 *Let $u \in C^N(\Omega)$. For almost every $a \in u(\Omega)$, the level set $L_a \subset \mathbb{R}^N$ is an $(N-1)$-dimensional submanifold of $\mathbb{R}^N$.*

In particular, if $u \in C^N(\Omega)$, for almost every $a \in u(\Omega)$, the level set L_a has zero N-dimensional Lebesgue measure. It is interesting to note that if there exists a value a, such that the level set L_a has nonzero Lebesgue measure, then the gradient of u must vanish at almost every point of L_a.

Lemma C.3.1 *Let Ω be an open set of $\mathbb{R}^N$, $N \geq 1$, $u \in W^{1,1}_{loc}(\Omega)$ and $a \in u(\Omega)$. Then,*
$$\nabla u(x) = 0, \qquad \text{for almost every } x \in L_a.$$

Proof. We may assume that $a = 0$ without loss of generality. For $\varepsilon > 0$, take a cutoff function $\chi_\varepsilon \in C_c(\mathbb{R})$, such that $0 \leq \chi_\varepsilon \leq 1$, $\chi_\varepsilon \equiv 1$ in $(-\varepsilon, \varepsilon)$ and $\chi_\varepsilon \equiv 0$ in $\mathbb{R} \setminus (-2\varepsilon, 2\varepsilon)$. Also let $\Phi_\varepsilon(t) = \int_0^t \chi_\varepsilon \, ds$, for all $t \in \mathbb{R}$. Then, given any function $\varphi \in C_c^1(\Omega)$,

$$\int_\Omega \varphi \chi_\varepsilon(u) \nabla u \, dx = - \int_\Omega \Phi_\varepsilon(u) \nabla \varphi \, dx. \tag{C.22}$$

Note that $|\Phi_\varepsilon(t)| \leq 2\varepsilon$ for all $t \in \mathbb{R}$. Hence, the right-hand side of (C.22) converges to zero as $\varepsilon \to 0$. Now, $\chi_\varepsilon(u)$ is bounded by 1 and converges pointwise to $\chi_{[u=0]}$. By dominated convergence, we conclude that

$$\int_{[u=0]} \varphi \, \nabla u \, dx = 0.$$

This being true for all $\varphi \in C_c^1(\Omega)$, the result follows. $\qquad \square$

C.3.1 Coarea formula

When evaluating integrals on a radially symmetric domain, for example, the unit ball $\Omega = B$, one is often led to using polar coordinates:

$$\int_\Omega \varphi \, dx = \int_0^1 \left(\int_{|x|=r} \varphi \, d\sigma \right) dr,$$

for $\varphi \in L^1(\Omega)$. Extending φ by zero outside $\Omega = B$ and thinking of spheres as the level sets of a given radial, decreasing function u, the above formula can be restated as follows:

$$\int_\Omega \varphi \, dx = \int_{-\infty}^{+\infty} \left(\int_{[u=t]} \frac{1}{|\nabla u|} \varphi \, d\sigma \right) dt. \tag{C.23}$$

The above formula remains valid more generally.

Theorem C.3.2 *Let Ω be an open set of $\mathbb{R}^N$, $N \geq 2$, $u \in C^2(\Omega)$ such that $|\nabla u| \neq 0$ in Ω, and $\varphi \in L^1(\Omega)$. Then, (C.23) holds.*

Proof. Fix a point $x_0 \in \Omega$ and let $\alpha = u(x_0)$. Without loss of generality, we may assume that φ is compactly supported in a neighborhood of x_0. We may also assume that $\frac{\partial u}{\partial x_N} \neq 0$ in that neighborhood. Now consider variables $t = (t_1, \ldots, t_{N-1})$ simply defined by $t_1 = x_1$, ..., $t_{N-1} = x_{N-1}$. By the implicit function theorem, the mapping $x = x(t) = (t, x_N(t))$ given implicitly by

$$u(t, x_N(t)) = \alpha \tag{C.24}$$

defines a representation of the level set $M := [u = \alpha]$ at x_0. The metric tensor of M is given by

$$g_{ij} = \frac{\partial x}{\partial t_i} \cdot \frac{\partial x}{\partial t_j} = \sum_{k=1}^{N} \frac{\partial x_k}{\partial t_i} \frac{\partial x_k}{\partial t_j} = \delta_{ij} + \frac{\partial x_N}{\partial t_i} \frac{\partial x_N}{\partial t_j}. \tag{C.25}$$

We claim that

$$\sqrt{g} = \frac{|\nabla u|}{\frac{\partial u}{\partial x_N}}. \tag{C.26}$$

For now, take (C.26) for granted. Then, letting D denote the domain of the representation x,

$$\int_{[u=\alpha]} \frac{1}{|\nabla u|} \varphi \, d\sigma = \int_D \frac{1}{\frac{\partial u}{\partial x_N}} \varphi \, dt_1 \ldots dt_{N-1}.$$

Integrating over α, it follows that

$$\int_{-\infty}^{+\infty} \left(\int_{[u=\alpha]} \frac{1}{|\nabla u|} \varphi \, d\sigma \right) d\alpha = \int_{-\infty}^{+\infty} \left(\int_D \frac{1}{\frac{\partial u}{\partial x_N}} \varphi \, dt_1 \ldots dt_{N-1} \right) d\alpha,$$

$$= \int_\Omega \varphi \, dx_1 \ldots dx_{N-1} dx_N,$$

where we used the change of variable $t_1 = x_1$, ..., $t_{N-1} = x_{N-1}$, $\alpha = u(t_1, \ldots, t_{N-1}, x_N)$ in the last equality.

It remains to prove (C.26). To this end, consider $v = (\frac{\partial x_N}{\partial t_1}, \ldots, \frac{\partial x_N}{\partial t_{N-1}})$. Using (C.25), we obtain

$$\sum_{j=1}^{N-1} g_{ij} v_j = v_i + \sum_{j=1}^{N-1} \frac{\partial x_N}{\partial t_i} \frac{\partial x_N}{\partial t_j} v_j = \left(1 + \sum_{j=1}^{N-1} \left(\frac{\partial x_N}{\partial t_j} \right)^2 \right) \frac{\partial x_N}{\partial t_i},$$

and so $\lambda = \left(1 + \sum_{j=1}^{N-1} \left(\frac{\partial x_N}{\partial t_j}\right)^2\right)$ is an eigenvalue of (g_{ij}). Differentiate (C.24) with respect to t_j:

$$\frac{\partial u}{\partial t_j} + \frac{\partial u}{\partial x_N}\frac{\partial x_N}{\partial t_j} = 0.$$

So, the eigenvalue λ can be rewritten as

$$\lambda = 1 + \sum_{j=1}^{N-1}\left(\frac{\frac{\partial u}{\partial t_j}}{\frac{\partial u}{\partial x_N}}\right)^2 = \left(\frac{|\nabla u|}{\frac{\partial u}{\partial x_N}}\right)^2.$$

Now take any vector w which is orthogonal to $v = (\frac{\partial x_N}{\partial t_1}, \ldots, \frac{\partial x_N}{\partial t_{N-1}})$, with respect to the standard inner product on $\mathbb{R}^{N-1}$. Clearly, w can be written as a linear combination of vectors of the form $\tilde{w} = \frac{\partial x_N}{\partial t_l}e_k - \frac{\partial x_N}{\partial t_k}e_l$, where (e_i) denotes the canonical basis of $\mathbb{R}^{N-1}$. Then,

$$\sum_{j=1}^{N-1} g_{ij}\tilde{w}_j = \sum_{j=1}^{N-1}\left(\delta_{ij} + \frac{\partial x_N}{\partial t_i}\frac{\partial x_N}{\partial t_j}\right)\left(\frac{\partial x_N}{\partial t_l}\delta_{kj} - \frac{\partial x_N}{\partial t_k}\delta_{lj}\right)$$

$$= \delta_{ik}\frac{\partial x_N}{\partial t_l} - \delta_{il}\frac{\partial x_N}{\partial t_k} + \frac{\partial x_N}{\partial t_i}\frac{\partial x_N}{\partial t_k}\frac{\partial x_N}{\partial t_l} - \frac{\partial x_N}{\partial t_i}\frac{\partial x_N}{\partial t_l}\frac{\partial x_N}{\partial t_k}$$

$$= \tilde{w}_i.$$

So, 1 is the only other eigenvalue of (g_{ij}), hence

$$g = \lambda = \left(\frac{|\nabla u|}{\frac{\partial u}{\partial x_N}}\right)^2,$$

as claimed. $\qquad\qquad\qquad\qquad\qquad\qquad\qquad\qquad\qquad\qquad\qquad\qquad\square$

C.4 Spectral theory of the Laplace operator on the sphere

Definition C.4.1 *A function F defined on $\mathbb{R}^N \setminus \{0\}$ is homogeneous of degree $k \in \mathbb{Z}$ if*

$$F(tx) = t^k F(x), \qquad \text{for any } t > 0, \, x \in \mathbb{R}^N \setminus \{0\}.$$

Definition C.4.2 *Given $N \geq 2$ and $k \in \mathbb{N}$, let P_k denote the space of polynomials (in N variables), which are homogeneous of degree k. Let A_k denote the subspace of P_k consisting of all harmonic polynomials. Then, the space $H_k = A_k\big|_{\mathscr{S}^{N-1}}$, obtained by restriction from A_k is called the space of spherical harmonics of degree k.*

Theorem C.4.1 *Let $N \geq 2$ and let $\mathscr{S}^{N-1}$ denote the unit sphere of $\mathbb{R}^N$. The eigenvalues of the Laplace-Beltrami operator $-\Delta_{\mathscr{S}^{N-1}}$ are the numbers*

$$\mu_k = k(k + N - 2), \qquad k \in \mathbb{N}.$$

Furthermore,

- *Let H_k denote the set of spherical harmonics of degree k. Then, H_k is the eigenspace associated to μ_k.*

- $L^2(\mathscr{S}^{N-1}) = \bigoplus_{k=0}^{+\infty} H_k$.

Theorem C.4.2

$$dim(H_k) = \frac{(2k + N - 2)(N + k - 3)!}{k!(N-2)!} = \mathcal{O}(k^{N-2}).$$

To prove the above results, we follow [200], [11] and begin with a series of elementary lemmata.

Lemma C.4.1 *Let $u, v \in C^2(\mathscr{S}^{N-1})$. Then,*

$$\int_{\mathscr{S}^{N-1}} u\Delta_{\mathscr{S}^{N-1}} v \, d\sigma = \int_{\mathscr{S}^{N-1}} v\Delta_{\mathscr{S}^{N-1}} u \, d\sigma.$$

Proof. To see this, let $U(x) = u(x/|x|)$ and $V(x) = v(x/|x|)$, for all $x \in \mathbb{R}^N \setminus \{0\}$. That is, U, V are extensions of u, v, which are constant along the normal lines of $\mathscr{S}^{N-1}$. In particular, $\Delta_{\mathscr{S}^{N-1}} u = \Delta U|_{\mathscr{S}^{N-1}}$ and a similar formula holds for v. In addition, since U, V are homogeneous of degree 0, $\Delta U, \Delta V$ are homogeneous of degree -2 and so we have

$$\int_{B_2 \setminus B_1} (V\Delta U - U\Delta V) \, dx = \left(\int_1^2 r^{N-3} dr\right) \int_{\mathscr{S}^{N-1}} (v\Delta_{\mathscr{S}^{N-1}} u - u\Delta_{\mathscr{S}^{N-1}} v) \, d\sigma.$$

Integrating by parts the left-hand side of the above expression, we obtain

$$\int_{B_2 \backslash B_1} (V \Delta U - U \Delta V) dx =$$

$$\int_{\partial B_2} \left(V \frac{\partial U}{\partial r} - U \frac{\partial V}{\partial r} \right) d\sigma - \int_{\partial B_1} \left(V \frac{\partial U}{\partial r} - U \frac{\partial V}{\partial r} \right) d\sigma.$$

Since U, V are homogeneous of degree 0, $\frac{\partial U}{\partial r}, \frac{\partial V}{\partial r}$ vanish, and the lemma follows. $\qquad\square$

Lemma C.4.2 *Given $U \in C^2(\mathbb{R}^N \backslash \{0\})$,*

$$\Delta U = \frac{\partial^2 U}{\partial r^2} + \frac{N-1}{r} \frac{\partial U}{\partial r} + \frac{1}{r^2} \Delta_{\mathscr{S}^{N-1}} U.$$

In particular, if $U \in P_k$, then

$$\Delta U = k(k+N-2)|x|^{k-2} u(x/|x|) + |x|^{k-2} \Delta_{\mathscr{S}^{N-1}} u(x/|x|),$$

where $U(x) = |x|^k u(x/|x|)$.

Proof. By the Stone-Weierstrass theorem, it suffices to prove the identity when U is a homogeneous polynomial of degree k, so that $U(x) = |x|^k u(x/|x|)$ for some $u \in C^2(\mathscr{S}^{N-1})$. Now, $\Delta(HG) = H\Delta G + 2\nabla H \cdot \nabla G + G\Delta H$; we apply this with $H(x) = |x|^k$ and $G(x) = u(x/|x|)$. With this H and G, ∇H is perpendicular to the sphere $|x| = $ constant, and ∇G is tangential to that sphere, so $\nabla H \cdot \nabla G = 0$. Since u is homogeneous of degree 0, Δu is homogeneous of degree -2, and so $|x|^{k-2} \Delta_{\mathscr{S}^{N-1}}(u(x/|x|)) = |x|^k \Delta u(x)$. Finally, a direct calculation shows that

$$\Delta |x|^k = k(k+N-2)|x|^{k-2},$$

so that, letting $r = |x|$ and $\omega = x/|x|$,

$$\Delta U(x) = k(k+N-2)|x|^{k-2} u(x/|x|) + |x|^{k-2} \Delta_{\mathscr{S}^{N-1}} u(x/|x|)$$

$$= \left(\frac{\partial^2}{\partial r^2} + \frac{N-1}{r} \frac{\partial}{\partial r} \right) r^k u(\omega) + \frac{1}{r^2} r^k \Delta_{\mathscr{S}^{N-1}} u(\omega)$$

$$= \frac{\partial^2 U}{\partial r^2} + \frac{N-1}{r} \frac{\partial U}{\partial r} + \frac{1}{r^2} \Delta_{\mathscr{S}^{N-1}} U,$$

as desired. $\qquad\square$

Lemma C.4.3 *For $k \geq 2$, any polynomial p of degree at most k can be written as*

$$p = q_1 + (1 - |x|^2)q_2, \tag{C.27}$$

where q_1 is a harmonic polynomial and q_2 is a polynomial of degree at most $k-2$.

Proof. It suffices to find a polynomial q_2 of degree at most $k - 2$ such that

$$\Delta((1 - |x|^2)q_2) = \Delta p.$$

To do this, let W denote the vector space of polynomials of degree at most $k - 2$ and define the linear map $T : W \rightarrow W$ by

$$T(q) = \Delta((1 - |x|^2)q).$$

Observe that T is one-to-one: if $T(q) = 0$, then $(1 - |x|^2)q$ is harmonic. In addition, $(1 - |x|^2)q = 0$ on $\mathscr{S}^{N-1}$. By the maximum principle, $(1 - |x|^2)q \equiv 0$ in B and so $q \equiv 0$. Hence, T is a one-to-one endomorphism of W, so it is also surjective and the lemma follows. $\qquad\square$

Proof of Theorem C.4.1. We begin by proving that any nonzero spherical harmonic $\varphi \in H_k$ is an eigenvector associated to μ_k. Indeed, take $p = |x|^k\varphi(x/|x|) \in A_k$. Applying Lemma C.4.2 to p, we have on $\mathscr{S}^{N-1}$, $0 = \Delta p = k(N + k - 2)\varphi + \Delta_{\mathscr{S}^{N-1}}\varphi$ and so φ is an eigenvector associated to μ_k. By Lemma C.4.1, if φ_k, φ_j are eigenvectors associated to $\mu_k \neq \mu_j$, then

$$0 = \int_{\mathscr{S}^{N-1}} \left(\varphi_k \Delta_{\mathscr{S}^{N-1}}\varphi_j - \varphi_j \Delta_{\mathscr{S}^{N-1}}\varphi_k \right) d\sigma = (\mu_k - \mu_j) \int_{\mathscr{S}^{N-1}} \varphi_j \varphi_k \, d\sigma$$

and so H_k is orthogonal to H_j in $L^2(\mathscr{S}^{N-1})$. It remains to prove that the space of spherical harmonics is dense in $L^2(\mathscr{S}^{N-1})$. To see this, by the density of continuous functions in $L^2(\mathscr{S}^{N-1})$ and the Stone-Weierstrass theorem, it suffices to prove that given any polynomial p, its restriction $p\big|_{\mathscr{S}^{N-1}}$ can be written as a linear combination of spherical harmonics. We work inductively on the degree k of p. If $k = 0$, then p is clearly harmonic and the claim follows. Now, assume by induction that any polynomial of degree at most $k-1$ is a sum of spherical harmonics, when restricted to the sphere. Take p a polynomial of degree k and let p_k denote its homogeneous part of degree k. Apply Lemma C.4.3 to p_k. Take the homogeneous part of degree k on both sides of (C.27). Then,

$$p_k = q_k - |x|^2 q_{k-2},$$

where q_k is a homogeneous harmonic polynomial of degree k and q_{k-2} is a homogeneous polynomial of degree $k - 2$. By the induction hypothesis, q_{k-2} is a sum of spherical harmonics when restricted to the sphere. This implies that p_k and p can also be decomposed in spherical harmonics on $\mathscr{S}^{N-1}$. □

Proof of Theorem C.4.2. Let $d_{k,N}$ denote the dimension of the space of homogeneous polynomials of degree k. Separating monomials of degree k into those divisible by x_N and those not divisible by x_N, we get $d_{k,N} = d_{k-1,N} + d_{k,N-1}$, and clearly $d_{k,1} = d_{0,N} = 1$. By induction on $k+N$, we deduce that

$$ d_{k,N} = \frac{(N+k-1)!}{k!(N-1)!}. $$

By Theorem C.4.1, any polynomial p of degree at most k can be written as a linear combination of spherical harmonics q_i, $i = 0..k$, when restricted to the sphere. Noting that $|x|^2 q = q$ on $\mathscr{S}^{N-1}$, we deduce that

$$ p\big|_{\mathscr{S}^{N-1}} = q_1\big|_{\mathscr{S}^{N-1}} + q_2\big|_{\mathscr{S}^{N-1}}, $$

where q_1 is a homogeneous polynomial of degree k and q_2 a homogeneous polynomial of degree $k-1$. Note that the decomposition is unique, since any homogeneous polynomial q of degree j must be either odd or even with j. We deduce that the space of restrictions of polynomials of degree at most k to the sphere, has dimension $d_{k,N} + d_{k-1,N}$. Finally, applying Theorem C.4.1, any polynomial of degree at most k can be uniquely decomposed in the form $p\big|_{\mathscr{S}^{N-1}} = q_1\big|_{\mathscr{S}^{N-1}} + q_2\big|_{\mathscr{S}^{N-1}}$, where this time $q_1\big|_{\mathscr{S}^{N-1}}$ is a spherical harmonic of degree k and q_2 is a polynomial of degree at most $k-1$. We deduce that

$$ \dim H_k = d_{k,N} + d_{k-1,N} - (d_{k-1,N} + d_{k-2,N}), $$

which proves Theorem C.4.2. □

References

[1] S. Agmon, A. Douglis, and L. Nirenberg, *Estimates near the boundary for solutions of elliptic partial differential equations satisfying general boundary conditions. I*, Comm. Pure Appl. Math. **12** (1959), 623–727.

[2] A. Aftalion and F. Pacella, *Qualitative properties of nodal solutions of semilinear elliptic equations in radially symmetric domains*, C. R. Math. Acad. Sci. Paris **339** (2004), no. 5, 339–344 (English, with English and French summaries).

[3] G. Alberti, L. Ambrosio, and X. Cabré, *On a long-standing conjecture of E De Giorgi: symmetry in 3D for general nonlinearities and a local minimality property*, Acta Appl. Math. **65** (2001), no. 1-3, 9–33. Special issue dedicated to Antonio Avantaggiati on the occasion of his 70th birthday.

[4] N. D. Alikakos and P. W. Bates, *On the singular limit in a phase field model of phase transitions*, Ann. Inst. H. Poincaré Anal. Non Linéaire **5** (1988), no. 2, 141–178 (English, with French summary).

[5] W. K. Allard, *On the first variation of a varifold*, Ann. of Math. (2) **95** (1972), 417–491.

[6] F. J. Almgren Jr., *Some interior regularity theorems for minimal surfaces and an extension of Bernstein's theorem*, Ann. of Math. (2) **84** (1966), 277–292.

[7] L. Ambrosio and X. Cabré, *Entire solutions of semilinear elliptic equations in $\mathbf{R}^3$ and a conjecture of De Giorgi*, J. Amer. Math. Soc. **13** (2000), no. 4, 725–739 (electronic).

[8] L. Ambrosio, G. De Philippis, and L. Martinazzi, *Gamma-convergence of nonlocal perimeter functionals*, Preprint available at http://cvgmt.sns.it/cgi/get.cgi/people/dephilipp/ (2010).

[9] G. Arioli, F. Gazzola, and H.-C. Grunau, *Entire solutions for a semilinear fourth order elliptic problem with exponential nonlinearity*, J. Differential Equations **230** (2006), no. 2, 743–770, DOI 10.1016/j.jde.2006.05.015.

[10] G. Arioli, F. Gazzola, H.-C. Grunau, and E. Mitidieri, *A semilinear fourth order elliptic problem with exponential nonlinearity*, SIAM J. Math. Anal. **36** (2005), no. 4, 1226–1258 (electronic), DOI 10.1137/S0036141002418534.

[11] S. Axler, P. Bourdon, and W. Ramey, *Harmonic function theory*, 2nd ed., Graduate Texts in Mathematics, vol. 137, Springer-Verlag, New York, 2001.

[12]   S. Baraket and F. Pacard, *Construction of singular limits for a semilinear elliptic equation in dimension* 2, Calc. Var. Partial Differential Equations **6** (1998), no. 1, 1–38.

[13]   J. Bebernes and D. Eberly, *Mathematical problems from combustion theory*, Applied Mathematical Sciences, vol. 83, Springer-Verlag, New York, 1989.

[14]   H. Bellout, *A criterion for blow-up of solutions to semilinear heat equations*, SIAM J. Math. Anal. **18** (1987), no. 3, 722–727, DOI 10.1137/0518055.

[15]   E. Berchio and F. Gazzola, *Some remarks on biharmonic elliptic problems with positive, increasing and convex nonlinearities*, Electron. J. Differential Equations (2005), No. 34, 20 pp. (electronic).

[16]   H. Berestycki, L. A. Caffarelli, and L. Nirenberg, *Symmetry for elliptic equations in a half space*, Boundary value problems for partial differential equations and applications, RMA Res. Notes Appl. Math., vol. 29, Masson, Paris, 1993, pp. 27–42.

[17]   ———, *Further qualitative properties for elliptic equations in unbounded domains*, Ann. Scuola Norm. Sup. Pisa Cl. Sci. (4) **25** (1997), no. 1-2, 69–94 (1998). Dedicated to Ennio De Giorgi.

[18]   H. Berestycki and L. Nirenberg, *On the method of moving planes and the sliding method*, Bol. Soc. Brasil. Mat. (N.S.) **22** (1991), no. 1, 1–37.

[19]   S. Bernstein, *Über ein geometrisches Theorem und seine Anwendung auf die partiellen Differentialgleichungen vom elliptischen Typus*, Math. Z. **26** (1927), no. 1, 551–558, DOI 10.1007/BF01475472 (German).

[20]   M.-F. Bidaut-Véron and L. Véron, *Nonlinear elliptic equations on compact Riemannian manifolds and asymptotics of Emden equations*, Invent. Math. **106** (1991), no. 3, 489–539, DOI 10.1007/BF01243922.

[21]   T. Boggio, *Sulle funzioni di Green d'ordine m*, Rend. Circ. Mat. Palermo (1905), 97–135.

[22]   E. Bombieri, E. De Giorgi, and E. Giusti, *Minimal cones and the Bernstein problem*, Invent. Math. **7** (1969), 243–268.

[23]   D. Bonheure, V. Bouchez, C. Grumiau, and J. Van Schaftingen, *Asymptotics and symmetries of least energy nodal solutions of Lane-Emden problems with slow growth*, Commun. Contemp. Math. **10** (2008), no. 4, 609–631, DOI 10.1142/S0219199708002910.

[24]   V. Bouchez, *Nonlinear boundary value problems: existence, estimates, symmetry*, PhD Thesis, Université catholique de Louvain **167** (2009).

[25]   H. Brezis, *Functional Analysis, Sobolev Spaces and Partial Differential Equations*, Universitext, Springer, 2011 (English).

[26]   ———, *Is there failure of the inverse function theorem?*, Morse theory, minimax theory and their applications to nonlinear differential equations, New Stud. Adv. Math., vol. 1, International Press, Somerville, MA, 2003, pp. 23–33.

[27]   H. Brezis and X. Cabré, *Some simple nonlinear PDE's without solutions*, Boll. Unione Mat. Ital. Sez. B Artic. Ric. Mat. (8) **1** (1998), no. 2, 223–262 (English, with Italian summary).

[28]   H. Brezis and T. Cazenave, *Nonlinear heat equation*, unpublished (English).

[29]   H. Brezis, T. Cazenave, Y. Martel, and A. Ramiandrisoa, *Blow up for $u_t - \Delta u = g(u)$ revisited*, Adv. Differential Equations **1** (1996), no. 1, 73–90.

[30]   H. Brezis and F. Merle, *Uniform estimates and blow-up behavior for solutions of $-\Delta u = V(x)e^u$ in two dimensions*, Comm. Partial Differential Equations **16** (1991), no. 8-9, 1223–1253.

[31]   H. Brézis and L. Nirenberg, *Positive solutions of nonlinear elliptic equations involving critical Sobolev exponents*, Comm. Pure Appl. Math. **36** (1983), no. 4, 437–477, DOI 10.1002/cpa.3160360405.

[32]   H. Brezis and J. L. Vázquez, *Blow-up solutions of some nonlinear elliptic problems*, Rev. Mat. Univ. Complut. Madrid **10** (1997), no. 2, 443–469.

[33]   C. C. Burch, *The Dini condition and regularity of weak solutions of elliptic equations*, J. Differential Equations **30** (1978), no. 3, 308–323, DOI 10.1016/0022-0396(78)90003-7.

[34]   X. Cabré, *Regularity of minimizers of semilinear elliptic problems up to dimension four*, arXiv:0909.4696v1.

[35]   ———, *Regularity of radial extremal solutions of semilinear elliptic equations*, Bol. Soc. Esp. Mat. Apl. SeMA (2006), no. 34, 92–98.

[36]   ———, *Topics in regularity and qualitative properties of solutions of nonlinear elliptic equations*, Discrete Contin. Dyn. Syst. **8** (2002), no. 2, 331–359. Current developments in partial differential equations (Temuco, 1999).

[37]   X. Cabré and S. Chanillo, *Stable solutions of semilinear elliptic problems in convex domains*, Selecta Math. (N.S.) **4** (1998), no. 1, 1–10, DOI 10.1007/s000290050022.

[38]   X. Cabré and A. Capella, *Regularity of radial minimizers and extremal solutions of semilinear elliptic equations*, J. Funct. Anal. **238** (2006), no. 2, 709–733.

[39]   ———, *Regularity of minimizers for three elliptic problems: minimal cones, harmonic maps, and semilinear equations*, Pure Appl. Math. Q. **3** (2007), no. 3, 801–825.

[40]   X. Cabré, A. Capella, and M. Sanchón, *Regularity of radial minimizers of reaction equations involving the p-Laplacian*, Calc. Var. Partial Differential Equations **34** (2009), no. 4, 475–494, DOI 10.1007/s00526-008-0192-3.

[41]   X. Cabré and E. Cinti, *Energy estimates and 1-D symmetry for nonlinear equations involving the half-Laplacian*, arXiv:1004.2866v1.

[42]   X. Cabré and M. Sanchón, *Semi-stable and extremal solutions of reaction equations involving the p-Laplacian*, Commun. Pure Appl. Anal. **6** (2007), no. 1, 43–67.

[43]   X. Cabré and J. Solà-Morales, *Layer solutions in a half-space for boundary reactions*, Comm. Pure Appl. Math. **58** (2005), no. 12, 1678–1732, DOI 10.1002/cpa.20093.

[44]   L. Caffarelli, R. Hardt, and L. Simon, *Minimal surfaces with isolated singularities*, Manuscripta Math. **48** (1984), no. 1-3, 1–18.

[45] L. Caffarelli, J.-M. Roquejoffre, and O. Savin, *Nonlocal minimal surfaces*, Comm. Pure Appl. Math. **63** (2010), no. 9, 1111–1144.

[46] L. Caffarelli and E. Valdinoci, *Uniform estimates and limiting arguments for nonlocal minimal surfaces*, Calc. Var. Partial Differential Equations, posted on 2010, DOI 10.1007/s00526-010-0359-6, (to appear in print).

[47] L. Caffarelli and L. Silvestre, *An extension problem related to the fractional Laplacian*, Comm. Partial Differential Equations **32** (2007), no. 7-9, 1245–1260, DOI 10.1080/03605300600987306.

[48] A. P. Calderon and A. Zygmund, *On the existence of certain singular integrals*, Acta Math. **88** (1952), 85–139.

[49] S. Campanato, *Equazioni ellittiche del II deg ordine espazi $\mathfrak{L}^{(2,\lambda)}$*, Ann. Mat. Pura Appl. (4) **69** (1965), 321–381 (Italian).

[50] A. Capella, J. Dávila, L. Dupaigne, and Y. Sire, *Regularity of radial extremal solutions for some nonlocal semilinear elliptic equations*, arXiv:1004.1906v1.

[51] H. Cartan, *Differential calculus*, Hermann, Paris, 1971. Exercises by C. Buttin, F. Rideau and J. L. Verley; Translated from the French.

[52] D. Castorina, P. Esposito, and B. Sciunzi, *Degenerate elliptic equations with singular nonlinearities*, Calc. Var. Partial Differential Equations **34** (2009), no. 3, 279–306, DOI 10.1007/s00526-008-0184-3.

[53] ———, *Low dimensional instability for semilinear and quasilinear problems in $\mathbb{R}^N$*, Commun. Pure Appl. Anal. **8** (2009), no. 6, 1779–1793, DOI 10.3934/cpaa.2009.8.1779.

[54] A. Castro, J. Cossio, and J. M. Neuberger, *A sign-changing solution for a superlinear Dirichlet problem*, Rocky Mountain J. Math. **27** (1997), no. 4, 1041–1053, DOI 10.1216/rmjm/1181071858.

[55] D. Cassani, J. M. do Ó, and N. Ghoussoub, *On a fourth order elliptic problem with a singular nonlinearity*, Adv. Nonlinear Stud. **9** (2009), no. 1, 177–197.

[56] T. Cazenave, M. Escobedo, and M. A. Pozio, *Some stability properties for minimal solutions of $-\Delta u = \lambda g(u)$*, Port. Math. (N.S.) **59** (2002), no. 4, 373–391.

[57] W. X. Chen and C. Li, *A priori estimates for solutions to nonlinear elliptic equations*, Arch. Rational Mech. Anal. **122** (1993), no. 2, 145–157, DOI 10.1007/BF00378165.

[58] ———, *Classification of solutions of some nonlinear elliptic equations*, Duke Math. J. **63** (1991), no. 3, 615–622.

[59] P. Clément, D. G. de Figueiredo, and E. Mitidieri, *Quasilinear elliptic equations with critical exponents*, Topol. Methods Nonlinear Anal. **7** (1996), no. 1, 133–170.

[60] C. Cowan, P. Esposito, and N. Ghoussoub, *Regularity of extremal solutions in fourth order nonlinear eigenvalue problems on general doamins*, DCDS **28** (2010), no. 3.

[61] C. Cowan, P. Esposito, N. Ghoussoub, and A. Moradifam, *The critical dimension for a fourth order elliptic problem with singular nonlinearity*, Arch. Rat. Mech. Anal. (2010).

References

[62] R. Courant and D. Hilbert, *Methods of mathematical physics. Vol. I*, Interscience Publishers, Inc., New York, 1953.

[63] M. G. Crandall and P. H. Rabinowitz, *Some continuation and variational methods for positive solutions of nonlinear elliptic eigenvalue problems*, Arch. Rational Mech. Anal. **58** (1975), no. 3, 207–218.

[64] M. Cwikel, *Weak type estimates for singular values and the number of bound states of Schrödinger operators*, Ann. Math. (2) **106** (1977), no. 1, 93–100.

[65] R. Dalmasso, *Symmetry properties in higher-order semilinear elliptic equations*, Nonlinear Anal. **24** (1995), no. 1, 1–7.

[66] L. Damascelli, A. Farina, B. Sciunzi, and E. Valdinoci, *Liouville results for m-Laplace equations of Lane-Emden-Fowler type*, Ann. Inst. H. Poincaré Anal. Non Linéaire **26** (2009), no. 4, 1099–1119, DOI 10.1016/j.anihpc.2008.06.001.

[67] L. Damascelli and B. Sciunzi, *Regularity, monotonicity and symmetry of positive solutions of m-Laplace equations*, J. Differential Equations **206** (2004), no. 2, 483–515, DOI 10.1016/j.jde.2004.05.012.

[68] E. N. Dancer, *Infinitely many turning points for some supercritical problems*, Ann. Mat. Pura Appl. (4) **178** (2000), 225–233, DOI 10.1007/BF02505896.

[69] _____, *Weakly nonlinear Dirichlet problems on long or thin domains*, Mem. Amer. Math. Soc. **105** (1993), no. 501, viii+66.

[70] _____, *Some notes on the method of moving planes*, Bull. Austral. Math. Soc. **46** (1992), no. 3, 425–434, DOI 10.1017/S0004972700012089.

[71] E. N. Dancer and A. Farina, *On the classification of solutions of $-\Delta u = e^u$ on $\mathbb{R}^N$: stability outside a compact set and applications*, Proc. Amer. Math. Soc. **137** (2009), no. 4, 1333–1338, DOI 10.1090/S0002-9939-08-09772-4.

[72] J. Dávila, M. del Pino, M. Musso, and J. Wei, *Fast and slow decay solutions for supercritical elliptic problems in exterior domains*, Calc. Var. Partial Differential Equations **32** (2008), no. 4, 453–480, DOI 10.1007/s00526-007-0154-1.

[73] J. Dávila and L. Dupaigne, *Perturbing singular solutions of the Gelfand problem*, Commun. Contemp. Math. **9** (2007), no. 5, 639–680.

[74] _____, *Hardy-type inequalities*, J. Eur. Math. Soc. (JEMS) **6** (2004), no. 3, 335–365.

[75] _____, *Comparison results for PDEs with a singular potential*, Proc. Roy. Soc. Edinburgh Sect. A **133** (2003), no. 1, 61–83.

[76] J. Dávila, L. Dupaigne, and A. Farina, *Partial regularity of finite Morse index solutions to the Lane-Emden equation*, to appear in J. Functional Analysis.

[77] J. Dávila, L. Dupaigne, I. Guerra, and M. Montenegro, *Stable solutions for the biplacian with exponential nonlinearity*, SIAM J. Math. Anal. **39** (2007), no. 2, 565–592 (electronic), DOI 10.1137/060665579.

[78] J. Dávila, L. Dupaigne, and M. Montenegro, *The extremal solution of a boundary reaction problem*, Commun. Pure Appl. Anal. **7** (2008), no. 4, 795–817, DOI 10.3934/cpaa.2008.7.795.

[79] J. Dávila, I. Flores, and I. Guerra, *Multiplicity of solutions for a fourth order problem with exponential nonlinearity*, J. Differential Equations **247** (2009), no. 11, 3136–3162, DOI 10.1016/j.jde.2009.07.023.

[80] _____, *Multiplicity of solutions for a fourth order problem with power-type nonlinearity*, Math. Annalen **348** (2010), no. 1, 143–193.

[81] E. De Giorgi, *Convergence problems for functionals and operators*, Analysis (Rome, 1978), Pitagora, Bologna, 1979, pp. 131–188.

[82] _____, *Una estensione del teorema di Bernstein*, Ann. Scuola Norm. Sup. Pisa (3) **19** (1965), 79–85 (Italian).

[83] D. G. de Figueiredo, P.-L. Lions, and R. D. Nussbaum, *A priori estimates and existence of positive solutions of semilinear elliptic equations*, J. Math. Pures Appl. (9) **61** (1982), no. 1, 41–63.

[84] M. del Pino, M. Kowalczyk, and J. Wei, *A counterexample to a conjecture by De Giorgi in large dimensions*, C. R. Math. Acad. Sci. Paris **346** (2008), no. 23-24, 1261–1266 (English, with English and French summaries).

[85] E. DiBenedetto, $C^{1+\alpha}$ *local regularity of weak solutions of degenerate elliptic equations*, Nonlinear Anal. **7** (1983), no. 8, 827–850, DOI 10.1016/0362-546X(83)90061-5.

[86] W. Ding, personal communication to W. X. Chen and C. Li.

[87] J. Dolbeault and R. Stańczy, *Non-existence and uniqueness results for supercritical semilinear elliptic equations*, Ann. Henri Poincaré **10** (2010), no. 7, 1311–1333, DOI 10.1007/s00023-009-0016-9.

[88] L. Dupaigne and A. Farina, *Stable solutions of* $-\Delta u = f(u)$ *in* $\mathbb{R}^N$, J. Eur. Math. Soc. (JEMS) **12** (2010), no. 4, 855–882, DOI 10.4171/JEMS/217.

[89] _____, *Liouville theorems for stable solutions of semilinear elliptic equations with convex nonlinearities*, Nonlinear Anal. **70** (2009), no. 8, 2882–2888, DOI 10.1016/j.na.2008.12.017.

[90] L. Dupaigne and Y. Sire, *A Liouville theorem for non local elliptic equations*, arXiv:0909.1650v1.

[91] P. Esposito, N. Ghoussoub, and Y. Guo, *Mathematical analysis of partial differential equations modeling electrostatic MEMS*, Courant Lecture Notes in Mathematics, vol. 20, Courant Institute of Mathematical Sciences, New York, 2010.

[92] M. J. Esteban and P.-L. Lions, *Existence and nonexistence results for semilinear elliptic problems in unbounded domains*, Proc. Roy. Soc. Edinburgh Sect. A **93** (1982/83), no. 1-2, 1–14.

[93] L. C. Evans, *Partial differential equations*, Graduate Studies in Mathematics, vol. 19, American Mathematical Society, Providence, RI, 1998.

[94] A. Farina, *Liouville-type results for solutions of* $-\Delta u = |u|^{p-1}u$ *on unbounded domains of* $\mathbb{R}^N$, C. R. Math. Acad. Sci. Paris **341** (2005), no. 7, 415–418 (English, with English and French summaries).

[95] _____, *Stable solutions of* $-\Delta u = e^u$ *on* $\mathbb{R}^N$, C. R. Math. Acad. Sci. Paris **345** (2007), no. 2, 63–66 (English, with English and French summaries).

[96] _____, *Liouville-type theorems for elliptic problems*, Handbook of Differential Equations: Stationary Partial Differential Equations., vol. 4, Edited by M.Chipot, Elsevier B.V., 2007.

[97] _____, *On the classification of solutions of the Lane-Emden equation on unbounded domains of* $\mathbb{R}^N$, J. Math. Pures Appl. (9) **87** (2007), no. 5, 537–561 (English, with English and French summaries).

[98] _____, *One-dimensional symmetry for solutions of quasilinear equations in* $\mathbb{R}^2$, Boll. Unione Mat. Ital. Sez. B Artic. Ric. Mat. (8) **6** (2003), no. 3, 685–692 (English, with English and Italian summaries).

[99] A. Farina, B. Sciunzi, and E. Valdinoci, *Bernstein and De Giorgi type problems: new results via a geometric approach*, Ann. Sc. Norm. Super. Pisa Cl. Sci. (5) **7** (2008), no. 4, 741–791.

[100] _____, *On a Poincaré type formula for solutions of singular and degenerate elliptic equations*, Manuscripta Mathematica **132** (2010), no. 3–4, 335–342.

[101] A. Farina, Y. Sire, and E. Valdinoci, *Stable solutions of elliptic equations on Riemannian manifolds*, arxiv:0809.3025v1.

[102] A. Farina and E. Valdinoci, *Flattening Results for Elliptic PDEs in Unbounded Domains with Applications to Overdetermined Problems*, Archive for Rational Mechanics and Analysis **195** (2010), no. 3, 1025–1058.

[103] _____, *Rigidity results for elliptic PDES with uniform limits: an abstract framework with applications*, to appear in Indiana Univ. Math. J.

[104] _____, *The state of the art for a conjecture of De Giorgi and related problems*, Recent progress on reaction-diffusion systems and viscosity solutions, World Scientific Publishing, Hackensack, NJ, 2009, pp. 74–96.

[105] A. Ferrero, H.-C. Grunau, and P. Karageorgis, *Supercritical biharmonic equations with power-type nonlinearity*, Ann. Mat. Pura Appl. (4) **188** (2009), no. 1, 171–185, DOI 10.1007/s10231-008-0070-9.

[106] A. Ferrero and G. Warnault, *On solutions of second and fourth order elliptic equations with power-type nonlinearities*, Nonlinear Anal. **70** (2009), no. 8, 2889–2902, DOI 10.1016/j.na.2008.12.041.

[107] M. Fila and H. Matano, *Connecting equilibria by blow-up solutions*, Discrete Contin. Dynam. Systems **6** (2000), no. 1, 155–164.

[108] M. Fila and P. Poláčik, *Global solutions of a semilinear parabolic equation*, Adv. Differential Equations **4** (1999), no. 2, 163–196.

[109] D. L. Finn, *Convexity of level curves for solutions to semilinear elliptic equations*, Commun. Pure Appl. Anal. **7** (2008), no. 6, 1335–1343, DOI 10.3934/cpaa.2008.7.1335.

[110] D. A. Frank-Kamenetskii, *Diffusion and heat exchange in chemical kinetics*, Princeton University Press, Princeton, NJ, 1955.

[111] H. Fujita, *On the nonlinear equations $\Delta u + e^u = 0$ and $\partial v/\partial t = \Delta v + e^v$*, Bull. Amer. Math. Soc. **75** (1969), 132–135.

[112] ———, *On the asymptotic stability of solutions of the equation $v_t = \Delta v + e^v$*, Proc. Internat. Conf. on Functional Analysis and Related Topics (Tokyo, 1969), Univ. of Tokyo Press, Tokyo, 1970, pp. 252–259.

[113] J. García Azorero and I. Peral Alonso, *On an Emden-Fowler type equation*, Nonlinear Anal. **18** (1992), no. 11, 1085–1097, DOI 10.1016/0362-546X(92)90197-M.

[114] J. García Azorero, I. Peral Alonso, and J.-P. Puel, *Quasilinear problems with exponential growth in the reaction term*, Nonlinear Anal. **22** (1994), no. 4, 481–498, DOI 10.1016/0362-546X(94)90169-4.

[115] C. F. Gauss, *Werke: herausgegeben von der Koniglichen Gesellschaft der Wissenschaften zu Gottingen*, Georg Olms Verlag, New York (1863).

[116] F. Gazzola and H.-C. Grunau, *Radial entire solutions for supercritical biharmonic equations*, Math. Ann. **334** (2006), no. 4, 905–936, DOI 10.1007/s00208-005-0748-x.

[117] F. Gazzola, H.-C. Grunau, and G. Sweers, *Polyharmonic boundary value problems* (Springer Berlin, ed.), Lecture notes in Mathematics, vol. 1991, 2010.

[118] F. Gazzola and A. Malchiodi, *Some remarks on the equation $-\Delta u = \lambda(1+u)^p$ for varying λ, p and varying domains*, Comm. Partial Differential Equations **27** (2002), no. 3-4, 809–845, DOI 10.1081/PDE-120002875.

[119] I. M. Gel′fand, *Some problems in the theory of quasilinear equations*, Amer. Math. Soc. Transl. (2) **29** (1963), 295–381.

[120] N. Ghoussoub and C. Gui, *On a conjecture of De Giorgi and some related problems*, Math. Ann. **311** (1998), no. 3, 481–491.

[121] M. Giaquinta, *Multiple integrals in the calculus of variations and nonlinear elliptic systems*, Ann. Math. Stud., vol. 105, Princeton University Press, Princeton, NJ, 1983.

[122] M. Giaquinta and S. Hildebrandt, *Calculus of variations. I*, Grundlehren der Mathematischen Wissenschaften [Fundamental Principles of Mathematical Sciences], vol. 310, Springer-Verlag, Berlin, 1996. The Lagrangian formalism.

[123] B. Gidas, W. M. Ni, and L. Nirenberg, *Symmetry and related properties via the maximum principle*, Comm. Math. Phys. **68** (1979), no. 3, 209–243.

[124] B. Gidas and J. Spruck, *A priori bounds for positive solutions of nonlinear elliptic equations*, Comm. Partial Differential Equations **6** (1981), no. 8, 883–901, DOI 10.1080/03605308108820196.

References

[125] _____, *Global and local behavior of positive solutions of nonlinear elliptic equations,* Comm. Pure Appl. Math. **34** (1981), no. 4, 525–598, DOI 10.1002/cpa.3160340406.

[126] D. Gilbarg and N. S. Trudinger, *Elliptic partial differential equations of second order,* Classics in Mathematics, Springer-Verlag, Berlin, 2001. Reprint of the 1998 edition.

[127] E. Giusti, *Minimal surfaces and functions of bounded variation,* Monographs in Mathematics, vol. 80, Birkhäuser Verlag, Basel, 1984.

[128] Z. Guo and J. Li, *The blow up locus of semilinear elliptic equations with supercritical exponents,* Calc. Var. Partial Differential Equations **15** (2002), no. 2, 133–153, DOI 10.1007/s00526-002-0078-8.

[129] E. Hopf, *A remark on linear elliptic differential equations of second order,* Proc. Amer. Math. Soc. **3** (1952), 791–793.

[130] _____, *Elementare Bemerkungen über die Lösungen partieller Differentialgleichungen zweiter Ordnung vom elliptischen Typus,* Sitzungsber. d. Preuss. Akad. d. Wiss. **19** (1927), 147–152.

[131] J. Jacobsen and K. Schmitt, *The Liouville-Bratu-Gelfand problem for radial operators,* J. Differential Equations **184** (2002), no. 1, 283–298.

[132] D. D. Joseph and T. S. Lundgren, *Quasilinear Dirichlet problems driven by positive sources,* Arch. Rational Mech. Anal. **49** (1972/73), 241–269.

[133] T. Kato, *Schrödinger operators with singular potentials,* Proceedings of the International Symposium on Partial Differential Equations and the Geometry of Normed Linear Spaces (Jerusalem, 1972), 1972, pp. 135–148 (1973).

[134] B. Kawohl, *Rearrangements and convexity of level sets in PDE,* Lecture Notes in Mathematics, vol. 1150, Springer-Verlag, Berlin, 1985.

[135] _____, *When are solutions to nonlinear elliptic boundary value problems convex?,* Comm. Partial Differential Equations **10** (1985), no. 10, 1213–1225, DOI 10.1080/03605308508820404.

[136] J. P. Keener and H. B. Keller, *Positive solutions of convex nonlinear eigenvalue problems,* J. Differential Equations **16** (1974), 103–125.

[137] H. B. Keller and D. S. Cohen, *Some positone problems suggested by nonlinear heat generation,* J. Math. Mech. **16** (1967), 1361–1376.

[138] N. J. Korevaar, *Convexity properties of solutions to elliptic PDEs,* Variational methods for free surface interfaces (Menlo Park, Calif., 1985), Springer, New York, 1987, pp. 115–121.

[139] T. Laetsch, *On the number of solutions of boundary value problems with convex nonlinearities,* J. Math. Anal. Appl. **35** (1971), 389–404.

[140] A. A. Lacey, *Mathematical analysis of thermal runaway for spatially inhomogeneous reactions,* SIAM J. Appl. Math. **43** (1983), no. 6, 1350–1366, DOI 10.1137/0143090.

[141] P. Li and S. T. Yau, *On the Schrödinger equation and the eigenvalue problem*, Comm. Math. Phys. **88** (1983), no. 3, 309–318.

[142] E. Lieb, *Bounds on the eigenvalues of the Laplace and Schroedinger operators*, Bull. Amer. Math. Soc. **82** (1976), no. 5, 751–753.

[143] G. M. Lieberman, *Boundary regularity for solutions of degenerate elliptic equations*, Nonlinear Anal. **12** (1988), no. 11, 1203–1219, DOI 10.1016/0362-546X(88)90053-3.

[144] C. S. Lin and W.-M. Ni, *A counterexample to the nodal domain conjecture and a related semilinear equation*, Proc. Amer. Math. Soc. **102** (1988), no. 2, 271–277, DOI 10.2307/2045874.

[145] J.-L. Lions, *Théorèmes de trace et d'interpolation. I*, Ann. Scuola Norm. Sup. Pisa (3) **13** (1959), 389–403.

[146] J.-L. Lions and E. Magenes, *Problèmes aux limites non homogènes et applications. Vol. 1*, Travaux et Recherches Mathématiques, No. 17, Dunod, Paris, 1968.

[147] J. Liouville, *Sur l'équation aux différences partielles $\frac{d^2 \log \lambda}{du\,dv} \pm \frac{\lambda}{2a^2} = 0.$*, J. Math. Pures Appl. **18** (1853), 71–72.

[148] H. Lou, *On singular sets of local solutions to p-Laplace equations*, Chin. Ann. Math. Ser. B **29** (2008), no. 5, 521–530, DOI 10.1007/s11401-007-0312-y.

[149] Y. Martel, *Uniqueness of weak extremal solutions of nonlinear elliptic problems*, Houston J. Math. **23** (1997), no. 1, 161–168.

[150] ———, *Complete blow up and global behaviour of solutions of $u_t - \Delta u = g(u)$*, Ann. Inst. H. Poincaré Anal. Non Linéaire **15** (1998), no. 6, 687–723 (English, with English and French summaries).

[151] A. Mellet and J. Vovelle, *Existence and regularity of extremal solutions for a mean-curvature equation*, J. Differential Equations **249** (2010), no. 1, 37–75.

[152] P. Mironescu and V. D. Rădulescu, *The study of a bifurcation problem associated to an asymptotically linear function*, Nonlinear Anal. **26** (1996), no. 4, 857–875.

[153] J. H. Michael and L. M. Simon, *Sobolev and mean-value inequalities on generalized submanifolds of R^n*, Comm. Pure Appl. Math. **26** (1973), 361–379.

[154] F. Mignot and J.-P. Puel, *Sur une classe de problèmes non linéaires avec non linéairité positive, croissante, convexe*, Comm. Partial Differential Equations **5** (1980), no. 8, 791–836 (French).

[155] ———, *Quelques résultats sur un problème elliptique avec non linéarité exponentielle*, Équations aux dérivées partielles et applications, Gauthier-Villars, Éd. Sci. Méd. Elsevier, Paris, 1998, pp. 683–704 (French, with French summary).

[156] L. Modica, *The gradient theory of phase transitions and the minimal interface criterion*, Arch. Rational Mech. Anal. **98** (1987), no. 2, 123–142.

[157] _____, *Monotonicity of the energy for entire solutions of semilinear elliptic equations*, Partial differential equations and the calculus of variations, Vol. II, Progr. Nonlinear Differential Equations Appl., vol. 2, Birkhäuser Boston, Boston, MA, 1989, pp. 843–850.

[158] _____, *A gradient bound and a Liouville theorem for nonlinear Poisson equations*, Comm. Pure Appl. Math. **38** (1985), no. 5, 679–684.

[159] M. Montenegro and A. C. Ponce, *The sub-supersolution method for weak solutions*, Proc. Amer. Math. Soc. **136** (2008), no. 7, 2429–2438.

[160] A. Moradifam, *The singular extremal solutions of the bi-Laplacian with exponential non-linearity*, Proc. Amer. Math. Soc. **138** (2010), no. 4, 1287–1293, DOI 10.1090/S0002-9939-09-10257-5.

[161] _____, *On the critical dimension of a fourth order elliptic problem with negative exponent*, J. Differential Equations **248** (2010), no. 3, 594–616, DOI 10.1016/j.jde.2009.09.011.

[162] C. G. Moreira and M. A. S. Ruas, *The curve selection lemma and the Morse-Sard theorem*, Manuscripta Math. **129** (2009), no. 3, 401–408, DOI 10.1007/s00229-009-0275-2.

[163] J.-M. Morel and L. Oswald, *A uniform formulation for Hopf maximum principle*, unpublished preprint (1985).

[164] A. P. Morse, *The behavior of a function on its critical set*, Ann. of Math. (2) **40** (1939), no. 1, 62–70.

[165] K. Nagasaki and T. Suzuki, *Spectral and related properties about the Emden-Fowler equation* $-\Delta u = \lambda e^u$ *on circular domains*, Math. Ann. **299** (1994), no. 1, 1–15, DOI 10.1007/BF01459770.

[166] G. Nedev, *Regularity of the extremal solution to semilinear elliptic equations*, C. R. Acad. Sci. Paris Sér. I Math. **330** (2000), no. 11, 997–1002 (English, with English and French summaries).

[167] _____, *Extremal solutions of semilinear elliptic equations*, Unpublished notes.

[168] M. Nicolesco, *Sur les fonctions de n variables, harmoniques d'ordre p*, Bull. Soc. Math. France **60** (1932), 129–151 (French).

[169] O. A. Oleĭnik, *On properties of solutions of certain boundary problems for equations of elliptic type*, Mat. Sbornik N.S. **30(72)** (1952), 695–702 (Russian).

[170] R. Osserman, *A survey of minimal surfaces*, 2nd ed., Dover Publications Inc., New York, 1986.

[171] _____, *Minimal varieties*, Bull. Amer. Math. Soc. **75** (1969), 1092–1120.

[172] F. Pacard, *A note on the regularity of weak solutions of* $-\Delta u = u^\alpha$ *in* $\mathbf{R}^n$, $n \geq 3$, Houston J. Math. **18** (1992), no. 4, 621–632.

[173] _____, *Partial regularity for weak solutions of a nonlinear elliptic equation*, Manuscripta Math. **79** (1993), no. 2, 161–172, DOI 10.1007/BF02568335.

313

[174] _____, *Convergence and partial regularity for weak solutions of some nonlinear elliptic equation: the supercritical case*, Ann. Inst. H. Poincaré Anal. Non Linéaire **11** (1994), no. 5, 537–551 (English, with English and French summaries).

[175] F. Pacella, *Symmetry results for solutions of semilinear elliptic equations with convex nonlinearities*, J. Funct. Anal. **192** (2002), no. 1, 271–282, DOI 10.1006/jfan.2001.3901.

[176] F. Pacella and M. Ramaswamy, *Symmetry of solutions of elliptic equations via maximum principles* (M. Chipot, ed.), Handbook of differential equations: stationary partial differential equations. Vol. VI, Elsevier/North-Holland, Amsterdam, 2008.

[177] F. Pacella and T. Weth, *Symmetry of solutions to semilinear elliptic equations via Morse index*, Proc. Amer. Math. Soc. **135** (2007), no. 6, 1753–1762 (electronic), DOI 10.1090/S0002-9939-07-08652-2.

[178] J. A. Pelesko and D. H. Bernstein, *Modeling MEMS and NEMS*, Chapman & Hall/CRC, Boca Raton, FL, 2003.

[179] M. Picone, *Maggiorazione degli integrali delle equazioni totalmente paraboliche alle derivate parziali del secondo ordine*, Ann. Mat. Pura Appl. **7** (1929), no. 1, 145–192 (Italian).

[180] _____, *Maggiorazioni degli integrali di equazioni lineari ellittico-paraboliche alle derivate parziali del secondo ordine*, Rend. Accad. Naz. Lincei. **5** (1927), no. 6, 138–143 (Italian).

[181] S. I. Pohožaev, *On the eigenfunctions of the equation* $\Delta u + \lambda f(u) = 0$, Dokl. Akad. Nauk SSSR **165** (1965), 36–39 (Russian).

[182] M. H. Protter and H. F. Weinberger, *Maximum principles in differential equations*, Prentice-Hall Inc., Englewood Cliffs, NJ, 1967.

[183] P. Pucci and J. Serrin, *The maximum principle*, Progress in Nonlinear Differential Equations and Their Applications, 73, Birkhäuser Verlag, Basel, 2007.

[184] P. Quittner and P. Souplet, *Superlinear parabolic problems*, Birkhäuser Advanced Texts: Basler Lehrbücher. [Birkhäuser Advanced Texts: Basel Textbooks], Birkhäuser Verlag, Basel, 2007. Blow-up, global existence and steady states.

[185] A. Ramiandrisoa, *Blow-up for two nonlinear problems*, Nonlinear Anal. **41** (2000), no. 7-8, Ser. A: Theory Methods, 825–854, DOI 10.1016/S0362-546X(98)00312-5.

[186] Y. Rébaï, *Solutions of semilinear elliptic equations with one isolated singularity*, Differential Integral Equations **12** (1999), no. 4, 563–581.

[187] G. V. Rozenbljum, *Distribution of the discrete spectrum of singular differential operators*, Dokl. Akad. Nauk SSSR **202** (1972), 1012–1015 (Russian).

[188] M. V. Safonov, *Boundary estimates for positive solutions to second order elliptic equations*, http://arxiv.org/abs/0810.0522v2.

[189] L. Saloff-Coste, *Aspects of Sobolev-type inequalities*, London Mathematical Society Lecture Note Series, vol. 289, Cambridge University Press, Cambridge, 2002.

[190] M. Sanchón, *Regularity of the extremal solution of some nonlinear elliptic problems involving the p-Laplacian*, Potential Anal. **27** (2007), no. 3, 217–224, DOI 10.1007/s11118-007-9053-5.

[191] _____, *Boundedness of the extremal solution of some p-Laplacian problems*, Nonlinear Anal. **67** (2007), no. 1, 281–294, DOI 10.1016/j.na.2006.05.010.

[192] A. Sard, *The measure of the critical values of differentiable maps*, Bull. Amer. Math. Soc. **48** (1942), 883–890.

[193] D. H. Sattinger, *Monotone methods in nonlinear elliptic and parabolic boundary value problems*, Indiana Univ. Math. J. **21** (1971/72), 979–1000.

[194] O. Savin, *Regularity of flat level sets in phase transitions*, Ann. of Math. (2) **169** (2009), no. 1, 41–78.

[195] O. Savin and E. Valdinoci, *Γ-convergence for nonlocal phase transitions*, arXiv:1007.1725v1.

[196] _____, *Density estimates for a variational model driven by the Gagliardo norm*, arXiv:1007.2114v1.

[197] R. Schaaf, *Uniqueness for semilinear elliptic problems: supercritical growth and domain geometry*, Adv. Differential Equations **5** (2000), no. 10-12, 1201–1220.

[198] J. Schauder, *Über lineare elliptische Differentialgleichungen zweiter Ordnung*, Math. Z. **38** (1934), no. 1, 257–282, DOI 10.1007/BF01170635 (German).

[199] K. Schmitt, *Positive solutions of semilinear elliptic boundary value problems*, Topological methods in differential equations and inclusions (Montreal, PQ, 1994), NATO Adv. Sci. Inst. Ser. C Math. Phys. Sci., vol. 472, Kluwer Academic Publishers, Dordrecht, 1995, pp. 447–500.

[200] R. T. Seeley, *Spherical harmonics*, Amer. Math. Monthly **73** (1966), no. 4, 115–121.

[201] J. Serrin, *Local behavior of solutions of quasi-linear equations*, Acta Math. **111** (1964), 247–302.

[202] H. S. Shapiro and M. Tegmark, *An elementary proof that the biharmonic Green function of an eccentric ellipse changes sign*, SIAM Rev. **36** (1994), no. 1, 99–101, DOI 10.1137/1036005.

[203] L. Simon, *The minimal surface equation*, Geometry, V, Encyclopaedia Math. Sci., vol. 90, Springer, Berlin, 1997, pp. 239–272.

[204] J. Simons, *Minimal varieties in riemannian manifolds*, Ann. of Math. (2) **88** (1968), 62–105.

[205] Y. Sire and E. Valdinoci, *Fractional Laplacian phase transitions and boundary reactions: a geometric inequality and a symmetry result*, J. Funct. Anal. **256** (2009), no. 6, 1842–1864, DOI 10.1016/j.jfa.2009.01.020.

[206] G. Stampacchia, *Equations elliptiques du second ordre à coefficients discontinus*, Séminaire de Mathématiques Supérieures, No. 16 (Été, 1965), Les Presses de l'Université de Montréal, Montreal, Que., 1966 (French).

[207] P. Sternberg and K. Zumbrun, *On the connectivity of boundaries of sets minimizing perimeter subject to a volume constraint*, Comm. Anal. Geom. **7** (1999), no. 1, 199–220.

[208] ———, *A Poincaré inequality with applications to volume-constrained area-minimizing surfaces*, J. Reine Angew. Math. **503** (1998), 63–85.

[209] L. Tartar, *An introduction to Sobolev spaces and interpolation spaces*, Lecture Notes of the Unione Matematica Italiana, vol. 3, Springer, Berlin, 2007.

[210] J. F. Toland, *On positive solutions of* $-\Delta u = F(x, u)$, Math. Z. **182** (1983), no. 3, 351–357.

[211] P. Tolksdorf, *Regularity for a more general class of quasilinear elliptic equations*, J. Differential Equations **51** (1984), no. 1, 126–150, DOI 10.1016/0022-0396(84)90105-0.

[212] N. S. Trudinger, *On Harnack type inequalities and their application to quasilinear elliptic equations*, Comm. Pure Appl. Math. **20** (1967), 721–747.

[213] E. Valdinoci, *From the long jump random walk to the fractional Laplacian*, Bol. Soc. Esp. Mat. Apl. SēMA (2009), no. 49, 33–44.

[214] ———, personal communication.

[215] J. L. Vazquez and E. Zuazua, *The Hardy inequality and the asymptotic behaviour of the heat equation with an inverse-square potential*, J. Funct. Anal. **173** (2000), no. 1, 103–153.

[216] S. Villegas, *Asymptotic behavior of stable radial solutions of semilinear elliptic equations in* $\mathbb{R}^N$, J. Math. Pures Appl. (9) **88** (2007), no. 3, 241–250 (English, with English and French summaries).

[217] ———, *Sharp estimates for semi-stable radial solutions of semilinear elliptic equations*, http://fr.arxiv.org/abs/0906.1443v1.

[218] G. Warnault, *Regularity of the extremal solution for a biharmonic problem with general nonlinearity*, Commun. Pure Appl. Anal. **8** (2009), no. 5, 1709–1723, DOI 10.3934/cpaa.2009.8.1709.

[219] ———, *Liouville theorems for stable radial solutions for the biharmonic operator*, Asymptotic Analysis (2010).

[220] K. Wang, *Partial regularity of stable solutions to the supercritical equations and its applications*, preprint.

[221] ———, *Partial regularity of stable solutions to Emden equation*, preprint.

[222] X.-J. Wang, *Schauder estimates for elliptic and parabolic equations*, Chinese Ann. Math. Ser. B **27** (2006), no. 6, 637–642, DOI 10.1007/s11401-006-0142-3.

References

[223] J. Wei and X. Xu, *Classification of solutions of higher-order conformally invariant equations*, Math. Ann. **313** (1999), no. 2, 207–228, DOI 10.1007/s002080050258.

[224] X.-F. Yang, *Nodal sets and Morse indices of solutions of super-linear elliptic PDEs*, J. Funct. Anal. **160** (1998), no. 1, 223–253, DOI 10.1006/jfan.1998.3301.

[225] S Zaremba, *Sur un problème mixte relatif à l'équation de Laplace*, Bull. Intern. de l' Acad. Sci. de Cracovie. Série A, Sci. Math. (1910), 313–344.

[226] Z. X. Zhao, *Green function for Schrödinger operator and conditioned Feynman-Kac gauge*, J. Math. Anal. Appl. **116** (1986), no. 2, 309–334.

Index

Milton Keynes UK
Ingram Content Group UK Ltd.
UKHW051947071024
449327UK00026B/2207

9 780367 382971